COMPUTATIONAL METHODS
FOR
DATA EVALUATION AND ASSIMILATION

COMPUTATIONAL METHODS
FOR
DATA EVALUATION AND ASSIMILATION

DAN GABRIEL CACUCI
IONEL MICHAEL NAVON
MIHAELA IONESCU-BUJOR

CRC Press
Taylor & Francis Group
Boca Raton London New York

CRC Press is an imprint of the
Taylor & Francis Group, an **informa** business

A CHAPMAN & HALL BOOK

CRC Press
Taylor & Francis Group
6000 Broken Sound Parkway NW, Suite 300
Boca Raton, FL 33487-2742

First issued in paperback 2019

© 2014 by Taylor & Francis Group, LLC
CRC Press is an imprint of Taylor & Francis Group, an Informa business

No claim to original U.S. Government works

ISBN-13: 978-1-58488-735-5 (hbk)
ISBN-13: 978-0-367-37961-2 (pbk)

Visit the Taylor & Francis Web site at
http://www.taylorandfrancis.com

and the CRC Press Web site at
http://www.crcpress.com

Contributors

Dan Gabriel Cacuci
University of South Carolina
Columbia, South Carolina

Ionel Michael Navon
Florida State University
Tallahassee, Florida

Mihaela Ionescu-Bujor
Karlsruher Institute of Technology
Karlsruhe, Campus North

Preface

This book is addressed to graduate and postgraduate students and researchers in the interdisciplinary methods of data assimilation, which refers to the integration of experimental and computational information. Since experiments and corresponding computations are encountered in many fields of scientific and engineering endeavors, the concepts presented in this book are illustrated using paradigm examples that range from the geophysical sciences to nuclear physics. In an attempt to keep the book as self-contained as possible, the mathematical concepts mostly from probability theory and functional analysis needed to follow the material presented in the book's five chapters, are summarized in the book's three appendices.

This book was finalized at the University of South Carolina. The authors wish to acknowledge the outstanding professional assistance of Dr. Madalina Corina Badea of the University of South Carolina, who has thoroughly reviewed the final version of the book, providing very valuable suggestions while improving its readability. Also acknowledged are the services of Dr. Erkan Arslan for his typing the word-version of this book into Latex. Last but not least, this book would have not have appeared without the continued patience, guidance, and understanding of Bob Stern (Executive Editor, Taylor and Francis Group), whom the authors appreciate immensely.

List of Figures

List of Tables

Contents

Introduction

Experience shows that it is practically impossible to measure exactly the true value of a physical quantity. This is because various imperfections occur at various stages involved in a measurement, including uncontrollable experimental errors, inaccurate standards, and other uncertainties arising in the data measurement and interpretation (reduction) process. Around any reported experimental value, therefore, there always exists a certain range of similar, more or less plausible, values that may also be true. In turn, this means that all inferences, predictions, engineering computations, and other applications of measured data are necessarily founded on weighted averages over all the possibly true values, with weights indicating the degree of plausibility of each value. These weights and weighted averages are what we call *probabilities* and *expectation values*. Consequently, the evaluation of scientific data is intrinsically intertwined with probability theory. The basic concepts underlying the evaluation of experimental data are presented in Chapter 1, which commences with a discussion, in Section 1.1, of the basic types of errors and probability distributions usually associated with them.

Section 1.2 presents the basic concepts of probability theory involved in the evaluation of uncertainty-afflicted scientific data. Since probabilities cannot be measured directly, they are either inferred from the results of observations or they are postulated and (partially) verified through accumulated experience. In scientific data evaluation, probabilities encode incomplete information. Persons possessing different information or knowledge assign different probabilities; furthermore, these probabilities are updated whenever new relevant information becomes available. It follows that probabilities cannot be considered as measurable physical properties. They are subjective in the sense that they depend on a person's knowledge. However, this does not mean that probabilities are arbitrary. They must obey the rules of logic, which demand, for instance, that rational persons with the same knowledge assign the same probabilities. The elementary conditions for logical consistency imply two fundamental rules from which all other mathematical relationships between probabilities can be derived. These two fundamental rules are the sum and the product rules, respectively. A powerful logical tool that follows immediately from the two forms of the product rule is Bayes' theorem (1763). The customary way to express Bayes' theorem in words is: "the posterior is proportional to the likelihood times the prior" where the "prior" distribution summarizes our knowledge extant prior to observing the (new) data, the "likelihood" distribution conveys the impact of the new information brought by the (new) data, while the "posterior" distribution contains the full information available for further inferences, predictions, and decision making. Thus, Bayes' theorem is a formal model for updating, or learning from observations.

A "measurable" or "physical" quantity is a property of phenomena, bodies, or substances that can be determined qualitatively and can be expressed quantitatively. Measurement is the process of experimentally finding the value of a physical quantity, with the help of devices called *measuring instruments*. It is important to note that: (1) the purpose of a measurement is to represent a property of an object by a number, so the result of a measurement must always be a number expressed in sanctioned units of measurements; (2) a measurement is always performed with the help of a measuring instrument; and (3) a measurement is always an experimental procedure. If it were known, the true value of a measurable quantity would ideally reflect, both qualitatively and quantitatively, the corresponding property of the object. The theory of measurement relies on the following postulates: (a) the true value of the measurable quantity exists; (b) the true value of the measurable quantity is constant relative to the conditions of the measurement; and (c) the true value cannot be found.

Since measuring instruments are imperfect, and since every measurement is an experimental procedure, the results of measurements cannot be accurate. This unavoidable imperfection of measurements is quantitatively characterized by measurement uncertainty or measurement error, which can be expressed in absolute or relative form. Consequently, repeated measurements of the same physical quantity can never yield identical results; even the most carefully measured scientific data in data banks will inevitably differ from the true values of the measured quantities. Consequently, nominal values for data, by themselves, are insufficient for applications. Quantitative uncertainties are also needed, along with the respective nominal values. Since the use of uncertain data may necessitate costly safety margins (in medicine, weather and climate prediction, or in the chemical, automotive, aerospace, or nuclear industries), working groups of the International Standards Organization have been developing uniform rules for reporting data uncertainties.

Combination of data from different sources involves a weighted propagation (via sensitivities, as will be seen subsequently) of all input uncertainties to uncertainties in the output values. Hence, data evaluation is intrinsically intertwined with uncertainty analysis, requiring reasoning from incomplete information, using probability theory for extracting "best estimate" values together with "best estimate" uncertainties from often sparse, incomplete, error-afflicted, and occasionally discrepant experimental data. A wide range of probability theory concepts and tools are employed in data evaluation and combination, from deductive statistics involving mainly frequencies and sample tallies to inductive inference for assimilating non-frequency data and a *priori knowledge*. Although grossly erroneous procedures and unintended mistakes (e.g., overlooking or miscalculating important corrections, equipment failure or improper calibration, bugs in computer codes, etc.) can produce defective data, such defective data will not be treated as "uncertainties" in this book. Nevertheless, data points that exhibit atypical behavior, which cannot be explained, need to be carefully scrutinized since outright rejection may not necessarily be appropriate. The terms "error" and "uncertainty" are interpreted in this book as being equivalent to the standard deviation of the probability distribution associated with the measurement process. This interpretation is consistent with the usual convention of considering "error" or "uncertainty" as an inherently positive number that quantifies the measurement dispersion of a specific observable parameter.

Legitimate errors are categorized either as random errors or as systematic errors. If the results of separate measurements of the same quantity differ from one another, and the respective differences cannot be predicted individually, then the error owing to this scatter of the results is called *random error*. Random errors can be identified by repeatedly measuring the same quantity under the same conditions. The scatter in results cannot be always tested in practice, particularly in large-scale modern experiments, where it may be impractical to provide sufficient repetition in order to satisfy the explicit needs for quantifying the random errors based on strict statistical requirements. Nevertheless, reasonable estimates of random errors can often be made, particularly when the nature of the underlying probability distribution can be inferred from previous experience. Furthermore, due to the influence of the central limit theorem, many sources of random error tend to be normally distributed. A significant feature of random errors is that repeated measurements (under fixed conditions) not only permit these errors to be better determined, but they also lead to error reduction, as assured by the law of large numbers. This feature is particularly important when high precision (i.e., small random errors) is required. A subtle issue regarding random errors stems, from the fact that such errors may contain correlated components: whether an error component is correlated or not within a particular data set depends upon the role that the associated random variable plays in the respective physical problem.

In contradistinction to a random error, a systematic error is defined as a measurement error that remains constant or changes in a regular fashion when the measurements of that quantity are repeated. Such errors arise because of inherent laws in the investigative process itself, and they lead to bias. Although systematic errors are difficult to distinguish from blunders, particularly when the impact of a blunder is small, the most reliable way to uncover systematic errors is by using a more accurate measuring instrument and/or by comparing a given result with a measurement of the same quantity, but performed by a different method. Each distinct approach leads to results that differ somewhat from those obtained in other ways. These differences exhibit a pattern (i.e., are "systematic") no matter how many times each approach is repeated, because the inherent systematic deficiencies of each method cannot be avoided by mere repetition. When the errors are truly systematic, statistical regularity will emerge from the ensemble of all measurements. Such a statistical regularity will not emerge when the data sets are afflicted with blunders, since blunders are generally one-time occurrences that can be detected if a particular procedure is repeated. Consequently, redundancy within a given investigative procedure is desirable not only to improve precision, but also to purge the results of blunders.

Probability theory is a branch of mathematical sciences that provides a model for describing the process of observation. The need for probability theory arises from the fact that most observations of natural phenomena do not lead to uniquely predictable results. Probability theory provides the tools for dealing with actual variations in the outcome of realistic observations and measurements. The challenging pursuit to develop a probability theory which is mathematically rigorous and also describes many phenomena observable in nature has generated over the years notable disputes over conceptual and logical issues. Modern probability theory is based on postulates constructed from three axioms attributed to Kolmogorov (1933), all of which are consistent with the notion of frequency of occurrence of events. The alternative approach, traceable to Laplace (1812), is based on the concept that probability is simply a way of providing a numerical scale to quantify our reasonable beliefs about a situation which we know only incompletely; this approach is consistent with Bayes' theorem, conditional probabilities, and inductive reasoning. Either approach to probability theory would completely describe a natural phenomenon if sufficient information were available to determine the underlying probability distribution exactly. In practice, though, such exact knowledge is seldom, if ever, available so the features of the probability distribution underlying the physical phenomenon under consideration must be estimated. Such estimations form the study object of statistics, which is defined as the branch of mathematical sciences that uses the results of observations and measurements to estimate, in a mathematically well-defined manner, the essential features of probability distributions. Both statistics and probability theory use certain generic terms for defining the objects or phenomena under study. A *system* is the object or phenomena under study. It represents the largest unit being considered. A system can refer to a nuclear reactor, corporation, chemical process, mechanical device, biological mechanism, society, economy, or any other conceivable object that is under study. The output or *response* of a system is a result that can be measured quantitatively or enumerated. The power of a nuclear reactor, the yield of a process, life span of cell, the atmospheric temperature and pressure, are all examples of system responses. A *model* is a mathematical idealization that is used as an approximation to represent the system and its output. Models can be quite simple or highly complex; regardless of its complexity, though, the model is an idealization of the system, so it cannot be exact. Usually, the more complex the system, the less exact the model, particularly since the ability to solve exactly mathematically highly complex expressions diminishes with increasing complexity. In other words, the simpler the model, the easier it is to analyze but the less precise the results.

Probabilities cannot be measured directly; they can be inferred from the results of observations or they can be postulated and (partially) verified through accumulated experience. In practice, though, certain random vectors tend to be more probable, so that most probability functions of practical interest tend to be localized. Therefore, the essential features regarding probability distributions of practical interest are measures of location and of dispersion of the observed results. Practice indicates that location is best described by the mean value, while dispersion of observed results appears to be best described by the variance, which is a second-order moment. In particular, the mean value can be interpreted as a locator of the center of gravity, while the variance is analogous to the moment of inertia (which linearly relates applied torque to induced angular acceleration in mechanics). For multivariate probability distributions, the collection of all second-order moments forms the so-called variance-covariance matrix, or, simply, the covariance matrix. If the probability function is known, then these moments can be calculated directly, through a process called *statistical deduction*. Otherwise, if the probability function is not known, then the respective moments must be estimated from experiments, through a process called *statistical inference*. The definitions, interpretations, and quantifications of the moments of a distribution, particularly the means and covariances, are discussed in Section 1.3. Particularly important is the method of propagation of errors or propagation of moments, which can be used to compute the error in a systems response (which can be either the result of an indirect measurement or the result of a computation), by propagating the uncertainties of the component system parameters using a form Taylor-series expansion of the response as a function of the underlying model parameters.

In practice, users of measured data seldom require knowledge of the complete posterior distribution, but usually request a "recommended value" for the respective quantity, accompanied by "error bars" or some suitably equivalent summary of the posterior distribution. Decision theory can provide such a summary, since it describes the penalty for bad estimates by a loss function. Since the true value is never known in practice, it is not possible to avoid a loss completely, but it is possible to minimize the expected loss, which is what an optimal estimate must accomplish. As will be shown in the first section of Chapter 2, in the practically most important case of "quadratic loss" involving a multivariate posterior distribution, the "recommended value" turns out to be the vector of mean values, while the "error bars" are provided by the corresponding covariance matrix. Thus, Section 2.1 also presents simple examples of estimating covariances and confidence intervals from experimental data.

Practical applications require not only mathematical relations between probabilities, but also rules for assigning numerical values to probabilities. As is well known, Bayesian statistics provides no fundamental rule for assigning the prior probability to a theory. The choice of the "most appropriate" prior distribution lies at the heart of applying Bayes' theorem to practical problems, and has caused considerable debates in the past, lasting over a century. Section 2.2 discusses the assignment of prior probability distributions under incomplete information. Of course, when complete prior information related to the problem under consideration is available and can be expressed in the form of a probability distribution, this information should certainly be used. In such cases, the repeated application of Bayes' theorem will serve to refine the knowledge about the respective problem. At the other extreme, when no specific information is available, it may be possible to construct prior distributions using concepts of group theory to reflect the possible invariance and/or symmetry properties of the problem under consideration, as discussed in Section 2.2.1. On the other hand, if repeatable trials are not feasible, but some information could nevertheless be inferred by some other means, information theory can be used in conjunction with the maximum entropy principle (the modern generalization of Bernoulli's principle of insufficient reason) to assign numerical values to probabilities, thus constructing a prior distribution, as will be shown in Section 2.2.2. The material presented in Section 2.2 is certainly not exhaustive regarding the use of group theory and symmetries for assigning priors (which continues to remain an area of active research), but is limited to presenting only the most commonly encountered priors in practice, and which are also encountered throughout this book.

Section 2.3 presents methods for evaluating unknown parameters from data which is consistent "within error bars," and is afflicted solely by random errors. Three common situations are considered, as follows: (i) evaluation of a location parameter when the scale parameters are known; (ii) both the scale and location parameters are unknown but need to be evaluated; and (iii) evaluation of a counting rate (a scale parameter) in the presence of background (noise). It is probably not too unfair to say that, although measurements without systematic errors are the exception rather than the rule, conventional (frequentist) sampling theory has not much to offer to practitioners in science and technology who are confronted with systematic errors and correlations. This is in marked contrast to the wealth of material on statistical errors, for which satisfactory techniques are available, based on counting (Poisson) statistics or on Gaussian models for the scatter of repeatedly measured data. Using Bayesian parameter estimation under quadratic loss, group-theoretical least informative priors, and probability assignment by entropy maximization, Section 2.4 addresses the practical situation of evaluating means and covariances from measurements affected by both random (uncorrelated) and systematic (correlated) errors issues. It is explained how common errors, the most frequent type of systematic error, invariably induce correlations, and how correlations are described by nondiagonal covariance matrices. As will be mathematically shown in this section, the random errors can be reduced by repeated measurements of the same quantity, but the systematic errors cannot be reduced this way. They remain as a "residual" uncertainty that could be reduced only by additional measurements, using different techniques and instrumentation, geometry, and so on. Eventually, additional measurements using different techniques would reduce the correlated (systematic) error just as repetitions of the same measurement using the same technique reduce the uncorrelated statistical uncertainty.

As discussed in Chapter 1, unrecognized or ill-corrected experimental effects, including background, dead time of the counting electronics, instrumental resolution, sample impurities, and calibration errors usually yield inconsistent experimental data. Although legitimately discrepant data may occur with a nonzero probability (e.g., for a Gaussian distribution, the probability that two equally precise measurements are outside of two standard deviations is about 15.7%), it is much more likely that apparently discrepant experiments actually indicate the presence of unrecognized errors. Section 2.5 illustrates the basic principles for evaluating discrepant data fraught by unrecognized, including systematic (common), errors. The marginal distributions for both recognized and unrecognized errors are obtained for both the Jeffreys least informative prior and for an exponential prior (when the unrecognized errors can be characterized by a known scale factor). This treatment of unrecognized systematic errors is an example of a two-stage "hierarchical" Bayesian method, involving a twofold application of Bayes' theorem to the sampling distribution that depends on parameters having a Gaussian prior, which in turn depended on a so-called "hyper-parameter," which had itself a "hyper-prior" distribution. Also the first-order expressions obtained for the various quantities are similar to the James-Stein estimators, which have sometimes lower risk than the estimates resulting from Bayesian estimation under quadratic loss (that minimize the square error averaged over all possible parameters, for the sample at hand). It is important to note, though, that the two-stage method Bayesian used in Section 2.5 yields results that are superior to the James-Stein estimators, especially for small samples. Moreover, the results presented in this section yield further improvements in a systematic and unambiguous way, without the discontinuities, questions of interpretation and restrictions associated with James-Stein estimators. This fact is particularly valuable and relevant for scientific data evaluation, where best values must often be inferred (for quadratic or any other loss) from just a single available sample.

Chapter 3 presents minimization algorithms, which are best suited for unconstrained and constrained minimization of large-scale systems such as time-dependent variational data assimilation in weather prediction and similar applications in the geophysical sciences. The operational implementation of "four-dimensional variational" data assimilation (customarily called 4–D VAR) hinges crucially upon the fast convergence of efficient gradient-based large-scale unconstrained minimization algorithms which are called to minimize a cost function that attempts to quantify the discrepancies between forecast and observations in a window of assimilation, subject to constraints imposed by the geophysical model. Data assimilation problems in oceanography and meteorology contain many typically of the order of ten million degrees of freedom. Consequently, conjugate-gradient (CG) methods and limited-memory quasi-Newton (LMQN) methods come into consideration since they typically require storage of only a few vectors, containing information from a few iterations, converge to local minima even from remote starting points, and can be efficiently implemented on multiprocessor machines.

Section 3.2 highlights the common as well as distinctive salient features of the minimization algorithms called (acronyms) CONMIN, E04DGF, and of the Limited Memory Broyden-Fletcher-Goldfarb-Shanno (L-BFGS) method, and Bert-Buckley-Variable-Storage-Conjugate-Gradient (BBVSCG) method. These methods all fall under the category of Limited-Memory Quasi-Newton (LMNQ) methods, combining the advantages of CG-algorithms (low storage requirements) with the advantages of quasi-Newton of (QN) methods (computational efficiency stemming from their superlinear convergence). The LMQN algorithms build several rank one or rank two matrix updates to the Hessian matrix, thereby avoiding the need to store the approximate Hessian matrix, as required by full QN methods. Like the CG-methods, the LMQN methods require only modest amount storage for generating the search directions. Currently, the L-BFGS algorithm is the widest used minimization algorithm at the operational numerical weather prediction centers which rely on large-scale 4–D VAR assimilation and prediction methodologies.

Section 3.3 presents Truncated Newton (T-N) methods, which attempt to retain the rapid (quadratic) convergence rate of classic Newton methods, while requiring evaluations of functions and gradients only, thereby reducing the storage and computational requirements to sizes that are feasible for large-scale minimization applications. The T-N methods are also called *Hessian-free methods*. When used together with a finite difference approximation to the Hessian-vector products, the T-N methods achieve a quasi-quadratic convergence rate. In recent implementation, these Hessian-vector products are computed most efficiently using adjoint methods. Thus, the T-N methods require forward and backward adjoint inner iterations within a CG-formalism. Therefore, although the T-N methods offer competitive alternatives for two-dimensional problems, they are not yet competitive for 3-D operational problems. This is because the high cost of the CG inner iterations offsets the advantage of its almost quadratic convergence rate.

Section 3.4 discusses the use of information provided by the Hessian matrix for large-scale optimization, highlighting, in particular, the way in which the eigenvalues of the Hessian matrix determine the convergence rate for unconstrained minimization. Section 3.5 discusses issues related to nonsmooth and nondifferentiable optimization, in view of the fact that precipitation and radiation parameterizations involve on/off processes. Methods of nondifferentiable optimization are needed to minimize nonsmooth functionals. Nonsmooth optimization methods are based on the assumptions that: (i) the functional to be minimized is locally Lipschitz continuous, and (ii) the functional and its arbitrary subgradients can be evaluated at each point. Nonsmooth optimization methods can be divided into two main classes: subgradient methods and bundle methods. The guiding principle underlying bundle methods is to gather the subgradient information from previous iterations into a bundle of subgradients. Although the additional computational cost for building the sub-gradient is several times larger than that for L-BFGS, bundle nonsmooth optimization methods may work advantageously for problems with discontinuities, where L-BFGS methods usually fail. Bundle nonsmooth optimization methods have not been tested yet on operational 4–D VAR systems, but investigations are in progress to assess the applicability of such methods to realistic large-scale problems.

Section 3.6 addresses two fundamental issues related to step-size searches in conjugate-gradient type methods: (i) How good is the search direction?; and (ii) What is the best choice for the length of the step along the search direction? Section 3.7 highlights the salient features of trust-region methods, which also seek global convergence while retaining fast local convergence of optimization algorithms. It is noted that the trust region methods follow a reverse sequence of operations, by first choosing a trial step length, and subsequently using a quadratic model to select the best step length.

Section 3.8 discusses scaling and preconditioning for linear and nonlinear problems. The goal of preconditioning is to improve the performance of conjugate gradient–type minimization methods, by reducing the number of iterations required to achieve a prescribed accuracy. Scaling can substantially improve the performance of minimization algorithms. An effective automatic scaling could also improve the condition number of the Hessian matrix for well-scaled problems, thus facilitating their solution. Scaling by variable transformations converts the variables from units that reflect the physical nature of the problem to units that display desirable properties for improving the efficiency of the minimization algorithms. On the other hand, badly scaled nonlinear problems can become extremely difficult to solve.

Several popular methods for performing nonlinear constrained optimization are discussed in Section 3.9, namely: (i) the penalty method; (ii) barrier methods; (iii) augmented Lagrangian methods; and (iv) sequential quadratic programming (SQP) methods. The penalty method replaces a constrained optimization problem by a series of unconstrained problems whose solutions should converge to the solution of the original constrained problem. The unconstrained problems minimize an objective function which is constructed by adding to the original objective function a term that comprises a penalty parameter multiplying a measure of the violation of the constraints. The measure of violation is nonzero when the constraints are not satisfied, and is zero in the region where the constraints are satisfied. The original problem can thus be solved by formulating a sequence of unconstrained subproblems. Barrier methods are an alternative class of algorithms for constrained optimization. These methods also use a penalty-like term added to the objective function, but the results of iterations within the barrier methods are forced by the barrier to remain interior to and away from the boundary of the feasible solution domain. Augmented Lagrangian methods turn a constrained minimization problem into the unconstrained minimization The SQP solves a sequence of sub-problems designed to minimize a quadratic model of the objective functional subject to linearization of the constraints. The SQP method can be used within either a linesearch or a trust region framework, and is very efficient for solving both small and large problems. To be practical, a SQP method must be able to converge on nonconvex problems, starting from remote points.

Nonlinear optimization may involve cost functions characterized by the presence of multiple minima. The aim of global optimization is to determine all of the critical points of a function, particularly if several local optima exist where the corresponding function values differ substantially from one another. Global optimization methods can be classified into two major categories, namely deterministic and stochastic methods. Deterministic methods attempt to compute all of the critical points with probability one (i.e., with absolute success). On the other hand, stochastic methods sacrifice the possibility of an absolute guarantee of success, attempting to minimize the function under consideration in a random sample of points from a set, which is assumed to be convex, compact, and to contain the global minimum as an interior point. Section 3.10 briefly discusses two stochastic global minimization methods: simulated annealing and genetic algorithms, which have recently been implemented in variational data assimilation in geophysical sciences applications. The simulated annealing algorithm exploits the analogy between the search for a global minimum and the annealing process (i.e., the way in which a metal cools and freezes) into a minimum energy crystalline structure. The genetic algorithms attempt to simulate the phenomenon of natural evolution, as each species searches for beneficial adaptations in an ever-changing environment. As species evolve, new attributes are encoded in the chromosomes of individual members. This information changes by random mutation, but the actual driving force behind evolutionary development is the combination and exchange of chromosomal material during breeding.

Simulated annealing algorithms are intrinsically sequential. On the other hand, genetic algorithms are particularly well suited for implementation on parallel computers. Evaluation of the objective function and constraints can be done simultaneously for the entire population; the production of the new population by mutation and crossover can also be parallelized. On highly parallel machines, therefore, a genetical algorithm (GA) can be expected to run nearly N times faster than on non-parallel machines, where N is the population size. Currently, however, the convergence rate of these global minimization methods does not outperform the convergence rate of LMQN methods for large-scale operational 4–D VAR models.

Chapter 4 discusses several basic principles of four-dimensional variational assimilation (4–D VAR). Initially, data assimilation methods were referred to as "objective analyses," in contradistinction to "subjective analyses," in which numerical weather predictions (NWP) forecasts were adjusted "by hand" by meteorologists, using their professional expertise. Subsequently, methods called "nudging" were introduced based on the simple idea of Newtonian relaxation. In nudging, the rightside of the model's dynamical equations is augmented with a term which is proportional to the difference between the calculated meteorological variable and the observation value. This term keeps the calculated state vector closer to the observations. Nudging can be interpreted as a simplified Kalman-Bucy filter with the gain matrix being prescribed rather than obtained from covariances. The nudging method is used in simple operational global-scale and meso-scale models for assimilating small-scale observations when lacking statistical data. The recent advances in nudging methods are briefly presented in Section 4.1.

Section 4.2 briefly mentions the "optimal interpolation" (OI) method, "three-dimensional variational data assimilation" (3–D VAR), and the physical space statistical analysis (PSAS) methods. These methods were introduced independently, but were shown to be formally equivalent; in particular, PSAS is a dual formulation of 3–D VAR.

Data assimilation requires the explicit specification of the error statistics for model forecast and the current observations, which are the primary quantities needed for producing an analysis. A correct specification of observation and background error covariances are essential for ensuring the quality of the analysis, because these covariances determine to what extent background fields will be corrected to match the observations. The essential parameters are the variances, but the correlations are also very important because they specify the manner in which the observed information will be smoothed in the model space if the resolution of the model does not match the density of the observations. Section 4.3 briefly outlines the prevailing operational practices employed for the practical estimation of observation error covariance matrices and background error covariance matrices.

The goal of the 4–D VAR formalism is to find the solution of a numerical forecast or numerical weather prediction (NWP) model that best fits sequences of observational fields distributed in space over a finite time interval. Section 4.4 discusses the basic framework of "four-dimensional variational" data assimilation (4–D VAR) methods utilizing optimal control theory (variational approach). The advance brought by the variational approaches is that the meteorological fields satisfy the dynamical equations of the forecast model while simultaneously minimizing a cost functional, which measures the differences between the computed and the observed fields, by solving a constrained minimization problem. The 4–D VAR formalism is first presented without taking the modeling errors into account; subsequently, the functional to be minimized is extended to include model errors. This section concludes with a discussion of the consistent optimality and transferable optimality properties of the 4–D VAR procedure.

Section 4.5 presents results of numerical experiments with unconstrained minimization methods for 4–D VAR using the shallow water equations, which are widely used in meteorology and oceanography for testing new algorithms since they contain most of the physical degrees of freedom (including gravity waves) present in the more sophisticated operational models. The numerical experiments were performed with four limited-memory quasi-Newton (LMQN) methods (CONMIN-CG, E04DGF, L-BFGS, and BBVSCG) and two truncated Newton (T-N) methods. The CONMIN-CG and BBVSCG algorithms failed after the first iteration, even when both gradient scaling and non-dimensional scaling were applied. The L-BFGS algorithm was successful only with gradient scaling. On the other hand, the E04DGF algorithm worked only with the non-dimensional shallow water equations model. This indicates that using additional scaling is essential for the success of LMQN minimization algorithms when applied to large-scale minimization problems. On the other hand, T-N methods appear to perform best for large-scale minimization problems, especially in conjunction with a suitable preconditioner. The importance of preconditioning increases with increasing dimensionality of the minimization problem under consideration. Furthermore, for the Shallow-Water Equations (SWE) numerical experiments, the T-N methods required far fewer iterations and function calls than the LMQN methods.

In the so-called strong constraint Variational Data Assimilation (VDA), or classical VDA, it is assumed that the forecast model perfectly represents the evolution of the actual atmosphere. The best fit model trajectory is obtained by adjusting only the initial conditions via the minimization of a cost functional that is subject to the model equations as strong constrains. However, numerical weather prediction (NWP) models are imperfect since subgrid processes are not included. Furthermore numerical discretizations produce additional dissipative and dispersion errors. Modeling errors also arise from the incomplete mathematical modeling of the boundary conditions and forcing terms, and from the simplified representation of physical processes and their interactions in the atmosphere. Usually, all of these modeling imperfections are collectively called *model error* (ME). Model error is formally introduced as a correction to the time derivatives of model variables. Section 4.6 highlights the treatment of MEs in VDA, taking into account numerical errors explicitly as additional terms in the cost functional to be minimized. However, taking MEs into account doubles (and can even triple) the size of the system to be optimized by comparison to minimizing the cost functional when the model errors are neglected.

Chapter 5 highlights specific difficulties in applying 4–D VAR to large-scale operational numerical weather prediction models. Recall that the objective of 4–D VAR is to find the optimal set of control variables, usually the initial conditions and/or boundary conditions, such that a cost function, comprising a weighted least square norm that quantifies the misfit between model forecast and observations, is minimized subject to the constraints of satisfying the geophysical model. In order to minimize this cost function, we need to know the gradient of this function with respect to the control variables. A straightforward way of computing this gradient is to perturb each control variable in turn and estimate the change in the cost function. But this method is impractical when the number of control variables is large a typical meteorological model comprises $O(10^7)$ control variables. Furthermore, the iterative minimization of the cost function requires several gradient estimations on the way to finding a local minimum. Often, these gradient estimations are insufficiently accurate to guarantee convergence of the minimization process. As discussed in Section 5.1, the convergence of the minimization process in 4–D VAR is particularly affected (negatively) by strong nonlinearities and on/off physical processes such as precipitation and clouds.

Section 5.2 highlights the highly efficient adjoint method for computing exactly the gradient of the cost function with respect to the control variables by integrating once the adjoint model backwards in time. Such a backward integration is of similar complexity to a single integration of the forward model. Another key advantage of adjoint variational data assimilation is the possibility to minimize the cost function using standard unconstrained minimization algorithms (usually iterative descent methods). However, for precipitation observations, highly nonlinear parameterization schemes must be linearized for developing the adjoint version of the model required by the minimization procedure for the cost function. Alternatively, simpler physics schemes, which are not a direct linearization of the full model physics (i.e., the "linear model" is not tangent linear to the nonlinear full model), can be coded for the linear model. Of course, it is desirable to have a linearized model that approximates as closely as possible the sensitivity of the full nonlinear model; otherwise the forecast model may not be in balance with its own analysis, producing so-called "model spin-up." Furthermore, multi-incremental approaches can exhibit discrete transitions affecting the stability of the overall minimization process.

On the one hand, nonlinear models have steadily evolved in complexity in order to improve forecast skill. For example, the prognostic cloud scheme introduced into the European Centre for Medium Range Weather Forecasts (ECMWF) model includes many highly nonlinear processes that are often controlled by threshold switches. On the other hand, even if it were possible to construct the tangent linear and, respectively, adjoint models corresponding to this complex cloud scheme, the validity of these models would be restricted due to these thresholds and their value would be questionable. Issues related to using non-smooth optimization methods to address discontinuities is an ongoing research topic. Standard solvers of elliptic equation often perform "fast Fourier transform" (FFT) and inverse FFT operations. Section 5.3 summarizes the adjoint coding of the FFT and of the inverse FFT, showing that the adjoint of the FFT is obtained by calling the inverse FFT routine and multiplying the output by the number of data sample points. Conversely, the adjoint of the inverse FFT is obtained by calling the FFT routine and dividing the output by the number of data sample points.

Section 5.4 indicates that the correctness of the adjoint code for interpolations and on/off processes can be verified by performing an additional integration of the nonlinear model with added bit vectors in order to determine the routes for the IF statements included in the physical processes. This additional integration of the nonlinear model with added bit vectors is needed for the verification of both the tangent linear model and the adjoint model. Section 5.5 discusses the construction of background covariance matrices, which play the following important roles: (i) spreading the information from the observations to neighboring domains; (ii) providing statistically consistent increments at the neighboring grid points and levels of the model; and (iii) ensuring that observations of one model variable (e.g., temperature) produce dynamically consistent increments in the other model variables (e.g., vorticity and divergence).

Section 5.6 revisits the characterization of model errors specifically for the 4–D VAR data assimilation procedure. Such errors are attributable to the dynamical model (e.g., poor representation of processes, omissions, or incorrect formulations of key processes, numerical approximations) and observations or measurements (e.g., sensor design, performance, noise, sample averaging, aliasing). For dynamically evolving systems, the model errors are expected to depend on time and, possibly, on the model state variables. Controlling modeling errors in addition to the model's initial conditions in the weak constraint 4–D VAR doubles the size of the optimization problem by comparison to the strong constraint 4–D VAR. Furthermore, if the stochastic component is included in the model error formulation, then the random realization would need to be saved at each model time step. Consequently, the size of the optimization problem would be tripled. The size of the model error control vector can be reduced by projecting it onto the subspace of eigenvectors corresponding to the leading eigenvalues of the adjoint-tangent linear operators.

Most data assimilation systems are not equipped to handle large, systematic corrections; they were designed to make small adjustments to the background fields that are consistent with the presumed multivariate and spatial structures of random errors. Statistics of "observed-minus-background" residuals provide a different, sometimes more informative, view on systematic errors afflicting the model or observations. Operational NWP centers routinely monitor time- and space-averaged background residuals associated with different components of the observing system, providing information on the quality of the input data as well as on the performance of the assimilation system. In general, small root-mean-square residuals imply that the system is able to accurately predict future observations. Nonzero mean residuals, however, indicate the presence of biases in the observations and/or their model-predicted equivalents. It is paramount to develop physically meaningful representations of model errors that can be clearly distinguished from possible observation errors. This issue is the subject of intensive ongoing research.

Section 5.7 discusses the incremental 4–D VAR algorithm, which was formulated in the mid-1990s, and decisively facilitated the adoption, application, and implementation of 4–D VAR data assimilation at major operational centers, thereby advancing the timely state of weather prediction. Prior to the development of the "incremental 4–D VAR algorithm," implementation of the full 4–D VAR algorithm in operational models was impractical, since a typical minimization requires between 10 and 100 evaluations of the gradient. The cost of the adjoint model is typically 3 times that of the forward model, and the analysis window in a typical operational model such as the ECMWF system is 12-hours. Thus, the cost of a 12 hour analysis was roughly equivalent to between 20 and 200 days of model integration (with 108 variables), making it computationally prohibitive for NWP centers which had to deliver timely forecasts to the public. In addition, the nonlinearity of the model and/or of the observation operator could produce multiple minima in the cost function, which impacted the convergence of the minimization algorithm.

The incremental 4–D VAR algorithm reduces the resolution of the model and eliminates most of the time-consuming physical packages, thereby enabling the 4–D VAR method to become computationally feasible. Furthermore, the incremental 4–D VAR algorithm removes the nonlinearities in the cost minimization by using a forward integration of the linear model instead of a nonlinear one. The minimization procedure is identical to the usual 4–D VAR algorithm except that the increment trajectory is obtained by integration of the linear model. The reference trajectory (which is needed for integrating the linear and adjoint models and which starts from the background integration) is not updated at every iteration. This simplified iterative procedure for minimizing the incremental cost function is called the *inner loop*, and is much cheaper computationally to implement by comparison to the full 4–D VAR algorithm. However, when the quadratic cost function is approximated in this way, the incremental 4–D VAR algorithm no longer converges to the solution of the original problem. Furthermore, the analysis increments are calculated at reduced resolution and must be interpolated to conform to the high-resolution model's grid. Consequently, after performing a user-defined number of inner loops, one outer loop is performed to update the high-resolution reference trajectory and the observation departures. After each outer loop update, it is possible to use progressively higher resolutions for the inner loop.

However, experiments show that the current implementations of the incremental 4–D VAR algorithm lead to divergent computational results after four outer loop iterations (e.g., this was the case when the incremental 4–D VAR was initially implemented into the ECMWF weather prediction system). Various numerical experiments indicate that convergence can be attained when the inner and outer loops use the same resolution and/or use the same time step. This feature is explained by the presence of gravity waves which propagate at different speeds in the linear and nonlinear models. These gravity waves are related to the shape of the leading eigenvector of the Hessian of the 4–D VAR cost function; this eigenvector is determined by the surface pressure observation and controls the convergence of the minimization algorithm. Chapter 5 concludes with a short discussion, in Section 5.8, of current research issues.

This book also comprises three appendices. Appendix A (Chapter 6) is intended to provide a quick reference to selected properties of distributions commonly used for data analysis, evaluation, and assimilation. Appendix B (Chapter 7) introduces and summarizes the most important properties of adjoint operators in conjunction with differential calculus in vector spaces, as used for data assimilation. Appendix C (Chapter 8) highlights the main issues arising when identifying and estimating model parameters from experimental data. The problem of parameter identification can be formulated mathematically as follows: "an unknown parameter is *identifiable* if it can be determined uniquely at all points of its domain by using the input-output relation of the system and the input-output data." Such a mapping is generally known as an "inverse problem," in contradistinction with the forward mapping, which maps the space of parameters to the space of outputs. The uniqueness of inverse mappings is difficult to establish. Appendix C also discusses briefly three methods for estimating parameters: the maximum likelihood method, the maximum total variation L_1-regularization method for estimating parameters with discontinuities, and the "extended Kalman filter" method.

1

Experimental Data Evaluation: Basic Concepts

CONTENTS

Experience shows that it is practically impossible to measure exactly the true value of a physical quantity. This is because various imperfections occur at various stages involved in a measurement, including uncontrollable experimental errors, inaccurate standards, and other uncertainties arising in the data measurement and interpretation (reduction) process. Around any reported experimental value, therefore, there always exists a certain range of similar, more or less plausible, values that may also be true. In turn, this means that all inferences, predictions, engineering computations, and other applications of measured data are necessarily founded on weighted averages over all the possibly true values, with weights indicating the degree of plausibility of each value. These weights and weighted averages are what we call *probabilities* and *expectation values*. Consequently, the evaluation of scientific data is intrinsically intertwined with probability theory. The basic types of errors and probability distributions usually associated with them will be presented in Section 1.1.

The interpretation of probabilities as degrees of plausibility or rational expectation, on a numerical scale ranging from 0 (impossibility) to 1 (certainty), dates back at least to Bernoulli [16] and Laplace [111]. In scientific

1

data evaluation, probabilities encode incomplete information. Persons possessing different information or knowledge assign different probabilities; furthermore, these probabilities are updated whenever new relevant information becomes available. It follows that probabilities cannot be considered as measurable physical properties. They are subjective in the sense that they depend on a person's knowledge. However, this does not mean that probabilities are arbitrary. They must obey the rules of logic, which demand, for instance, that rational persons with the same knowledge assign the same probabilities. The elementary conditions for logical consistency imply two fundamental rules from which all other mathematical relationships between probabilities can be derived. These two fundamental rules are the sum and the product rules, respectively (see, e.g., refs. [39] and [202]). A powerful logical tool that follows immediately from the two forms of the product rule is Bayes' theorem (1763). In data analysis and evaluation, Bayes' theorem [10] shows how, under given experimental conditions C, the "direct" probabilities of experimental errors (i.e., likelihoods of possible observed data B given the true values A) are related to the "inverse" probabilities of possible true values A given the observed data B. The customary way to express Bayes' theorem in words is: *"posterior* $\propto$ *likelihood* $\times$ *prior"* where the "prior" distribution summarizes our knowledge extant prior to observing the (new) data, the "likelihood" distribution conveys the impact of the new information brought by the (new) data, while the "posterior" distribution contains the full information available for further inference, prediction, and decision making. Thus, Bayes' theorem is a formal model for updating, or learning from observations. Section 1.2 presents the basic concepts of probability theory involved in the evaluation of uncertainty-afflicted scientific data.

Since probabilities cannot be measured directly, they are either inferred from the results of observations or they are postulated and (partially) verified through accumulated experience. In practice, though, certain random vectors tend to be more probable, so that most probability functions of practical interest tend to be localized. Therefore, the essential features regarding probability distributions of practical interest are measures of *location* and of *dispersion*. These measures are provided by the *expectation* and *moments* of the respective probability function. If the probability function is known, then these moments can be calculated directly, through a process called *statisti-*

cal deduction. Otherwise, if the probability function is not known, then the respective moments must be estimated from experiments, through a process called *statistical inference.* Since measurements do not yield true values, it is necessary to introduce surrogate parameters to describe the "location" and "dispersion" for the observed results. Practice indicates that *location is best described by the mean value,* while *dispersion* of observed results appears to be *best described* by the *variance,* which is a second-order moment. In particular, the mean value can be interpreted as a locator of the center of gravity, while the variance is analogous to the moment of inertia (which linearly relates applied torque to induced angular acceleration in mechanics). For multivariate probability distributions involving n random variables, the collection of all second-order moments forms an $n \times n$ matrix, called the *variance-covariance matrix,* or, simply, the covariance matrix. Section 1.3 highlights the important properties of, and methods for evaluating covariances.

1.1 Experimental Data Uncertainties

A "measurable" or "physical" quantity is a property of phenomena, bodies, or substances that can be defined qualitatively and can be expressed quantitatively. *Measurement* is the process of experimentally finding the value of a physical quantity, with the help of devices called *measuring instruments.* A measurement has three features:

1. The result of a measurement must always be a number expressed in sanctioned units of measurements. The purpose of a measurement is to represent a property of an object by a number.

2. A measurement is always performed with the help of a measuring instrument; measurement is impossible without measuring instruments.

3. A measurement is always an experimental procedure.

If it were known, the true value of a measurable quantity would ideally reflect, both qualitatively and quantitatively, the corresponding property of the object. The theory of measurement relies on the following postulates:

1. The true value of the measurable quantity exists;

2. The true value of the measurable quantity is constant relative to the conditions of the measurement; and

3. The true value cannot be found.

Since measuring instruments are imperfect, and since every measurement is an experimental procedure, the results of measurements cannot be absolutely accurate. This unavoidable imperfection of measurements is quantitatively characterized by *measurement uncertainty* or *measurement error*, which can be expressed in absolute or relative form. Consequently, repeated measurements of the same physical quantity can never yield identical results. An indication of the probability of measurements to bracket the true value can be inferred by considering the "Principle of Insufficient Reason," already used by Bernoulli [16] and Laplace [111]. For two measurements, for example, this principle would assign equal probabilities to the two alternatives, namely, "measured value too low" and "measured value too high," unless there is a reason to consider one or the other as more likely. Hence, if the two measurements were independent, the four alternatives "low-low," "low-high," "high-low," and "high-high" would be equally probable, so the two measured values would bracket the true value with a probability of 1/2. Even after a third measurement, the probability would still be (as high as) 1/4 that the true value is below or above all of the measured values.

Thus, since even the most carefully measured scientific data in data banks will inevitably differ from the true values of the measured quantities, it is apparent that nominal values for data, by themselves, are insufficient for applications; the quantitative uncertainties are also needed, along with the respective nominal values. Therefore, in response to user requests (e.g., from nuclear science and technology, fission and fusion reactor development, astrophysics, radiotherapy), nuclear data evaluators, for example, are now also including "covariance files" comprising information about uncertainties and correlations when updating the extensive computer libraries of recommended nuclear data. Since the use of uncertain data may necessitate costly safety margins (in medicine, weather and climate prediction, or in the chemical, automotive, aerospace, or nuclear industries), working groups of the International Standards Organization have been developing uniform rules for reporting data uncertainties.

Combination of data from different sources involves a weighted propagation (via sensitivities, as will be seen subsequently) of all input uncertainties to uncertainties in the output values finally recommended in data banks. Hence, data evaluation is intrinsically intertwined with uncertainty analysis, requiring reasoning from incomplete information, using probability theory for extracting "best" values together with "best" uncertainties from often sparse, incomplete, error-afflicted, and occasionally discrepant experimental data. A wide range of probability-theory concepts and tools are employed in data evaluation and combination, from deductive statistics involving mainly frequencies and sample tallies to inductive inference for assimilating non-frequency data and *a priori* knowledge.

In practice, grossly erroneous procedures and unintended mistakes (e.g., overlooking or miscalculating important corrections, equipment failure or improper calibration, bugs in computer codes, etc.) can produce defective data. In this book, such defective data will *not* be treated as "uncertainties." Nevertheless, data points that exhibit atypical behavior, which cannot be explained, need to be carefully scrutinized since outright rejection may not necessarily be appropriate: for example, a differential cross section point on an excitation curve might appear to be anomalous, but could be nevertheless correct due to the effect of a previously unidentified resonance. Such data points need to be well understood before subjecting the data to extensive statistical analysis for quantifying the associated uncertainties.

The terms "error" and "uncertainty" are interpreted in this book as being equivalent to the standard deviation of the probability distribution associated with the measurement process. This interpretation is consistent with the usual convention of considering "error" or "uncertainty" as an inherently positive number that quantifies the measurement dispersion of a specific observable parameter. Legitimate errors are categorized either as *random errors* or as *systematic errors*. If the results of separate measurements of the same quantity differ from one another, and the respective differences cannot be predicted individually, then the error owing to this scatter of the results is called *random error*. Random errors can be identified by repeatedly measuring the same quantity under the same conditions. The scatter in results cannot be always tested in practice, particularly in large-scale modern experiments, where it may be impractical to provide sufficient repetition in order to satisfy the explicit needs for quantifying the random errors based on strict statistical requirements. Nevertheless, reasonable estimates of random errors can often be made, particularly when the nature of the underlying probability distribution can be inferred from previous experience. Furthermore, due to the influence of the central limit theorem, many sources of random error tend to be normally distributed. A significant feature of random errors is that repeated measurements (under fixed conditions) not only permit these errors to be better determined, but they also lead to error reduction, as assured by the law of large numbers. This feature is particularly important when high precision (i.e., small random errors) is required. A subtle issue regarding random errors stems from the fact that such errors may contain correlated components: whether an error component is correlated or not within a particular data set depends upon the role that the associated random variable plays in the respective physical problem.

In contradistinction to a random error, a *systematic error* is defined as a measurement error that remains constant or changes in a regular fashion when the measurements of that quantity are repeated. Such errors arise because of inherent flaws in the investigative process itself, and they lead to bias. Although systematic errors are difficult to distinguish from blunders, particularly when the impact of a blunder is small, the most reliable way to uncover systematic errors is by using a more accurate measuring instrument and/or by comparing a given result with a measurement of the same quantity, but performed by a different method. Each distinct approach leads to results that differ somewhat from those obtained in other ways. These differences exhibit a pattern (i.e., are "systematic") no matter how many times each approach is repeated, because the inherent systematic deficiencies of each method cannot be avoided by mere repetition. When the errors are truly systematic, statistical regularity will emerge from the ensemble of all measurements. Such a statistical regularity will not emerge when the data sets are afflicted with blunders, since blunders are generally one-time occurrences that can be detected if a particular procedure is repeated. Consequently, redundancy within a given investigative procedure is desirable not only to improve precision, but also to purge the results of blunders.

Systematic errors can stem from many sources; typical systematic errors in the nuclear field, for example, can arise from cross sections used for neutron fluence determination, sample material standards, detector calibrations, shortcomings in deriving corrections (e.g., neutron multiple scattering), and nuclear decay properties. Once the sources of systematic error have been identified, it is necessary to estimate their respective magnitudes, corresponding to a consistent level of confidence. This is a very difficult task, since the applicable probability distribution laws are often unknown, and only an estimate of the ranges of possibilities for the variables in question may be available. The issue of confidence is important because the various error components must ultimately be combined to generate covariance matrices, and if the specific errors conform to widely different confidence levels, their combination may lead to misleading results.

Ideally, systematic errors should be estimated by theoretical analysis of the measurement conditions, based on the known properties of the measuring instruments and the quantity being measured. Although the systematic errors are reduced by introducing appropriate corrections, it is impossible to eliminate them completely. Ultimately, a residual error will always remain, and this residual error will then constitute the new systematic component of the measurement error. Good accuracy (i.e., small systematic error) can be obtained only by successfully reducing the magnitude of significant systematic errors. Important mathematical means for dealing with random and systematic errors will be presented in Chapter 2.

The quality of measurements that reflects the closeness of the results of measurements of the same quantity performed under the same conditions is called the *repeatability* of measurements. Good repeatability indicates that the random errors are small. On the other hand, the quality of measurements that reflects the closeness of the results of measurements of the same quantity performed under different conditions (e.g., in different laboratories, at different locations, and/or using different equipment) is called the *reproducibility* of measurements. Good reproducibility indicates that both the random and systematic errors are small.

Since the true value of the measured quantity is unknown, a measurement error cannot be found by using its definition as an algorithm. The measurement error must be evaluated by identifying its underlying sources and reasons, and by performing computations based on the estimates of all components of the respective measurement inaccuracy. The smallest of the measurement errors are customarily referred to as *elementary errors* (of a measurement), and are defined as those components of the overall measurement error that are associated with a single source of inaccuracy for the respective measurement. In turn, the total measurement error is computed by using the estimates of the component elementary errors. Even though it is sometimes possible to correct, partially, certain elementary errors (e.g., systematic ones), no amount or combination of corrections can produce an accurate measurement result; there always remains a residual error. In particular, the corrections themselves cannot be absolutely accurate, and, even after they are implemented, there remain residuals of the corresponding errors which cannot be eliminated and which later assume the role of elementary errors.

Since a measurement error can only be calculated indirectly, based on models and experimental data, it is important to identify and classify the underlying elementary errors. This identification and classification is subsequently used to develop mathematical models for the respective elementary errors. Finally, the resulting *overall measurement error* is obtained by synthesizing the mathematical models of the underlying elementary errors.

In the course of developing mathematical models for elementary errors, it has become customary to distinguish four types of elementary errors, namely absolutely constant errors, conditionally constant errors, purely random errors, and quasi-random errors. Thus, *absolutely constant errors* are defined as elementary errors that remain the same (i.e., are constant) in repeated measurements performed under the same conditions, for all measuring instruments of the same type. For example, an absolutely constant error arises from inaccuracies in the formula used to determine the quantity being measured, once the limits of the respective inaccuracies have been established. Typical situations of this kind arise in indirect measurements of quantities determined by linearized or truncated simplifications of nonlinear formulas (e.g., analog/digital instruments where the effects of electro-motive forces are linearized). Based on their properties, absolutely constant elementary errors are purely systematic errors, since each such error has a constant value in every measurement, but this constant is nevertheless unknown. Only the limits of these errors are known. Therefore, absolutely constant errors are often modeled mathematically by a determinate (as opposed to random) quantity whose magnitude lies within an interval of known limits.

Conditionally constant errors are, by definition, elementary errors that have definite limits (just like the absolutely constant errors) but (as opposed to the absolutely constant errors) such errors can vary within their limits due both to the non-repeatability and the non-reproducibility of the results. A typical example of such an error is the measurement error due to the intrinsic error of the measuring instrument, which can vary randomly between fixed limits. Usually, the conditionally constant error is mathematically modeled by a random quantity with a uniform probability distribution within prescribed limits. This mathematical model is chosen because the uniform distribution has the highest uncertainty (in the sense of information theory) among distributions with fixed limits. Note, in this regard, that the round-off error also has known limits, and this error has traditionally been regarded in mathematics as a random quantity with a uniform probability distribution.

Purely random errors appear in measurements due to noise or other random phenomena produced by the measuring device. In principle, the form of the distribution function for random errors can be found using data from each multiple measurement. In practice, however, the number of measurements performed in each experiment is insufficient for determining the actual form of the distribution function. Therefore, a purely random error is usually modeled mathematically by using a normal distribution characterized by a standard deviation that is computed from the experimental data.

Quasi-random errors occur when measuring a quantity defined as the average of nonrandom quantities that differ from one another such that their aggregate behavior can be regarded as a collection of random quantities. In contrast to the case of purely random errors, though, the parameters of the probability distribution for quasi-random errors cannot be unequivocally determined from experimental data. Therefore, a quasi-random error is modeled by a probability distribution with parameters (e.g., standard deviation) determined by expert opinion.

The term *uncertainty* is customarily used to express the inaccuracy of measurement results when the numerical value of the respective inaccuracy is accompanied by a corresponding confidence probability. In this respect, we also note that the second edition of the International Vocabulary of Basic and General Terms in Metrology (2nd edition, ISO 1993) defines the term *uncertainty* as "*an interval having a stated level of confidence.*"

The term *error* is customarily used for all components of uncertainty, while the term *limits of error* is used in cases in which the measurement inaccuracy is caused by the intrinsic error of the measuring instrument, and when a corresponding level of confidence cannot be stated. Measurements must be reproducible, since otherwise they lose their objective character and become meaningless. The *limits of measurement error or uncertainty* estimated by the experimenter provide a measure of the *nonreproducibility of a measurement permitted by the experimentalist*. The validity of the uncertainty computed for every measurement is based on the validity of the estimates of errors underlying the respective measurement. A correctly estimated measurement uncertainty permits comparisons of the respective result with results obtained by other experimenters.

When Q_m is the result of a measurement and Δ_U and Δ_L are the upper and lower limits of the error in the measurement, then *the result of a measurement and the respective measurement error* can be written in the form

$$(Q_m, \Delta_U, \Delta_L) , \quad \text{or} \quad (Q_m \pm \Delta) \tag{1.1}$$

when $|\Delta_L| = |\Delta_U| = \Delta$. When the inaccuracy of a measurement is expressed as uncertainty, then the corresponding confidence probability must also be given, usually in parentheses following the value of the uncertainty. For example, if a temperature measurement yields the value 316.24 K and the uncertainty for this result, say $\pm 0.2\ K$, was calculated for a confidence probability of 0.95, then this result should be written in the form

$$T(0.95) = (316.2 \pm 0.2)K \tag{1.2}$$

If the confidence probability is not indicated in the measurement result, e.g., if the result is written as

$$T = (316.2 \pm 0.2)K \qquad (1.3)$$

then the inaccuracy is assumed to have been estimated without the use of probability methods. Although an error estimate obtained without the use of probability methods can be very reliable, it cannot be associated with a probability of one or some other value. In other words, when a probabilistic model was not employed to obtain the error estimate, the respective probability cannot be estimated and, therefore, should not be indicated.

In many cases, it is of interest to know not only the limiting values of the total measurement error but also the characteristics of the random and systematic error components separately, in order to analyze discrepancies between results of measurements of the same quantity performed under different conditions. Knowing the error components separately is particularly important when the result of a measurement is to be used for computations together with other data that are not absolutely precise. Furthermore, the main sources of errors should also be described together with estimates of their contributions to the total measurement uncertainty. For a random error, for example, it is of interest to indicate the form and parameters of the distribution function underlying the observations, the method employed for testing the hypothesis regarding the form of the distribution function, the significance level used in the testing, and so on.

The number of significant figures retained in the number expressing the result of a measurement must correspond to the accuracy of the measurement. This means that the uncertainty of a measurement can be equal to one or two units (and should not exceed 5 units) in the last figure of the number expressing the result of the measurement. Since measurement uncertainty determines only the vagueness of the results, the uncertainty is customarily expressed in its final form by a number with one or two significant figures. Two figures are retained for the most precise measurements. At least two additional significant digits should be retained during intermediate computations, in order to keep the round-off error below the value of the final error. The numerical value of the result of a measurement must be represented such that the last decimal digit is of the same rank as its uncertainty. Including a larger number of digits will not reduce the uncertainty of the result; however, using a smaller number of digits (by further rounding off the number) would increase the uncertainty and would make the result less accurate, thereby offsetting the care and effort invested in the measurement. The rules for rounding off and for recording the results of measurements have been established by convention, and are listed below:

1. If the decimal fraction in the numerical value of the result of a measurement terminates in 0's, then the 0's are dropped only up to the digit that corresponds to the rank of the numerical value of the error. If the digit being discarded is equal to 5 and the digits to its right are unknown or are equal to 0, then the last retained digit is not changed if it is even and it is increased by 1 if it is odd. (Example: if three significant digits are retained, the number 100.5 is rounded off to 100.0 and the number 101.5 is rounded off to 102.0).

2. The last digit retained is not changed if the adjacent digit being discarded is less than 5. Extra digits in integers are replaced by 0's, while extra digits in decimal fractions are dropped. (Example: the numerical value of the result of a measurement 1.5333 with an error in the limits ± 0.04 should be rounded off to 1.53; however, if the error limits are ± 0.001, the same number should be rounded off to 1.533.)

3. The last digit retained is increased by 1 if the adjacent digit being discarded is greater than 5, or if it is 5 and there are digits other than 0 to its right. (Example: if three significant digits are retained, the number 2.6351 is rounded off to 2.64.)

1.2   Uncertainties and Probabilities

1.2.1   Axiomatic, Frequency, and Subjective Probability

Probability theory is a branch of mathematical sciences that provides a model for describing the process of observation. The need for probability theory arises from the fact that most observations of natural phenomena do not lead to uniquely predictable results. Probability theory provides the tools for dealing with actual variations in the outcome of realistic observations and measurements. The challenging pursuit to develop a probability theory which is mathematically rigorous and also describes many phenomena observable in nature has generated over the years notable disputes over conceptual and logical issues. Modern probability theory is based on postulates constructed from three axioms attributed to Kolmogorov [108], all of which are consistent with the notion of frequency of occurrence of events. The alternative approach, traceable to Laplace [111], is based on the concept that probability is simply a way of providing a numerical scale to quantify our reasonable beliefs about a situation which we know only incompletely. This approach is consistent with Bayes' theorem, conditional probabilities, and inductive reasoning. Either approach to probability theory would completely describe a natural phenomenon if sufficient information were available to determine the underlying probability distribution exactly. In practice, though, such exact knowledge is seldom, if ever, available so the features of the probability distribution underlying the physical phenomenon under consideration must be *estimated*. Such estimations form the study object of *statistics*, which is defined as the branch of mathematical sciences that uses the results of observations and measurements to estimate, in a mathematically well-defined manner, the essential features of probability distributions.

Both statistics and probability theory use certain generic terms for defining the objects or phenomena under study. A *system* is the object or phenomenon under study. It represents the largest unit being considered. A system can refer to a nuclear reactor, corporation, chemical process, mechanical device, biological mechanism, society, economy, or any other conceivable object that is under study. The *output* or *response* of a system is a result that can be measured quantitatively or enumerated. The power of a nuclear reactor, the yield of a process, life span of cell, the atmospheric temperature and pressure, are all examples of system outputs. A *model* is a mathematical idealization that is used as an approximation to represent the system and its output. Models can be quite simple or highly complex; they can be expressed in terms of a single variable, many variables, or sets of nonlinear integro-differential equations. Regardless of its complexity, the model is an idealization of the system, so it cannot be exact: usually, the more complex the system, the less exact the model, particularly since the ability to solve exactly mathematically highly complex expressions diminishes with increasing complexity. In other words, the simpler the model, the easier it is to analyze but the less precise the results.

A *statistical model* comprises mathematical formulations that express the various outputs of a system in terms of probabilities. Usually, a statistical model is used when the system's output cannot be expressed as a fixed function of the input variables. Statistical models are particularly useful for representing the behavior of a system based on a limited number of measurements, and for summarizing and/or analyzing a set of data obtained experimentally or numerically. For example, there exist families of probability distributions, covering a wide range of shapes, which can be used for representing experimentally obtained data sets. The resulting statistical model can then be used to perform extrapolations or interpolations, for determining the probability of occurrence of some event in some specific interval of values, and so on. Statistical models are constructed based on the physical properties of the system. When the physical properties and, hence, the principles of operation of the system are well understood, the model will be derived from these underlying principles. Most often, though, the basic physical properties of the system under consideration are known incompletely. In simple cases, and when sufficient data are available, it is possible to select a general yet flexible family of statistical models and fit one member of this family to the observations. The validity of such fits is then verified by performing a sensitivity analysis to determine the sensitivity of the system's output to the selection of a specific model to represent the system. In situations when the physical properties of the system under investigation are ill understood, but measurements are nevertheless available, the model for the system is selected on a trial-and-error basis, usually based not only on objective criteria but also on subjective expert opinion. Finally, when no data are available, it is possible to study the conceptual model on a computer to generate numerically synthetic observations. The previously mentioned statistical methods can then be applied to analyze the synthetic observations generated numerically, just as if they had been physical observations.

The group study to be measured or counted, or the conceptual entity for which predictions are to be made, is called *population*. A *model parameter* is a quantity that expresses a characteristic of the system; a parameter could be constant or variable. A *sample* is a subset of the population selected for study. An *experiment* is a sequence of a limited (or, occasionally, unlimited) number of trials. An *event* is an outcome of an experiment. The set of all events (i.e., the set that represents all outcomes of an experiment) is called the *event space* or sample space, of the experiment; the respective events are also referred to as *sample points*.

When a process operates on an event space such that the individual outcomes of the observations cannot be controlled, the respective process is called a *random process* and the respective events are called *random events*. Although the exact outcome of any single random trial is unpredictable, a sample of random trials of reasonable size is expected to yield a pattern of outcomes. In other words, randomness implies lack of deterministic regularity but existence of statistical regularity. Thus, random phenomena must be distinguished from totally unpredictable phenomena, for which no pattern can be construed, even for exceedingly large samples. Observations of physical phenomena tend to fall somewhere between total unpredictability, at the one extreme, and statistically well-behaved processes, at the other extreme. For this reason, experimentalists must invest a considerable effort to identify and eliminate, as much as possible, statistically unpredictable effects (e.g., malfunctioning equipment), in order to perform experiments under controlled conditions conducive to generating statistically meaningful results.

The quantitative analysis of statistical models relies on concepts of probability theory. The basic concepts of probability theory can be introduced in several ways, ranging from the intuitive notion of frequency of occurrence to the *axiomatic development* initiated by Kolmogorov [108], and including the subjective *inductive reasoning* ideas based on "degree of belief" as originally formulated by Laplace and Bayes. For the purposes of this book (namely data evaluation, assimilation, model adjustment/calibration, sensitivity and uncertainty analysis of models and data), all three interpretations of probability will be employed in order to take advantage of their respective strengths. From a mathematical point of view, the concepts of probability theory are optimally introduced by using Kolmogorov's axiomatic approach, in which probability is postulated in terms of abstract functions operating on well-defined event spaces. This axiomatic approach avoids both the mathematical ambiguities inherent to the concept of relative frequencies and the pitfalls of inadvertently misusing the concept of inductive reasoning. Thus, consider that S is the sample space consisting of a certain number of events, the interpretation of which is momentarily left unspecified. Assigned to each subset A of S, there exists a real number $P(A)$, called a *probability*, defined by the following three axioms (Kolmogorov [108]):

AXIOM I (EXISTENCE): For every subset A in S, the respective probability exists and is nonnegative, i.e., $P(A) \geq 0$.

AXIOM II (ADDITIVITY): For any two subsets A and B that are disjoint (i.e., $A \cap B = \emptyset$), the probability assigned to the union of A and B is the sum of the two corresponding probabilities, i.e., $P(A \cup B) = P(A) + P(B)$.

AXIOM III (NORMALIZATION): The probability assigned to the entire sample space is one, i.e., $P(S) = 1$; in other words, the certain event has unit probability.

Note that the statements of the three axioms mentioned above are not entirely rigorous from the standpoint of pure mathematics, but have been deliberately simplified somewhat, in order to suit the scope of this book. A mathematically more precise definition of probability requires that the set of subsets to which probabilities are assigned constitute a so-called σ-field.

Several useful properties of probabilities can be readily derived from the axiomatic definition introduced above:

a) $P(\bar{A}) = 1 - P(A)$, where $\bar{A}$ is the complement of A;

b) $0 \leq P(A) \leq 1$;

c) $P(\emptyset) = 0$, *but* $P(A) = 0$ *does NOT mean that* $A = \emptyset$;

d) if $A \subset B$, then $P(A) \leq P(B)$;

e) $P(A \cup B) = P(A) + P(B) - P(A \cap B)$. \hfill (1.4)

The last relation above can be extended to any *finite sequence* of events $(E_1, E_2, \ldots, E_k)$ in a sample space S, in which case it becomes

$$P(\bigcup_{i=1}^{n} E_i) = \sum_{i=1}^{n} P(E_i) - \sum_{i \neq j} P(E_i \cap E_j) + \sum_{i \neq j \neq k} P(E_i \cap E_j \cap E_k)$$

$$+ \ldots + (-1)^{n+1} P\left(\bigcap_{i=1}^{n} E_i\right). \hfill (1.5)$$

When the events $(E_1, E_2, \ldots, E_n)$ form a *finite sequence of mutually exclusive* events, Eq. (1.5) reduces to

$$P\left(\bigcup_{i=1}^{n} E_i\right) = \sum_{i=1}^{n} P(E_i). \hfill (1.6)$$

When extended to the infinite case $\to \infty$, Eq. (1.6) actually expresses one of the defining properties of the probability measure, which makes it possible to introduce the concept of *probability function* defined on an event space S. Specifically, a function P is a probability function defined on S if and only if it possesses the following properties:

a) $P(\emptyset) = 0$, where $\emptyset$ is the null set.

b) $P(S) = 1$.

c) $0 \leq P(E) \leq 1$ where E is any event in S.

d) $P\left(\bigcup_{i=1}^{\infty} E_i\right) = \sum_{i=1}^{\infty} P(E_i)$, where $(E_1, E_2, \ldots)$ is any sequence of mutually exclusive events in S.

The concept of probability defined thus far cannot address conditions (implied or explicit) such as "What is the probability that event E_2 occurs if it is known that event E_1 has actually occurred?" In order to address such conditions, an additional concept, namely the concept of *conditional probability*, must be introduced. For this purpose, consider that E_1 and E_2 are any two events in an event space S, with $P(E_1) > 0$. Then, the conditional probability $P(E_2|E_1)$ that event E_2 occurs given the occurrence of E_1 is defined as

$$P(E_2|E_1) \equiv \frac{P(E_1 \cap E_2)}{P(E_1)}. \qquad (1.7)$$

From the definition introduced in Eq. (1.7), it follows that

$$P(E_1 \cap E_2) = P(E_2|E_1) P(E_1). \qquad (1.8)$$

Conditional probabilities also satisfy the axioms of probability, as can be seen from the definition introduced in Eq. (1.7). In particular, since $P(E_1) = P(E_1|S)$, the foregoing definition implies that the usual probability $P(E_1)$ can itself be regarded as a conditional probability, i.e., $P(E_1)$ is the conditional probability for E_1 given S.

The definition of conditional probability can be readily extended to more than two events by partitioning the sample space S into k mutually exclusive events $(E_1, E_2, \ldots, E_k)$, and by considering that $E \subset S$ is an arbitrary event in S with $P(E) > 0$. Then the probability, $P(E_i|E)$, that event E_i occurs given the occurrence of E is expressed as

$$P(E_i|E) = \frac{P(E|E_i) P(E_i)}{\sum\limits_{j=1}^{k} P(E|E_j) P(E_j)}, \qquad (i = 1, 2, \ldots, k), \qquad (1.9)$$

where $P(E|E_i)$ denotes the probability that event E occurs, given the occurrence of E_i. Equation (1.9) is known as *Bayes' theorem*, and is of fundamental importance to practical applications of probability theory to the evaluation of scientific data. Two events, E_1 and E_2, are said to be *statistically* independent if $P(E_2|E_1) = P(E_2)$. This means that the occurrence (or nonoccurrence) of E_1 does not affect the occurrence of E_2. Note that if E_1 and E_2 are statistically independent, then $P(E_1 E_2) = P(E_1) P(E_2)$, $P(E_1|E_2) = P(E_1)$, and also conversely. Actually, the following statements are equivalent:

$$E_1 \text{ and } E_2 \text{ are statistically independent; or } P(E_1 | E_2) = P(E_1). \quad (1.10)$$

The numerical representation of the elementary events E in a set S is accomplished by introducing a function, say, X, which operates on all events $E \subset S$ in such a way as to establish a particular correspondence between E and the real number $x = X(E)$. The function X defined in this way is called a *random variable*. It is very important to note that X itself is the random variable, rather than the individual values $x = X(E)$, which X generates by operating on E. Thus, although X has the characteristics of a "function" rather than those of a "variable," the usual convention in probability theory and statistics is to call X a "random variable" rather than a "random function." There also exist "functions of random variables," which may, or may not, be random themselves.

Random events $E \subset S$ that can be completely characterized by a *single-dimensional random variable* X are called *single-variable events*. The qualifier "single-dimensional" is omitted when it is apparent from the respective context. The concept of single-dimensional random variable, introduced to represent numerically a single-variable event, can be extended to a multivariable event $E \subset S$ by considering that S is a sample space in k-dimensions. If each random variable X_i $(i = 1, 2, \ldots, k)$ is a real-valued function defined on a domain N_i (representing the ith dimension of S), then $(X_1, \ldots, X_k)$ is a *multivariate random variable* or *random vector*. Furthermore, consider that each domain N_i $(i = 1, 2, \ldots, k)$ is a discrete set, either finite or denumerably infinite (N_i is usually the set of nonnegative integers or a subset thereof). Then, a *probability function*, $p(x_1, \ldots, x_k)$, of the discrete random vector $(X_1, \ldots, X_k)$ is defined by requiring $p(x_1, \ldots, x_k)$ to satisfy, for each value x_i taken on by X_i $(i = 1, 2, \ldots, k)$, the following properties:

$$\text{(i)} \quad p(x_1, x_2, \ldots, x_k) = P\{X_1 = x_1, X_2 = x_2, \ldots, X_k = x_k\} \quad (1.11)$$

and

$$\text{(ii)} \quad P\{A\} = \sum_{(x_1, \ldots, x_k) \in A} p(x_1, \ldots, x_k), \quad (1.12)$$

for any subset A of N, where N is the k-dimensional set whose i^{th} component is N_i $(i = 1, 2, \ldots, k)$ with $P\{N\} = 1$. Consider that A is the set of all random vectors $(X_1, \ldots, X_k)$ such that $\mathbf{X}_i \leq x_i$ $(i = 1, 2, \ldots, k)$. Then

$$P\{A\} = P\{X_1 \leq x_1, X_2 \leq x_2, \ldots, X_k \leq x_k\}, \tag{1.13}$$

is called the *cumulative distribution function* (CDF) of $(X_1, \ldots, X_k)$. The usual notation for the CDF of $(X_1, \ldots, X_k)$ is $F(x_1, \ldots, x_k)$. Note that the cumulative distribution function is not a random variable; rather, it is a real numerical-valued function whose arguments represent compound events.

To define the probability density function of a *continuous random vector* $(X_1, \ldots, X_k)$, consider that $(X_1, \ldots, X_k)$ is a random vector whose i^{th} component, X_i, is defined on the real line $(-\infty, \infty)$ or on a subset thereof. Suppose $p(x_1, \ldots, x_k) > 0$ is a function such that for all $x_i \in [a_i, b_i]$, $(i = 1, 2, \ldots, k)$, the following properties hold:

(i) $P\{a_1 < X_1 < b_1, \ldots, a_k < X_k < b_k\} = \displaystyle\int_{a_k}^{b_k} \ldots \int_{a_1}^{b_1} p(x_1, \ldots, x_k) \, dx_1 \ldots dx_k$

$$\tag{1.14}$$

and if A is any subset of k-dimensional intervals,

(ii) $P\{A\} = \displaystyle\int_{(x_1, \ldots, x_k) \in A} \ldots \int p(x_1, x_2, \ldots, x_k) \, dx_1 \, dx_2 \ldots dx_k,$ (1.15)

then $p(x_1, \ldots, x_k)$ is said to be a *joint probability density function* (PDF) of the continuous random vector $(X_1, \ldots, X_k)$, if it is normalized to unity over its domain.

Consider that the set $(x_1, \ldots, x_k)$ represents a collection of random variables with a multivariate joint probability density $p(x_1, \ldots, x_k) > 0$; then, the *marginal probability density* of x_i, denoted by $p_i(x_i)$, is defined as

$$p_i(x_i) = \int_{-\infty}^{\infty} dx_1 \ldots \int_{-\infty}^{\infty} dx_{i-1} \int_{-\infty}^{\infty} dx_{i+1} \ldots \int_{-\infty}^{\infty} p(x_1, x_2, \ldots, x_k) \, dx_k \tag{1.16}$$

In addition, the *conditional probability density function* (PDF) $p(x_1, \ldots, x_{i-1}, x_{i+1}, \ldots, x_k | x_i)$ can be defined whenever $p_i(x_i) \neq 0$, by means of the expression

$$p(x_1, \ldots, x_{i-1}, x_{i+1}, \ldots, x_k | x_i) = p(x_1, \ldots, x_k) / p_i(x_i). \qquad (1.17)$$

The meaning of marginal and conditional probability can be illustrated by considering bivariate distributions. Thus, if (X, Y) is a discrete random vector whose joint probability function is $p(x, y)$, where X is defined over N_1 and Y is defined over N_2, then the *marginal probability distribution function* (PDF) of X, $p_x(x)$, is defined as

$$p_x(x) = \sum_{y \in N_2} p(x, y). \qquad (1.18)$$

On the other hand, if (X, Y) is a continuous random vector whose joint PDF is $p(x, y)$, where X and Y are each defined over the domain $(-\infty, \infty)$, then the marginal PDF of X, $p_x(x)$, is defined by the integral

$$p_x(x) = \int_{-\infty}^{\infty} p(x, y) \, dy. \qquad (1.19)$$

Consider that (X, Y) is a random vector (continuous or discrete) whose joint PDF is $p(x, y)$. The *conditional PDF* of y given $X = x$ (fixed), denoted by $h(y|x)$, is defined as

$$h(y|x) = \frac{p(x, y)}{p_x(x)}, \qquad (1.20)$$

where the domain of y may depend on x, and where $p_x(x)$ is the marginal PDF of X, with $p_x(x) > 0$.

Similarly, the conditional PDF, say $g(x|y)$, for x given y is

$$g(x|y) = \frac{p(x, y)}{p_y(y)} = \frac{p(x, y)}{\int p(x', y) \, dx'}. \qquad (1.21)$$

Combining equations (1.20) and (1.21) gives the following relationship between $g(x|y)$ and $h(y|x)$,

$$g(x|y) = \frac{h(y|x) \, p_x(x)}{p_y(y)}, \qquad (1.22)$$

which expresses *Bayes' theorem for the case of continuous variables.*

Consider that (X, Y) is a random vector whose joint PDF is $p(x, y)$. The random variables X and Y are called *stochastically independent* if and only if

$$p(x, y) = p_x(x) \, p_y(y), \tag{1.23}$$

over the entire domain of (X, Y) (i.e., for all x and y). From this definition and from Eq. (1.21), it follows that X and Y are independent if and only if $g(x | y) = p_x(x)$ over the entire domain of (X, Y); this definition of stochastic independence can be generalized to random vectors.

Consider that $x = (x_1, \ldots, x_n)$ and $y = (y_1, \ldots, y_n)$ are two distinct vector random variables that describe the same events. Consider, further, that the respective multivariate probability densities $p_x(x)$ and $p_y(y)$ are such that the mappings $y_i = y_i(x_1, \ldots, x_n)$, $(i = 1, 2, \ldots, n)$, are continuous, one-to-one, and all partial derivatives $\partial y_i / \partial x_j$, $(i, j = 1, \ldots, n)$, exist. Then, the transformation from one *PDF* to the other is given by the relationship

$$p_y(y) \, |dy| = p_x(x) \, |dx|, \quad \text{or} \quad p_x(x) = |J| \, p_y(y), \tag{1.24}$$

where $|J| \equiv \det |\partial y_i / \partial x_j|$, $(i, j = 1, \ldots, n)$, is the Jacobian of the respective transformation. An alternative way of writing the above relation is

$$p_y(y) = p_x\left(f^{-1}(y)\right) \left| \frac{\partial f^{-1}(y)}{\partial y} \right|; \quad \frac{\partial f^{-1}(y)}{\partial y} \triangleq \left[\left\{ \frac{\partial f(x)}{\partial x} \right\}_{x = f^{-1}(y)} \right]^{-1}. \tag{1.25}$$

If the inverse function $f^{-1}(y)$ is not unique, i.e., then the above relation becomes, more generally:

$$p_y(y) = \sum_{i=1}^{m} p_x\left(f_i^{-1}(y)\right) \left| \frac{\partial f_i^{-1}(y)}{\partial y} \right|; \quad f_i^{-1}(y) = x_i^{-1}. \tag{1.26}$$

For example, the distribution $p_y(y)$ of a quadratic function $y = ax^2 + bx + c$, where a, b, c are known constants and the distribution $p_x(x)$ of x is known, is found by using Eq. (1.26) to obtain

$$p_y\left(y\right) = \left\{ p_x \left[\frac{-b + \left(b^2 - 4ac + 4ay\right)^{1/2}}{2a}\right] + p_x \left[\frac{-b - \left(b^2 - 4ac + 4ay\right)^{1/2}}{2a}\right]\right\}$$
$$\times \left(b^2 - 4ac + 4ay\right)^{-1/2}.$$

The dimensionality of a random number is defined by the dimensionality of the particular set of values $x = X(E)$, generated by the function X when operating on the events E. Although the cardinalities (i.e., dimensions) of event sets and corresponding random variables are usually equal, it is important to note that the explicit dimensionality of E is irrelevant in this context. By definition, *finite random variables* are those for which the set of values x obtainable from X operating on E is in one-to-one correspondence with a finite set of integers. *Infinite but discrete random variables* are those for which the set of values x obtainable from X operating on E is in one-to-one correspondence with the infinite set of all integers. *Unaccountable or nondenumerable random variables* are those for which the set of values x obtainable from X operating on E is in one-to-one correspondence with the infinite set of all real numbers.

As discussed above, random variables are actually well-behaved functions that operate on event spaces to yield numerical values that characterize, in turn, the respective events. Random variables can also serve as arguments of other functions, whenever these functions are well behaved in the sense that such functions are bounded and are devoid of singularities except, perhaps, for a finite number of jump-discontinuities.

The main interpretations of probability commonly encountered in data and model analysis are that of *relative frequency* (which is used, in particular, for assigning statistical errors to measurements) and that of *subjective probability* (which is used, in particular, to quantify systematic uncertainties). These two interpretations will be discussed in more detail below.

When probability is interpreted as a *limiting relative frequency*, the elements of the set S correspond to the possible outcomes of a measurement, assumed to be (at least hypothetically) repeatable. A subset E of S corresponds to the occurrence of any of the outcomes in the subset. Such a subset is called an *event*, which is said to occur if the outcome of a measurement is in the subset. A subset of S consisting of only one element denotes an *elementary outcome*. In turn, the probability of an elementary outcome E is defined as the fraction of times that E occurs in the limit when the measurement is repeated infinitely many of times, namely:

$$P(E) = \lim_{n \to \infty} \frac{\text{number of occurences of outcome } E \text{ in } n \text{ measurements}}{n}.$$

(1.27)

The probability for the occurrence of any one of several outcomes (i.e., for a non-elementary subset E) is determined from the probabilities for individual elementary outcomes by using the addition rule provided by the axioms of probability. These individual probabilities correspond, in turn, to relative frequencies of occurrence. The "relative frequency" interpretation is consistent with the axioms of probability since the fraction of occurrence is always greater than or equal to zero, the frequency of any outcome is the sum of the individual frequencies of the individual outcomes (as long as the set of individual outcomes is disjoint), and the measurement must, by definition, eventually yield some outcome [i.e., $P(S) = 1$]. Correspondingly, the conditional probability $P(E_2|E_1)$ represents the number of cases where both E_2 and E_1 occur divided by the number of cases in which E_1 occurs, regardless of whether E_2 occurs. In other words, $P(E_2|E_1)$ gives the frequency of E_2 with the subset E_1 taken as the sample space.

The interpretation of probability as a relative frequency is straightforward when studying physical laws, since such laws are assumed to act the same way in repeated experiments, implying that the validity of the assigned probability values can be tested experimentally. This point of view is appropriate, for example, in nuclear and particle physics, where repeated collisions of particles constitute repetitions of an experiment. Note, though, that the probabilities based on such an interpretation can never be determined experimentally with perfect precision. Hence, the fundamental tasks of classical statistics are to interpret the experimental data by: (a) estimating the probabilities (assumed to have some definite but unknown values) of occurrence of events of interest, given a finite amount of experimental data, and (b) testing the extent to which a particular model or theory that predicts the probabilities estimated in (a) is compatible with the observed data.

The concept of probability as a relative frequency becomes questionable when attempting to assign probabilities for very rare (or even uniquely occurring) phenomena such as a core meltdown in a nuclear reactor or the big bang. For such rare events, the frequency interpretation of probabilities might perhaps be rescued by imagining a large number of similar universes, in some fraction of which the rare event under consideration would occur. However, such a scenario is pure utopia, even in principle; therefore, the frequency interpretation of probability must be abandoned in practice when discussing extremely rare events. In such cases, probability must be considered as a mental construct to assist us in expressing a degree of belief about the single universe in which we live; this mental construct provides the premises of the Bayesian interpretation of probability, which will be discussed next.

Complementary to the frequency interpretation of probability is the so-called *subjective* (also called *Bayesian*) *probability*. In this interpretation, the elements of the sample space are considered to correspond to *hypotheses* or *propositions*, i.e., statements that are either true or false; the sample space is often called the *hypothesis space*. Then, the probability associated with a cause or hypothesis A is interpreted as a measure of degree of belief, namely:

$$P(A) \equiv \text{ a priori measure of the rational degree of belief}$$

that A is the correct cause or hypothesis. $\qquad$ (1.28)

The sample space S must be constructed so that the elementary hypotheses are mutually exclusive, i.e., only one of them is true. A subset consisting of more than one hypothesis is true if any of the hypotheses in the subset is true. This means that the union of sets corresponds to the Boolean OR operation, while the intersection of sets corresponds to the Boolean AND operation. One of the hypotheses must necessarily be true, implying that $P(S) = 1$.

Since the statement "a measurement will yield a given outcome for a certain fraction of the time" can be regarded as a hypothesis, it follows that the framework of subjective probability includes the relative frequency interpretation. Furthermore, subjective probability can be associated with (for example) the value of an unknown constant; this association reflects one's confidence that the value of the respective probability is contained within a certain fixed interval. This is in contrast with the frequency interpretation of probability, where the "probability for an unknown constant" is not meaningful, since if we repeat an experiment depending on a physical parameter whose exact value is not certain, then its value is either never or always in a given fixed interval. Thus, in the frequency interpretation, the "probability for an unknown constant" would be either zero or one, but we do not know which. For example, the mass of a physical quantity (e.g., neutron) may not be known exactly, but there is considerable evidence, in practice, that it lays between some upper and lower limits of a given interval. In the frequency interpretation, the statement "the probability that the mass of the neutron lies within a given interval" is meaningless. By contrast, though, a subjective probability of 90% that the neutron mass is contained within the given interval is a meaningful reflection of one's state of knowledge.

The use of subjective probability is closely related to Bayes' theorem and forms the basis of *Bayesian* (as opposed to classical) *statistics*. For example, in Bayesian statistics for the particular case of two subsets, the subset E_2 appearing in the definition of conditional probability is interpreted as the hypothesis that "a certain theory is true," while the subset E_1 designates the hypothesis that "an experiment will yield a particular result (i.e., data)." In this interpretation, Bayes' theorem takes on the form

$$P\left(theory\,|data\right) \propto P\left(data\,|theory\right) \cdot P\left(theory\right) . \qquad (1.29)$$

In the above expression of proportionality, $P\left(theory\right)$ represents the *prior probability* that the theory is true, while the *likelihood* $P\left(data|theory\right)$ expresses the probability of observing the data that were actually obtained under the assumption that the theory is true. The *posterior probability*, that the theory is correct after seeing the result of the experiment, is given by $P\left(theory\,|data\right)$. Note that the prior probability for the data, $P\left(data\right)$, does not appear explicitly, so the above relation expresses a proportionality rather than an equality. Furthermore, Bayesian statistics provides no fundamental rule for assigning the prior probability to a theory. However, once a prior probability has been assigned, Bayesian statistics indicates how one's degree of belief should change after obtaining additional information (e.g., experimental data).

The choice of the "most appropriate" prior distribution lies at the heart of applying Bayes' theorem to practical problems, and has caused considerable debates over the years. Thus, when prior information related to the problem under consideration is available and can be expressed in the form of a probability distribution, this information should certainly be used. In such cases, the repeated application of Bayes' theorem will serve to refine the knowledge about the respective problem. When only scant information is available, the maximum entropy principle (as described in statistical mechanics and information theory) is the recommended choice for constructing a prior distribution. Finally, in the extreme case when no information is available, the general recommendation is to use a continuous uniform distribution as the prior. In any case, the proper repeated use of Bayes' theorem ensures that the impact of the choice of priors on the final result diminishes as additional information (e.g., measurements) containing consistent data is successively incorporated. These issues will be revisited and set in a proper mathematical framework in Chapter 2.

As can be deduced from the previously mentioned *"axioms of probability,"* the fundamental relationships of probability theory are the sum and product rules

$$P(A|B) + P(\overline{A}|B) = 1. \tag{1.30}$$

$$P(AB|C) = P(A|BC)\,P(B|C) = P(B|AC)\,P(A|C). \tag{1.31}$$

where A, B, and C represent propositions (e.g., "the dice shows six"); AB signifies "both A and B are true," $\overline{A}$ signifies "A is false," and $P(A|B)$ denotes the "probability of A given B." The above notation indicates that all probability assignments are conditional, based on assumptions, on empirical and/or theoretical information. The two forms of the product rule reflect the symmetry $AB = BA$. As argued already by Bernoulli [16] and Laplace [111], probabilities can be interpreted as degrees of plausibility or rational expectation on a numerical scale ranging from 0 (impossibility) to 1 (certainty); intermediate values reflect intermediate degrees of plausibility. The sum rule indicates that, under all circumstances B, the more likely is A the less likely is $\overline{A}$; the unit sum of both probabilities indicates that one of these alternatives must be certainly true. The product rule indicates that, under all circumstances C, the probability that both A and B are true is equal to the probability of A given B, times the probability that, in fact, B is true. Since A and B enter symmetrically one can also take the probability of B given A and multiply it by the probability of A. The interpretation of the Ps as degrees of plausibility has been occasionally criticized in the past based on the following arguments: (i) probability must only mean "relative frequency in a random experiment" (e.g., coin tossing, in the limit of very many repetitions); and (ii) if causes (stochastic laws and their parameters) are given, then probabilities of effects (observations) can be assigned, but not vice versa (i.e., probabilities cannot be assigned to various possible causes if observations are given); (iii) since physical constants are not random variables that take on given values with certain frequencies, they should not have probability distributions associated with them.

The above arguments would be too restrictive for data evaluation, since they would not permit statements like "according to measured data, the value of a physical constant has the probability P of lying between given limits." The practical problem of estimating values of physical quantities (e.g., natural constants, half-lives, reaction cross sections) from error-affected, uncertain and incomplete data is not a random experiment that can be repeated at will, but is a problem to be addressed by inductive inference (i.e., reasoning in the face of uncertainty). Therefore, the Bernoulli-Laplace concept of probability is more appropriate for data analysis and evaluation (as opposed to a purely statistical viewpoint), since, for example, a probability distribution associated with a measured value for a physical constant does not imply that the constant varies; it merely indicates how plausible various possible values are for the true but unknown value of the respective physical constant. Under the general assumptions that $P\left(\overline{A}|B\right)$ depends on $P\left(A|B\right)$, and that $P\left(AB|C\right)$ depends on $P\left(A|BC\right)$ and $P\left(B|C\right)$, with probabilities interpreted as degrees of plausibility between 0 and 1, Cox [39] proved that any consistent scheme of logical inference must be equivalent to probability theory as derived from the basic sum and product rules expressed by Eqs. (1.30) and (1.31). Schrödinger [183] arrived independently at the same conclusions, and Rényi [168] proved these conclusions under the most general conditions, without needing Cox's assumptions of (twice-) differentiability of probability functions. These works have paved the way to the current consensus that:

1. Probabilities are not frequencies. They can be applied equally well to non-repetitive situations as to repeated trials.

2. All internally consistent schemes of logical inference (including fuzzy logic or artificial intelligence) must be equivalent to probability theory.

1.2.2 Bayes' Theorem for Assimilating New Information

Equation (1.31) can be written in the form

$$P(A|BC) = \frac{P(B|AC)P(A|C)}{P(B|C)}. \tag{1.32}$$

which expresses *Bayes' theorem* (1763) in its simplest form. In the course of evaluating experimental data, Bayes' theorem is used to compute *the updated or a posteriori probability* ("posterior"), $P(A|BC)$, which is proportional to the product $P(B|AC)P(A|C)$, where $P(B|AC)$ is the *likelihood function* and $P(A|C)$ is the *a priori probability* (or "prior," for short). In practice, the experimental data, B, to be evaluated and/or assimilated depends on the value of an unknown physical A and on other circumstances C. The likelihood function $P(B|AC)$ is provided by the theoretical model employed by the evaluator to assess how likely the data B would be under the circumstances C if the unknown quantity were in fact A. The prior reflects the information available about the data A before it actually became available. The likelihood function reflects the impact of the data, and the posterior contains the complete information available for further inference and prediction. For continuous variates A and B, the finite probabilities $P(B|AC)$ and $P(A|C)$ are replaced by infinitesimal probabilities $p(B|AC)dB$ and $p(A|C)dA$, where $p(B|AC)$ and $p(A|C)$ denote the respective probability densities.

The generalization of Eq. (1.32) to N distinct, mutually exclusive alternatives A_ν, $\nu = 1, 2, \ldots, N$, takes the form

$$P(A_\nu|BC) = \frac{P(B|A_\nu C)P(A_\nu|C)}{\sum_\nu P(B|A_\nu C)P(A_\nu|C)}, \quad \nu = 1, 2, \ldots, N. \tag{1.33}$$

The above probability is normalized to unity as demanded by the sum rule.

For continuous variates, Eq. (1.33) takes the form

$$p(A|BC)dA = \frac{p(B|AC)p(A|C)dA}{\int p(B|AC)p(A|C)dA}, \quad A_{\min} \leq A \leq A_{\max}. \quad (1.34)$$

The above forms of Bayes' theorem provide the foundation for data evaluation and assimilation, indicating how prior knowledge (e.g., a data file) is to be updated with new evidence (new data). Since the denominator in Bayes' theorem is simply a normalization constant, the formal rule for updating observations can be briefly stated as *"the posterior is proportional to the likelihood times the prior."* It should be understood that the terms "posterior" and "prior" have a logical rather than a temporal connotation, implying "with" and, respectively, "without" having assimilated the new data.

As an illustration of the use of Bayes' theorem for data evaluation and assimilation, consider the determination of the decay constant λ of some short-lived radioisotope from decays registered at times $t_1, t_2, \ldots, t_n$. In this case, the unknown value of the decay constant λ corresponds to A, the data $t_1, \ldots, t_n$ corresponds to B, while all other information about the situation such as applicability of the exponential decay law, purity of the sample, reliability of the recording apparatus, sufficiently long observation time for all observable decays to be recorded, and so on, corresponds to C. The statistical model for the experiment is represented by the so-called "sampling distribution," i.e., by the probability with which various alternatives can be "reasonably" expected after sampling once, given the parameters of the model. In this example, the statistical model is the probability that, given λ, one particular decay (e.g., the ith one), is recorded in the time interval dt_i at t_i, namely

$$p(t_i|\lambda)\, dt_i = \exp(-\lambda t_i)\, \lambda dt_i, \quad 0 < t_i < \infty. \quad (1.35)$$

Writing probability distributions in the above form, with the probability density p multiplied by the corresponding differential (i.e., as an infinitesimal probability), and with the range of the random variable explicitly stated, emphasizes the fact that all probability distributions are ultimately used for computing expectation values, as integrands subject to change of variables. Furthermore, the above notation omits explicit reference to the background information C.

Using the above models together with the product rule, it follows that the joint probability of observing the mutually independent data $t_1, \ldots, t_n$, given λ, is

$$p(t_1, \ldots, t_n | \lambda) \, dt_1, \ldots, dt_n = \exp\left(-\lambda \sum_{i=1}^{n} t_i\right) \lambda^n \, dt_1, \ldots, dt_n. \qquad (1.36)$$

The above expression corresponds to the likelihood $p(A|B)\,dB$. In this example, the likelihood function does not depend on the individual sample values, since they appear only in the form $\sum_i t_i \equiv n\bar{t}$. Hence, for a given n, the sample average $\bar{t}$ carries all the information contained in the data; customarily, $\bar{t}$ is called a "sufficient statistic," or an "ancillary statistic."

The prior is $p(\lambda)\,d\lambda$, and its expression will be discussed and specified later. According to Bayes' theorem, the posterior probability is proportional to the product of the likelihood function and the prior probability. Hence, multiplying the likelihood function $p(t_1, \ldots, t_n | \lambda)$ (which conveys the impact of the data) by the prior $p(\lambda)\,d\lambda$ (which summarizes the information that was already known without the data) leads to

$$p(\lambda | t_1, \ldots, t_n) \, d\lambda \, \propto \, \exp\left(-\lambda \sum_{i=1}^{n} t_i\right) \lambda^n \, p(\lambda)\,d\lambda, \qquad 0 < \lambda < \infty. \quad (1.37)$$

where $\propto$ denotes proportionality. When all values of λ between 0 and ∞ are considered as being equally probable a priori, the prior becomes $p(\lambda)\,d\lambda \propto d\lambda$, and the above expression simplifies to

$$p\left(\lambda | n\bar{t}\right) d\lambda \, \sim \, e^{-\lambda n\bar{t}} \lambda^n d\lambda, \qquad 0 < \lambda < \infty. \qquad (1.38)$$

Normalizing the above distribution to unity leads to the following properly normalized posterior distribution:

$$p\left(\lambda | n\bar{t}\right) d\lambda = \Gamma(n+1)^{-1} e^{-x} x^n \, dx, \qquad 0 < x \equiv \lambda n\bar{t} < \infty, \qquad (1.39)$$

where $\Gamma(n+1) \equiv \int_0^{\infty} e^{-x} x^n \, dx$ is the customary gamma function. The above distribution is the "gamma distribution" or the "chi-square distribution with $\nu \equiv 2n+2$ degrees of freedom"; it represents the complete information about λ that can be obtained from the data and the assumed prior. As the sample size

n increases, the posterior becomes more and more concentrated, indicating that the more data are available, the better defined is λ.

1.3   Moments, Means, and Covariances

Probabilities cannot be measured directly; they can be inferred from the results of observations or they can be postulated and (partially) verified through accumulated experience. In practice, though, certain random vectors tend to be more probable, so that most probability functions of practical interest tend to be localized. Therefore, the essential features regarding probability distributions of practical interest are measures of *location* and of *dispersion*. These measures are provided by the *expectation* and *moments* of the respective probability function. If the probability function is known, then these moments can be calculated directly, through a process called *statistical deduction*. Otherwise, if the probability function is not known, then the respective moments must be estimated from experiments, through a process called *statistical inference*.

Consider that $\mathbf{x} = (x_1, \ldots, x_n)$ is a collection of random variables that represent the events in a space E, and consider that S_x represents the n-dimensional space formed by all possible values of $\mathbf{x}$. The space S_x may encompass the entire range of real numbers (i.e., $-\infty < x_i < \infty$, $i = 1, \ldots, n$) or a subset thereof. Furthermore, consider a real-valued function, $g(\mathbf{x})$, and a probability density, $p(\mathbf{x})$, both defined on S_x. Then, the *expectation* of $g(\mathbf{x})$, denoted as $E[g(\mathbf{x})]$, is defined as:

$$E[g(\mathbf{x})] \equiv \int_{S_x} g(\mathbf{x})\, p(\mathbf{x})\, d\mathbf{x}, \tag{1.40}$$

if the condition of *absolute convergence*, namely

$$E(|g|) = \int_{S_x} |g(\mathbf{x})|\, p(\mathbf{x})\, d\mathbf{x} < \infty, \tag{1.41}$$

is satisfied. When $\mathbf{x}$ is discrete with the domain N, the expectation of g is defined as

$$E\left[g\left(\mathbf{x}\right)\right] = \sum_{\mathbf{x} \in N} g\left(\mathbf{x}\right) p\left(\mathbf{x}\right), \tag{1.42}$$

provided that $E\left(\left|g\left(\mathbf{x}\right)\right|\right) < \infty$.

In particular, the *moment of order k about a point c* is defined for a univariate probability function as

$$E\left[\left(x - c\right)^{k}\right] \equiv \int_{S_x} \left(x - c\right)^{k} p\left(x\right) \, dx \,, \tag{1.43}$$

where S_x denotes the set of values of x for which $p\left(x\right)$ is defined, and the integral above is absolutely convergent.

For a multivariate probability function, given a collection of n random variables $(x_1, \ldots, x_n)$ and a set of constants $(c_1, \ldots, c_n)$, the *mixed moment of order k* is defined as

$$E\left[\left(x_1 - c_1\right)^{k_1}, \ldots, \left(x_n - c_n\right)^{k_n}\right] \equiv$$
$$\int_{S_{x_1}} dx_1 \ldots \int_{S_{x_n}} dx_n (x_1 - c_1)^{k_1} \ldots (x_n - c_n)^{k_n} p\left(x_1 \ldots x_n\right) . \tag{1.44}$$

The *zeroth-order moment* is obtained by setting $k = 0$ (for univariate probability) in Eq. (1.43) or by setting $k_1 = \ldots = k_n = 0$ (for multivariate probability) in Eq. (1.44), respectively. *Since probability functions are required to be normalized, the zeroth moment is always equal to unity.*

In particular, when $c = 0$ (for a univariate probability) or when $c_1 = \ldots = c_n = 0$ (for a multivariate probability), the quantities defined as $\nu_k \equiv E\left(x^k\right)$, and, respectively, $\nu_{k_1 \ldots k_n} \equiv E\left(x_1^{k_1} \ldots x_n^{k_n}\right)$ are called the *moments about the origin* (often also called *raw* or *crude moments*). If $\sum_{i=1,n} k_i = k$, then the moments of the form $\nu_{k_1 \ldots k_n}$ are called the *mixed raw moments of order k*. For $k = 1$, these moments are called *mean values*, and are denoted as $m_o = \nu_1 = E\left(x\right)$ for univariate probability, and $m_{oi} \equiv \nu_{o \ldots 1 \ldots o} \equiv E\left(x_i\right)$, $(i = 1, \ldots, n)$, for multivariate probability, respectively. Note that a "0" in the jth subscript position signifies that $k_j = 0$ for the particular raw moment in question, while a "1" in the ith subscript position indicates that $k_i = 1$ for the respective moment.

The moments about the mean or *central moments* are defined as

$$\mu_k \equiv E\left[(x - m_o)^k\right], \quad \text{for univariate probability,} \qquad (1.45)$$

and

$$\mu_{k_1 \ldots k_n} \equiv E\left[(x_1 - m_{o1})^{k_1} \ldots (x_n - m_{on})^{k_n}\right],$$

$$\text{for multivariate probability.} \qquad (1.46)$$

Furthermore, if $\sum_{i=1,n} k_i = k$, then the above moments are called the *mixed central moments* of order k. Note that the central moments vanish whenever one particular $k_i = 1$ and all other $k_j = 0$, i.e., $\mu_{o\ldots 1 \ldots o} = 0$. Note also that all even-power central moments of univariate probability functions (i.e., μ_k, for $k = even$) are nonnegative.

1.3.1   Means and Covariances

The central moments for $k = 2$ play very important roles in statistical theory, and are therefore assigned special names. Thus, for univariate probability, the second moment, $\mu_2 \equiv E\left[(x - m_o)^2\right]$, is called *variance*, and is usually denoted as var(x) or σ^2. The positive square root of the variance is called the *standard deviation*, denoted as σ, and defined as

$$\sigma \equiv [\text{var}(x)]^{1/2} \equiv \mu_2^{1/2} \equiv \left\{E\left[(x - m_o)^2\right]\right\}^{1/2}. \qquad (1.47)$$

The terminology and notation used for univariate probability are also used for multivariate probability. Thus, for example, the standard deviation of the ith component is defined as:

$$\mu_{o\ldots 2 \ldots o} \equiv \text{var}(x_i) \equiv \sigma_i^2 \equiv E\left[(x_i - m_o)^2\right]. \qquad (1.48)$$

To simplify the notation, the subscripts accompanying the moments ν and μ for multivariate probability functions are usually dropped in favor of a simpler alternative notation. For example, μ_{ii} is often employed to denote var(x_i), and μ_{ij} signifies $E\left[(x_i - m_{oi})(x_j - m_{oj})\right]$, $(i, j = 1, \ldots, n)$.

The raw and central moments are related to each other through the important relationship

$$\mu_k = \sum_{i=0}^{k} C_i^k (-1)^i \nu_{k-i} \nu_1^i \quad (k \geq 1), \tag{1.49}$$

where $C_i^k = k!/[(k-i)!\,i!]$ is the *binomial coefficient*. This formula is very useful for estimating central moments from sampling data, since, in practice, it is more convenient to estimate the raw moments directly from the data, and then derive the central moments by using the above equation. Since measurements rarely yield true values, it is necessary to introduce surrogate parameters to measure location and dispersion for the observed results. Practice indicates that *location* is *best described by the mean value*, while *dispersion of observed results appears to be best described by the variance, or standard deviation*. In particular, the mean value can be interpreted as a locator of the center of gravity, while the variance is analogous to the moment of inertia (which linearly relates applied torque to induced angular acceleration in mechanics). Also very useful for the study of errors is the *Minimum Variance Theorem*, which states that: *if c is a real constant and x is a random variable, then* $\mathrm{var}(x) \leq E\left[(x-c)^2\right]$.

Henceforth, when we speak of *errors in physical observations*, they *are to be interpreted as standard deviations*, unless explicitly indicated otherwise. In short, errors are simply the measures of dispersion in the underlying probability functions that govern observational processes. *The fractional relative error or coefficient of variation*, f_x, is defined by $f_x = \sigma/|E(x)|$, when $E(x) \neq 0$. The reciprocal, $(1/f_x)$, is commonly called (particularly in engineering applications) the *signal-to-noise ratio*. Finally, the term percent error refers to the quantity $100 f_x$.

When the probability function is known and the respective mean and variance (or standard deviation) exist, they can be computed directly from their definitions. However, when the actual distribution is not known, it is considerably more difficult to interpret the knowledge of the mean and standard deviation in terms of confidence that they are representative of the distribution of measurements. The difficulty can be illustrated by considering a *confidence indicator associated with the probability function p*, $C_p(k\sigma)$, defined by means of the integral

$$C_p(k\sigma) \equiv \int_{m_o-k\sigma}^{m_o+k\sigma} p(x)\, dx,$$
(1.50)

where σ is the standard deviation and $k \geq 1$ is an integer. Since the probability density integrated over the entire underlying domain is normalized to unity, it follows that $C_p(k\sigma) < 1$ for all k. However, $C_p(k\sigma) \approx 1$ whenever $k \gg 1$. Thus, $C_p(k\sigma)$ can vary substantially in magnitude for different types of probability functions p, even for fixed values of σ and k. This result indicates that although the variance or standard deviation are useful parameters for measuring dispersion (error), knowledge of them alone does not provide an unambiguous measure of confidence in a result, unless the probability family to which the distribution in question belongs is *a priori* known. Consequently, when an experiment involves several observational processes, each governed by a distinct law of probability, it is difficult to interpret overall errors (which consist of several components) in terms of confidence. In practice, though, the consequences are mitigated by a very important theorem of statistics, called the *Central Limit Theorem*, which will be discussed in the sequel, following Eq. (1.58).

For multivariate probability, the second-order central moments comprise not only the variances $\mu_{ii} = \text{var}(x_i) = E\left[(x_i - m_{oi})^2\right]$, $(i = 1,\ldots,n)$, but also the moments $\mu_{ij} = E\left[(x_i - m_{oi})(x_j - m_{oj})\right]$, $(i,j = 1,\ldots,n)$. These moments are called *covariances*, and the notation $\text{cov}(x_i, x_j) \equiv \mu_{ij}$ is often used. The collection of all second-order moments of a multivariate probability function involving n random variables forms an $n \times n$ matrix, denoted in this section as $\mathbf{V}_x$, and called the *variance-covariance matrix*, or, simply, the *covariance matrix*. Since $\mu_{ij} = \mu_{ji}$ for all i and j, covariance matrices are symmetric.

When $\mu_{ii} > 0$, $(i = 1,\ldots,n)$, it is often convenient to use the quantities ρ_{ij} defined by the relationship

$$\rho_{ij} \equiv \mu_{ij}/(\mu_{ii}\,\mu_{jj})^{1/2}, \quad (i,j = 1,\ldots,n),$$
(1.51)

and called *correlation parameters* or, simply, *correlations*. The matrix obtained by using the correlations, ρ_{ij}, is called the *correlation matrix*, and will be denoted as $\mathbf{C}_x$ in this section.

Using the *Cauchy-Schwartz inequality*, it can be shown that the elements of $\mathbf{V}_x$ always satisfy the relationship

$$|\mu_{ij}| \leq (\mu_{ii}\mu_{jj})^{1/2}, \quad (i, j = 1, n), \tag{1.52}$$

while the elements ρ_{ij} of the correlation matrix $\mathbf{C}_x$ satisfy the relationship

$$-1 \leq \rho_{ij} \leq 1. \tag{1.53}$$

In the context of covariance matrices, the *Cauchy-Schwartz inequality* provides an *indicator of data consistency* that is very useful to verify practical procedures for processing experimental information. Occasionally, practical procedures may generate covariance matrices with negative eigenvalues (thus violating the condition of positive-definitiveness), or with coefficients that would violate the Cauchy-Schwartz inequality; such matrices would, of course, be unsuitable for representing physical uncertainty. Although the mathematical definition of the variance only indicates that it must be nonnegative, the variance for physical quantities should in practice be positive, because it provides a mathematical basis for the representation of physical uncertainty. Since zero variance means no error, probability functions for which some of the random variables have zero variance are not realistic choices for the representation of physical phenomena, since such probability functions would indicate that some parameters were without error, which is never the case in practice. Furthermore, $\mu_{ii} < 0$ would imply an imaginary standard deviation (since $\sigma_i = \mu_{ii}^{1/2}$), which is clearly unacceptable. The reason for mentioning these points here is because, in practice, the elements of covariance matrices are very seldom obtained from direct evaluation of expectations, but are obtained by a variety of other methods, many of them ad hoc. Practical considerations also lead to the requirement that $|\rho_{ij}| < 1$, for $i \neq j$, but a presentation of the arguments underlying this requirement is beyond the purpose of this book. These and other constraints on covariance and correlation matrices lead to the conclusion *that matrices which properly represent physical uncertainties are positive definite.*

Since covariance matrices are symmetric, the $n \times n$ covariance matrix contains no more than $n + [n(n-1)/2]$ distinct elements, namely the off-diagonal covariances and the n variances along the diagonal. Often, therefore, only the diagonal and upper or lower triangular part of covariance and correlation matrices are listed in the literature. A formula often used in practical computations of covariances is obtained by rewriting the respective definition in the form

$$\text{cov}(x_i, x_j) = E(x_i x_j) - m_{oi} m_{oj}. \tag{1.54}$$

Note that if any two random variables, x_i and x_j, in a collection of n random variables are independent, then $\text{cov}(x_i, x_j) = 0$. Note also that the converse of this statement is false: $\text{cov}(x_i, x_j) = 0$ *does not necessarily imply that x_i and x_j are independent.*

A very useful tool for practical applications is the so-called *scaling and translation theorem*, which states that if x_i and x_j are any two members of a collection of n random variables, then the following relations hold for the random variables $y_i = a_i x_i + b_i$ and $y_j = a_j x_j + b_j$:

$$E(y_i) = a_i E(x_i) + b_i, \text{var}(y_i) = a_i^2 \text{var}(x_i), \quad (i = 1, \dots, n);$$
$$\text{cov}(y_i, y_j) = a_i a_j \text{cov}(x_i, x_j), \quad (i, j = 1, \dots, n, \ i \neq j).$$

The constants a_i and a_j are called *scaling parameters*, while the constants b_i and b_j are called *translation parameters*. The above relationships show that mean values are affected by both scaling and translation, while the variances and covariances are only affected by scaling. In particular, the above relationships can be used to establish the following theorem regarding the relationship between ordinary random variables x_i and their standard random variable counterparts $u_i = (x_i - m_{oi})/\sigma_i$, $(i = 1, \dots, n)$: *the covariance matrix for the standard random variables $u_i = (x_i - m_{oi})/\sigma_i$ is the same as the correlation matrix for the random variables x_i.*

The *determinant*, $\det(\mathbf{V}_x)$, of the variance matrix is often referred to as the *generalized variance*, since it degenerates to a simple variance for univariate distributions. The probability distribution is called *nondegenerate* when $\det(\mathbf{V}_x) \neq 0$; when $\det(\mathbf{V}_x) = 0$, however, the distribution is called *degenerate*. Degeneracy is an indication that the information content of the set $\mathbf{x}$ of random variable is less than rank n, or that the probability function is confined to a hyperspace of dimension lower than n. Of course, the determinant of a covariance matrix vanishes if and only if there exist (one or more) linear relationships among the random variables of the set $\mathbf{x}$.

Due to the above mentioned properties of the positive definite matrix $\mathbf{V}_x$, it also follows that

$$\det(\mathbf{V}_x) \leq \prod_{i=1,n} \text{var}(x_i) = \prod_{i=1,n} \sigma_i^2. \tag{1.55}$$

The equality in the above relation is reached only when $\mathbf{V}_x$ is diagonal, i.e., when $\text{cov}(x_i, x_j) = 0$, $(i, j = 1, n, \ i \neq j)$; in this case, $\det(\mathbf{V}_x)$ attains its maximum value, equal to the product of the respective variances. The determinant $\det(\mathbf{V}_x)$ is related to the determinant of the correlation matrix, $\det(\mathbf{C}_x)$, by the relationship

$$\det(\mathbf{V}_x) = \det(\mathbf{C}_x) \prod_{i=1,n} \sigma_i^2. \tag{1.56}$$

From Eqs. (1.55) and (1.56), it follows that $\det(\mathbf{C}_x) \leq 1$. It further follows that $\det(\mathbf{C}_x)$ attains its maximum value of unity only when $\text{cov}(x_i, x_j) = 0$, $(i, j = 1, n, \ i \neq j)$. In practice, $\det(\mathbf{C}_x)$ is used as a *measure of degeneracy* of the multivariate probability function. In particular, the quantity $[\det(\mathbf{C}_x)]^{1/2}$ is called the *scatter coefficient for the probability function*. Note that $\det(\mathbf{C}_x) = 0$ when $\rho_{ij} = \rho_{ji} = 1$ for at least one pair (x_i, x_j), with $i \neq j$.

Two random variables, x_i and x_j, with $i \neq j$, are called *fully correlated* if $\text{cor}(x_i, x_j) = 1$; this situation arises if and only if the corresponding standard random variables u_i and u_j are identical, i.e., $u_i = u_j$. On the other hand, if $\text{cor}(x_i, x_j) = -1$, then x_i and x_j are fully *anti-correlated*, which can happen if and only if $u_i = -u_j$. Therefore, the statistical properties of fully correlated or fully anti-correlated random variables are identical, so that only one of them needs to be considered, a fact reflected by the practice of referring to such random variables as being redundant.

In addition to covariance matrices, $\mathbf{V}_x$, and their corresponding correlation matrices, $\mathbf{C}_x$, a third matrix, called the *relative covariance matrix or fractional error matrix*, can also be defined when the elements of the covariance matrix satisfy the condition that $m_{oi} \neq 0$, $(i = 1, \ldots, n)$. This matrix is usually denoted as $\mathbf{R}_x$, and its elements $(\mathbf{R}_x)_{ij} = \eta_{ij}$ are defined as

$$\eta_{ij} = \mu_{ij}/(m_{oi} m_{oj}) \ . \tag{1.57}$$

Moments of first- and second-order (i.e., means and covariance matrices) provide information only regarding the location and dispersion of probability distributions. Additional information on the nature of probability distributions is carried by the higher-order moments, although moments beyond fourth-order are seldom examined in practice. The nature of such information can be intuitively understood by considering the third- and fourth-order moments of univariate probability functions. For this purpose, it is easiest to consider the *respective reduced central moments*, α_k, defined in terms of central moments and the standard deviation by the relationship $\alpha_k \equiv \mu_k/\sigma^k$.

The reduced central moment α_3 is called the *skewness* of the probability distribution, because it measures quantitatively the departure of the probability distribution from symmetry (a symmetric distribution is characterized by the value $\alpha_3 = 0$). Thus, if $\alpha_3 < 0$, the distribution is skewed toward the left (i.e., it favors lower values of x relative to the mean), while $\alpha_3 > 0$ indicates a distribution skewed toward the right (i.e., it favors higher values of x relative to the mean). The reduced central moment α_4 measures the degree of sharpness in the peaking of a probability distribution and it is called *kurtosis*. Kurtosis is always nonnegative. The standard for comparison of kurtosis is the normal distribution (Gaussian) for which $\alpha_4 = 3$. Distributions with $\alpha_4 < 3$ are called *platykurtic distributions*. Those with $\alpha_4 = 3$ are called *mesokurtic distributions*. Finally, distributions with $\alpha_4 > 3$ are called *leptokurtic distributions*.

Very often in practice, the details of the distribution are unknown, and only the mean and standard deviations can be estimated from the limited amount of information available. Even under such circumstances, it is still possible to make statements regarding confidence by relying on *Chebyshev's theorem*, which can be stated as follows: consider that m_o and $\sigma > 0$ denote the mean value and standard deviation, respectively, of an otherwise unknown multivariate probability density p involving the random variable x. Furthermore, consider that P represents cumulative probability, C_p represents confidence, and $k \geq 1$ is a real constant (not necessarily an integer). Then, *Chebyshev's theorem states that the following relationship holds*:

$$C_p(k\sigma) = P(|x - m_o| \leq k\sigma) \geq 1 - (1/k^2) . \tag{1.58}$$

Chebyshev's theorem is a weak law of statistics since it *provides an upper bound on the probability of a particular deviation* ε. The actual probability of such a deviation (if the probability function were known in detail so that it could be precisely calculated) would always be smaller (implying greater confidence) than Chebyshev's limit. This important point is illustrated in Table (1.1), which compares probabilities for observing particular deviations (denoted as ε) from the respective means. This table clearly underscores the fact that normally distributed random variables are much more sharply localized with respect to the mean than indicated by Chebyshev's theorem.

TABLE 1.1
Probability of Occurrence

Deviation ε	Normal distribution	Chebyshev's limit
$> 1\sigma$	< 0.3173	< 1.0
$> 2\sigma$	< 0.0455	< 0.25
$> 3\sigma$	< 0.00270	< 0.1111
$> 4\sigma$	< 0.0000634	< 0.0625
$> 5\sigma$	$< 5.73 * 10 - 7$	< 0.04
$> 6\sigma$	$< 2.0 * 10 - 9$	< 0.02778

Central Limit Theorem: Consider that $(x_1, x_2, \ldots, x_n)$ is a random sample of the parent random variable x, with mean value m_o, var $(x) = \sigma^2$, and sample average $\xi_n = (\sum_{i=1}^{n} x_i)/n$. Furthermore, define $z_n \equiv (\xi_n - m_o)/(\sigma/n^{1/2})$ to be the reduced random-variable equivalent of ξ_n. Then, the central limit theorem states that z_n, ξ_n, and $n\xi_n$ are all asymptotically normal in the limit as $n \to \infty$. The least restrictive necessary and sufficient condition for the validity of the central limit theorem is the *Lindeberg condition*, which states that if the sequence of random variables $(x_1, x_2, \ldots, x_n)$ is uniformly bounded (i.e., if there exists a positive real constant C such that $|x_i| < C$ for each x_i and all possible n) and the sequence is not degenerate, then the central limit theorem holds. *In practice, the Lindeberg condition requires that the mean values and variances exist for each of these variables, and that the overall variance in the sum* ξ_n *of these random variables be not dominated by just a few of the components.* Application of the central limit theorem to correlated random variables is still an open field of research in mathematical statistics.

Rather than specify the conditions under which the central limit theorem holds exactly in the limit $n \to \infty$, *in practice it is more important to know the extent to which the Gaussian approximation is valid for finite n.* This is difficult to quantify exactly, but the rule of thumb is that the central limit theorem holds as long as the sum is made of a large number of small contributions. Discrepancies arise if, for example, the distributions of the individual terms have long tails, so that occasional large values make up a large part of the sum. Such contributions lead to "non-Gaussian" tails in the sum, which can significantly alter the probability to find values with large departures from the mean. In such cases, the main assumption underlying the central limit theorem, namely the assumption that the measured value of a quantity is a normally distributed variable centered about the mean value, breaks down. Since this assumption is often used when constructing a confidence interval, such intervals can be significantly underestimated if non-Gaussian tails are present. In particular, the relationship between the confidence level and the size of the interval will differ from the Gaussian prescription (i.e., 68.3% for a "1σ" interval, 95.4% for "$2\,\sigma$," etc.). A better understanding of the non-Gaussian tails can often be obtained from a detailed Monte Carlo simulation of the individual variables making up the sum.

For example, the central limit theorem cannot be used for calculating the angle by which a charged particle is deflected upon traversing a layer of matter. Although the total angle can be regarded as the sum of a small number of deflections caused by multiple Coulomb scattering collisions with nuclei in the substance being traversed, and although there are many such collisions, the total angle cannot be calculated by using the central limit theorem. This is because the distribution for individual deflections has a long tail extending to large angles, which invalidates the main assumption underlying the central limit theorem.

1.3.2 A Geometric Model for Covariance Matrices

For a set of n variates, $x_1, x_2, \ldots, x_n$, the square symmetrical matrix $\mathbf{C}$ of order $n \times n$, with elements

$$\mathbf{C} \triangleq (C_{ij})_{n \times n} \triangleq \text{cov}(x_i, x_j); \quad C_{ii} \triangleq \text{var}(x_i), \tag{1.59}$$

is, by definition, the *variance-covariance* (or, in short, *covariance*) matrix associated with the variates $x_1, x_2, \ldots, x_n$. The role of the covariances C_{ij} can be further illustrated by considering a joint distribution of n variates, $x_1, x_2, \ldots, x_n$, normalized to unity, of the form

$$f(\mathbf{x}) = A \exp\left[-1/2\, \mathbf{Q}(\mathbf{x})\right], \quad \mathbf{x} \triangleq (x_1, x_2, \ldots, x_n), \tag{1.60}$$

where $\mathbf{Q}(\mathbf{x})$ is a positive quadratic form expressed as

$$\mathbf{Q}(\mathbf{x}) = (\mathbf{x} - \mathbf{c})^T \mathbf{B}(\mathbf{x} - \mathbf{c}), \tag{1.61}$$

where $\mathbf{c}$ is a n-dimensional vector of constants, $\mathbf{B}$ is a $n \times n$ (positive) symmetric matrix with constant coefficients, the "T" denotes "transposition," and the normalization coefficient A is defined as

$$A = \sqrt{|\mathbf{B}|/(2\pi)^n}\,. \tag{1.62}$$

By differentiating the normalization relation $\int f(\mathbf{x})\, d^n x = 1$ with respect to $\mathbf{c}$ it follows that

$$0 = \frac{\partial}{\partial \mathbf{c}} \int f(\mathbf{x})\, d^n x = -\frac{1}{2} \int \frac{\partial \mathbf{Q}(\mathbf{x})}{\partial \mathbf{c}} f(\mathbf{x})\, d^n x = \mathbf{B}\langle \mathbf{x} - \mathbf{c}\rangle,$$

where the angular brackets $\langle , \rangle$ denote integration of the respective quantitiy over the distribution $f(x)$, i.e., $\langle g(\mathbf{x})\rangle \equiv \int g(\mathbf{x})f(\mathbf{x})d^n x$.

Since $|\mathbf{B}| > 0$, it follows that $\mathbf{c} = \langle \mathbf{x}\rangle$, i.e., the components

$$c_i = \mu_{x_i}, \quad i = 1, 2, \ldots, n, \tag{1.63}$$

of the vector $\mathbf{c}$ are the means of the variates x_i, $i = 1, \ldots, n$. Differentiating the normalization relation $\int f(\mathbf{x})\, d^n x = 1$ with respect to the elements of $\mathbf{B}$

yields

$$0 = \frac{\partial}{\partial B_{ij}} \int f(\mathbf{x}) \, d^n x = \frac{1}{2} \left(\frac{1}{|\mathbf{B}|} \frac{\partial |\mathbf{B}|}{\partial B_{ij}} - \int \frac{\partial Q(\mathbf{x})}{\partial B_{ij}} f(\mathbf{x}) \, d^n x \right).$$

Since the derivative of the determinant of a matrix with respect to an element of this matrix is the cofactor of the respective element, the above equation implies that

$$\frac{cofac(B_{ij})}{|\mathbf{B}|} = \int (x_i - \mu_i)(x_j - \mu_j) f(\mathbf{x}) \, d^n x. \tag{1.64}$$

Furthermore, since $\mathbf{B}$ is symmetric, the left side of the above equation is just the element (i, j) of the inverse of $\mathbf{B}$ or, in matrix notation

$$\mathbf{B}^{-1} = \left\langle (\mathbf{x} - \mu)(\mathbf{x} - \mu)^T \right\rangle \triangleq \mathbf{C}. \tag{1.65}$$

The right side of Eq. (1.65) is the average of the outer product of the vector $(\mathbf{x} - \mu)$ with itself, which is by definition the covariance matrix $\mathbf{C}$ associated with the variates $\mathbf{x}$. Summarizing the information contained in Eqs. (1.60) through Eq. (1.65) indicates that n variates, $x_1, x_2, \ldots, x_n$, characterized by the vector of mean-values μ and the covariance matrix $\mathbf{C}$, are effectively distributed according to the multivariate normal distribution

$$f(\mathbf{x}) = \frac{1}{\sqrt{(2\pi)^n |\mathbf{C}|}} \exp \left[-1/2 \, (\mathbf{x} - \mu)^T \mathbf{C}^{-1} (\mathbf{x} - \mu) \right] \tag{1.66}$$

Thus, the variance σ^2 of a single variate generalizes to the uncertainty (variance) matrix $\mathbf{C}$ containing the variances of the individual variates along its diagonal elements, and the covariances of every pair of variates as its off diagonal elements.

Since variances (or standard deviations) provide a measure for the dispersion of the probability distribution with respect to its mean, they can be visualized as providing a measure of the region of (n-dimensional) random-variable space where most of the probability is concentrated. The intuitive understanding of the meaning of correlation is less straightforward. Perhaps the simplest way to appreciate intuitively the meaning of correlation is to consider a bivariate distribution for the random variables x_1 and x_2. For simplicity, suppose that $-\infty \leq x_1 \leq \infty$ and $-\infty \leq x_2 \leq \infty$, and that p is a probability density for x_1 and x_2. Considering, without loss of generality, that x_1 and x_2 correspond to orthogonal coordinates of a two-dimensional Cartesian plane. Then, the surface $x_3 = p(x_1, x_2)$ will appear as a "hill" in the third (Cartesian) coordinate x_3; note that x_3 is not a random variable. The surface $x_3 = p(x_1, x_2)$ is "centered," in the (x_1, x_2)-plane, on the point (m_{o1}, m_{o2}), while the lateral extents of this surface are measured by the standard deviations σ_1 and σ_2; furthermore, the surface $x_3 = p(x_1, x_2)$ would be symmetric when $\sigma_1 = \sigma_2$. Since p is normalized and since $p(x_1, x_2) \geq 0$ for all (x_1, x_2), the surface $p(x_1, x_2)$ will have at least one maximum in the direction x_3, say $(x_3)_{\max}$. Slicing through the surface $x_3 = p(x_1, x_2)$ with a horizontal plane $x_3 = h$, where $0 < h < (x_3)_{\max}$, and projecting the resulting planar figure onto the (x_1, x_2)-plane yields elliptical shapes that can be inscribed in the rectangle $(x_1 = 2l\sigma_1, x_2 = 2l\sigma_2)$, with $0 < l < \infty$. The covariance $\text{cov}(x_i, x_j)$, or equivalently, *the correlation parameter $\rho_{12} = \rho$ indicates the orientation of the probability distribution* in the random variable space (x_1, x_2). This is particularly clear when $\sigma_1 \neq \sigma_2$, in which case the surface $x_3 = p(x_1, x_2)$ is not symmetric. Thus, a nonzero correlation implies that the surface $x_3 = p(x_1, x_2)$ is *tilted* relative to the x_1 and x_2 axes. On the other hand, a zero correlation implies that the surface $x_3 = p(x_1, x_2)$ is somehow aligned with respect to these axes. Since the intrinsic shape of the surface $x_3 = p(x_1, x_2)$ is governed by the underlying probability but not by the choice of variables, it is possible to perform an orthogonal transformation such as to align the surface relative to a new, transformed coordinate system. The transformed random variables generated by an orthogonal transformation are called *orthogonal random variables*, and they are independent from one another. Thus, consider a region Ω of the three-dimensional Euclidean space. Points in this space can be described by Cartesian coordinates (x_1, x_2, x_3) or, equivalently, by the vector representation

$$\mathbf{r} = \sum_{i=1}^{3} x_i \mathbf{u}_i, \tag{1.67}$$

where the $\mathbf{u}_i$ are fixed, orthogonal, unit basis vectors obeying the relations

$$\mathbf{u}_i^T \mathbf{u}_j = \delta_{ij}, \quad where \quad \delta_{ii} = 1, \quad \delta_{ij} = 0, \quad i \neq j. \tag{1.68}$$

The difference vector between two points, $\mathbf{r}$ and $\mathbf{r}'$, on a trajectory in Ω is

$$d\mathbf{r} = \mathbf{r}' - \mathbf{r} = \sum_{i=1}^{3} dx_i \mathbf{u}_i, \tag{1.69}$$

since the basis vectors are fixed in the orthogonal coordinate system representation. The *distance, dL*, between the points $\mathbf{r}$ and $\mathbf{r}'$ is given by

$$(dL)^2 = d\mathbf{r}^T d\mathbf{r} = \sum_{i=1}^{3}\sum_{j=1}^{3} dx_i \left(\mathbf{u}_i^T \mathbf{u}_j\right) dx_j = \sum_{i=1}^{3}(dx_i)^2. \tag{1.70}$$

Now consider *curvilinear coordinates* (y_1, y_2, y_3) as an alternative coordinate system representation for the region Ω. A typical example of an alternative coordinate system would be spherical coordinates. The curvilinear coordinates are related to the Cartesian coordinates by the transformation equations $y_i = y_i(\mathbf{x})$, where each function y_i is continuous and differentiable with respect to each element of $\mathbf{x}$, for $i = 1, 2, 3$. In terms of the curvilinear coordinates (y_1, y_2, y_3), the difference vector $d\mathbf{r}$ becomes

$$d\mathbf{r} = \sum_{i=1}^{3} dy_i \mathbf{g}_i, \quad \mathbf{g}_i \triangleq \partial \mathbf{r}/\partial y_i, \quad i = 1, 2, 3. \tag{1.71}$$

The vectors $\mathbf{g}_i$ are *effective basis vectors* for the alternative curvilinear coordinate system; they are neither orthogonal nor fixed unit vectors (as the vectors $\mathbf{u}_i$ in the Cartesian system). The distance dL is a scalar and is therefore an invariant quantity, independent of the coordinate representation. In terms of these alternative coordinates, the expression of dL becomes

$$(dL)^2 = d\mathbf{r}^T d\mathbf{r} = \sum_{i=1}^{3}\sum_{j=1}^{3} dy_i \left(\mathbf{g}_i^T \mathbf{g}_j\right) dy_j. \tag{1.72}$$

Equations (1.70) and (1.72) are examples of *differential quadratic forms*, which can be generally represented in the unified form

$$(dL)^2 = \sum_{i=1}^{3} \sum_{j=1}^{3} dq_i G_{ij} dq_j = d\mathbf{q}^T \mathbf{G} d\mathbf{q}, \tag{1.73}$$

where $\mathbf{G}$ is a symmetric matrix and q_i represents either x_i or y_i. Although $\mathbf{G}$ is diagonal for Cartesian coordinates (since $G_{ij} = \delta_{ij}$), it is not diagonal for general curvilinear coordinates. The coefficients G_{ij} are called *metric coefficients*, and the matrix $\mathbf{G}$ is usually referred to simply as the *metric matrix* for the coordinate representation used to describe points in the space Ω. The metric matrix enables the computation of the invariant quantity "distance" in terms of a particular coordinate representation for the respective space.

Since the vectors $\mathbf{g}_i$ need not be unit vectors, they can be expressed in terms of a set of unit vectors $\mathbf{v}_i$, parallel to the corresponding $\mathbf{g}_i$ in the form

$$\mathbf{g}_i = s_i \mathbf{v}_i, \quad i = 1, 2, 3. \tag{1.74}$$

Inserting Eq. (1.74) into Eq. (1.73) recasts the latter into the form

$$(dL)^2 = \sum_{i=1}^{3} \sum_{j=1}^{3} (s_i dq_i)\, C_{ij}\, (s_j dq_j) = (\mathbf{S} d\mathbf{q})^T \mathbf{C}\, (\mathbf{S} d\mathbf{q}), \tag{1.75}$$

where $\mathbf{S}$ is a diagonal matrix (with elements $s_i \delta_{ij}$), called the *dispersion matrix*, and $\mathbf{C}$ is a matrix with elements C_{ij} satisfying the conditions

$$C_{ij} \triangleq \mathbf{v}_i^T \mathbf{v}_j; \quad -1 \le C_{ij} \le 1, \quad i, j = 1, 2, 3. \tag{1.76}$$

Equation (1.75) indicates that $\mathbf{C}$ plays the role of a correlation matrix. Since the vectors $\mathbf{v}_i$ are unit vectors, it follows that

$$\mathbf{v}_i^T \mathbf{v}_j = \cos \theta_{ij}, \tag{1.77}$$

where θ_{ij} is the angle between the vectors. Hence, the partial covariances C_{ij} and the angles $\cos \theta_{ij}$ are equivalent, which provides a geometric interpretation of correlation; full correlation (anti-correlation) can be interpreted in terms of parallel (anti-parallel) vectors, while the lack of correlation can be visualized in terms of perpendicular vectors. This geometrical analogy can be summarized as follows:

1. The parameters of a physical problem correspond to the coordinates in a vector space Ω;

2. The space Ω represents the range of possibilities for these parameters;

3. Correlations arise because of the basic nonorthogonality of the set of parameters used to represent the physical problem;

4. Points along a curve in Ω correspond to various experimental outcomes for different values of the parameters (assuming, for simplicity, that the outcome of the experiment is a scalar quantity).

5. The "distance" between two nearby points, which differ only due to small positive or negative increments of the coordinates (deviations in the parameters) corresponds to the uncertainty in the experimental results; this uncertainty is computed using a differential quadratic form, with knowledge of the dispersion matrix and the correlation matrices.

The geometric model of error presented in the foregoing can be extended to include a wider range of possibilities, by considering that the set $\mathbf{x} = (x_1, \ldots, x_n)$ of random variables represents fundamental parameters of a physical problem, and that the components of the vector $\mathbf{y} = (y_1, \ldots, y_m)$ represent derived quantities, i.e., $y_k = y_k(\mathbf{x})$, for $k = 1, \ldots, m$. For example, in a typical nuclear experiment, the components of $\mathbf{x}$ might represent measured quantities such as reaction yields, sample masses, neutron fluences, neutron energies, etc., while the quantities $\mathbf{y}$ could represent computed reaction cross sections. Suppose that we know the covariance matrix, $\mathbf{C}_x$, of the parameters $\mathbf{x}$ and need to determine the covariance matrix, $\mathbf{C}_y$, of the derived quantities $\mathbf{y}$. For this purpose, we construct vectors $\mathbf{e}_{xi}$ in the error space corresponding to the variable x_i, such that

$$\mathbf{e}_{x\,i}^T \mathbf{e}_{xj} = C_{xij}, \quad i, j = 1, \ldots, n. \tag{1.78}$$

In particular, the amplitude $|\mathbf{e}_{xi}|$ equals the standard deviation, $\sigma_{x\,i}$, of x_i, so that

$$\mathbf{e}_{xi}^T \mathbf{e}_{xj} = \sigma_{xi}\sigma_{xj}\cos\alpha_{ij}, \quad i, j = 1, \ldots, n. \tag{1.79}$$

The parameter α_{ij} is the angle between $\mathbf{e}_{xi}$ and $\mathbf{e}_{xj}$ and, as was discussed in the foregoing, there is a one-to-one correspondence between the error correlation factor C_{xij} and $\cos\alpha_{ij}$.

The error in y_k can be represented as a vector by assuming that errors add like vectors, according to the law of linear superposition, i.e.,

$$\mathbf{e}_{yk} = \sum_{i=1}^{n} s_{ik}\mathbf{e}_{xi}, \quad k = 1, \ldots, m, \tag{1.80}$$

where the quantities s_{ik} remain to be determined by requiring that the treatment of the uncertainties of $\mathbf{y}$ should be consistent with the treatment for the uncertainties of $\mathbf{x}$. This requirement implies that

$$\mathbf{e}_{yk}^T \mathbf{e}_{yq} = \sigma_{yk}\sigma_{yq}\cos\beta_{kq} = C_{ykq}, \quad k, q = 1, \ldots, m, \tag{1.81}$$

so that the standard deviation σ_{yk} equals $|\mathbf{e}_{yk}|$, β_{kq} represents the angle between $\mathbf{e}_{yk}$ and $\mathbf{e}_{yq}$, while $\cos\beta_{kq}$ is equivalent to the correlation factor C_{ykq}. Comparing Eqs. (1.78) through (1.81) indicates that Eq. (1.80) is valid if

$$s_{ik} = (\partial y_k/\partial x_i)_\mu, \quad \mu = (\langle x_1 \rangle, \ldots, \langle x_n \rangle), \quad i = 1, \ldots, n; \quad k = 1, \ldots, m \tag{1.82}$$

The quantity s_{ik} is customarily called the *sensitivity* of (the derived quantity) y_k to the parameter x_i. In terms of the covariance matrix, $\mathbf{C}_x$, of the parameters $\mathbf{x}$ the explicit expression of the covariance matrix, $\mathbf{C}_y$, associated with the derived quantities $\mathbf{y}$ can be explicitly written in matrix form as

$$\mathbf{C}_y = \mathbf{S}\mathbf{C}_x\mathbf{S}^T, \quad \mathbf{S} \triangleq (s_{ik})_{n\times m}. \tag{1.83}$$

The above expression is colloquially known as the "*sandwich rule*" for "propagation of input parameter uncertainties" to compute the uncertainties in "responses" (i.e., derived quantities). The above "sandwich rule" holds rigorously only if the components of the derived quantity $\mathbf{y}$ are linear functions of the components of the parameters $\mathbf{x}$, i.e.,

$$y_k \left(\mu + \delta \mathbf{x} \right) = y_k \left(\mu \right) + \left(\partial y_k / \partial x_i \right)_\mu \delta \mathbf{x}. \tag{1.84}$$

The "geometrical model of errors" provides a useful mechanism for employing graphical analysis in the examination of errors, since error correlations can be visualized in terms of the cosines of angles between the various error vectors. The sandwich rule given in Eq. (1.83) is often used in practice, particularly for large-scale models involving many parameters, since information beyond first-order terms is often unavailable or impractical to use, which imposes (de facto) the assumption that responses are linearly dependent on parameters.

When information beyond first order is available, the respective response is expanded in a multivariate Taylor series, retaining derivatives beyond the first order sensitivities, by means of the *method of propagation of errors or propagation of moments* (see, e.g., ref. [29]), which can be used when the response is either the result of an indirect measurement or the result of a computation. The response, denoted by R, is considered to be a real-valued function of k system parameters, denoted as $(\alpha_1, \ldots, \alpha_k)$, with mean values $\left(\alpha_1^0, \ldots, \alpha_k^0 \right)$, i.e.,

$$R = R \left(\alpha_1, \ldots, \alpha_k \right) = R \left(\alpha_1^0 + \delta \alpha_1, \ldots, \alpha_k^0 + \delta \alpha_k \right). \tag{1.85}$$

Expanding $R \left(\alpha_1^0 + \delta \alpha_1, \ldots, \alpha_k^0 + \delta \alpha_k \right)$ in a Taylor series around the nominal values $\alpha^0 = \left(\alpha_1^0, \ldots, \alpha_k^0 \right)$ and retaining the terms up to the nth order in the variations $\delta\alpha_i \equiv (\alpha_i - \alpha_i^0)$ around α_i^0 gives:

$$R(\alpha_1,\ldots,\alpha_k) \equiv R\left(\alpha_1^0 + \delta\alpha_1,\ldots,\alpha_k^0 + \delta\alpha_k\right)$$

$$= R(\alpha^0) + \sum_{i_1=1}^{k} \left(\frac{\partial R}{\partial\alpha_{i_1}}\right)_{\alpha^0} \delta\alpha_{i_1}$$

$$+ \frac{1}{2}\sum_{i_1,i_2=1}^{k} \left(\frac{\partial^2 R}{\partial\alpha_{i_1}\partial\alpha_{i_2}}\right)_{\alpha^0} \delta\alpha_{i_1}\delta\alpha_{i_2}$$

$$+ \frac{1}{3!}\sum_{i_1,i_2,i_3=1}^{k} \left(\frac{\partial^3 R}{\partial\alpha_{i_1}\partial\alpha_{i_2}\partial\alpha_{i_3}}\right)_{\alpha^0} \delta\alpha_{i_1}\delta\alpha_{i_2}\delta\alpha_{i_3} + \cdots$$

$$+ \frac{1}{n!}\sum_{i_1,i_2,\ldots,i_n=1}^{k} \left(\frac{\partial^n R}{\partial\alpha_{i_1}\partial\alpha_{i_2}\ldots\partial\alpha_{i_n}}\right)_{\alpha^0} \delta\alpha_{i_1}\ldots\delta\alpha_{i_n}.$$

$$(1.86)$$

Recalling the definition of the moments for multivariate probability, cf. Eq. (1.46), the above Taylor-series expansion can now be used to determine the mean (expectation), variance, skewness, and kurtosis, of $R(\alpha_1,\ldots,\alpha_k)$ by formal integrations over the (unknown) joint probability density function $p(\alpha_1,\ldots,\alpha_k)$ of the parameters $(\alpha_1,\ldots,\alpha_k)$. For *uncorrelated parameters* $(\alpha_1,\ldots,\alpha_k)$, the expressions of the various moments of the response $R(\alpha_1,\ldots,\alpha_k)$ are as follows:

$$E(R) = R\left(\alpha_1^0,\ldots,\alpha_k^0\right) + \frac{1}{2}\sum_{i=1}^{k}\left\{\frac{\partial^2 R}{\partial\alpha_i^2}\right\}_{\alpha^0}\mu_2(\alpha_i)$$

$$+ \frac{1}{6}\sum_{i=1}^{k}\left\{\frac{\partial^3 R}{\partial\alpha_i^3}\right\}_{\alpha^0}\mu_3(\alpha_i) + \frac{1}{24}\sum_{i=1}^{k}\left\{\frac{\partial^4 R}{\partial\alpha_i^4}\right\}_{\alpha^0}\mu_4(\alpha_i)$$

$$+ \frac{1}{24}\sum_{i=1}^{k-1}\sum_{j=i+1}^{k}\left\{\frac{\partial^4 R}{\partial\alpha_i^2\partial\alpha_j^2}\right\}_{\alpha^0}\mu_2(\alpha_i)\mu_2(\alpha_j) ; \qquad (1.87)$$

$$\mu_2\left(R\right) = \sum_{i=1}^{k}\left\{\left(\frac{\partial R}{\partial\alpha_i}\right)^2\right\}_{\alpha^o}\mu_2\left(\alpha_i\right) + \sum_{i=1}^{k}\left\{\frac{\partial R}{\partial\alpha_i}\frac{\partial^2 R}{\partial\alpha_i^2}\right\}_{\alpha^o}\mu_3\left(\alpha_i\right)$$

$$+\frac{1}{3}\sum_{i=1}^{k}\left\{\frac{\partial R}{\partial\alpha_i}\frac{\partial^3 R}{\partial\alpha_i^3}\right\}_{\alpha^o}\mu_4\left(\alpha_i\right)$$

$$+\frac{1}{4}\sum_{i=1}^{k}\left\{\left(\frac{\partial^2 R}{\partial\alpha_i^2}\right)^2\right\}_{\alpha^o}\left[\mu_4\left(\alpha_i\right) - \left(\mu_2\left(\alpha_i\right)\right)^2\right] ; \tag{1.88}$$

$$\mu_3\left(R\right) = \sum_{i=1}^{k}\left\{\left(\frac{\partial R}{\partial\alpha_i}\right)^3\right\}_{\alpha^o}\mu_3\left(\alpha_i\right)$$

$$+\frac{3}{2}\sum_{i=1}^{k}\left\{\left(\frac{\partial R}{\partial\alpha_i}\right)^2\frac{\partial^2 R}{\partial\alpha_i^2}\right\}_{\alpha^o}\left[\mu_4\left(\alpha_i\right) - \left(\mu_2\left(\alpha_i\right)\right)^2\right] ; \tag{1.89}$$

$$\mu_4\left(R\right) = \sum_{i=1}^{k}\left\{\left(\frac{\partial R}{\partial\alpha_i}\right)^4\right\}_{\alpha^o}\left[\mu_4\left(\alpha_i\right) - 3(\mu_2\left(\alpha_i\right))^2\right] + 3[\mu_2\left(R\right)]^2. \tag{1.90}$$

In Eqs. (1.87) through (1.90), the quantities $\mu_l\left(R\right)$, $\left(l = 1,\ldots,4\right)$, denote the respective central moments of the response $R\left(\alpha_1,\ldots,\alpha_k\right)$, while the quantities $\mu_k\left(\alpha_i\right)$, $\left(i = 1,\ldots,k;\ k = 1,\ldots,4\right)$ denote the respective central moments of the parameters $\left(\alpha_1,\ldots,\alpha_k\right)$. Note that $E\left(R\right) \neq R^0$ when the response $R\left(\alpha_1,\ldots,\alpha_k\right)$ is a nonlinear function of the parameters $\left(\alpha_1,\ldots,\alpha_k\right)$.

It is important to emphasize that the "propagation of moments" equations, i.e., Eqs. (1.87) through (1.90), are used not only for processing experimental data obtained from indirect measurements, but are also used for performing statistical analysis of computational models. In the latter case, the "propagation of errors" equations provide a systematic way of obtaining the uncertainties in computed results, arising not only from uncertainties in the parameters that enter the respective computational model but also from the numerical approximations themselves. Clearly, the propagation of errors method can be used only if the exact response derivatives to parameters are also available, in addition to the respective parameter uncertainties.

1.3.3   Computing Covariances: Simple Examples

Consider the computation of the joint distribution of two variates u and v, defined as $u = x + z$ and $v = y + z$, where x, y, and z are three independent, normally distributed variates with zero means. The variances, and covariance of u and v can be readily obtained as

$$\langle \delta u \delta v \rangle = \langle (\delta x + \delta z)(\delta y + \delta z) \rangle = \ldots = \left\langle (\delta z)^2 \right\rangle, \left\langle (\delta u)^2 \right\rangle$$

$$= \left\langle (\delta x)^2 \right\rangle \left\langle (\delta z)^2 \right\rangle, and$$

$$\left\langle (\delta v)^2 \right\rangle = \left\langle (\delta y)^2 \right\rangle \left\langle (\delta z)^2 \right\rangle.$$

Hence, the uncertainty matrix associated with u and v is

$$\mathbf{C} = \begin{pmatrix} \sigma_x^2 & 0 \\ 0 & \sigma_y^2 \end{pmatrix} + \sigma_z^2 \begin{pmatrix} 1 & 1 \\ 1 & 1 \end{pmatrix}.$$

The determinant of this uncertainty matrix is $|\mathbf{C}| = \sigma_x^2 \sigma_y^2 + \sigma_y^2 \sigma_z^2 + \sigma_z^2 \sigma_x^2$. Hence, the inverse of $\mathbf{C}$ is

$$\mathbf{C}^{-1} = \frac{1}{|\mathbf{C}|} \left[\begin{pmatrix} \sigma_y^2 & 0 \\ 0 & \sigma_x^2 \end{pmatrix} + \sigma_z^2 \begin{pmatrix} 1 & -1 \\ -1 & 1 \end{pmatrix} \right].$$

It follows that the joint distribution is

$$f(\mathbf{w}) = \frac{1}{2\pi |\mathbf{C}|} \exp \left[-1/2 \, \mathbf{w}^T \mathbf{C}^{-1} \mathbf{w} \right],$$

where $\mathbf{w}$ denotes the column vector (u, v).

As the next example, consider a variate $y(x) = ax + b$ which depends linearly on a parameter x, but the coefficients a and b are unknown and need to be evaluated, together with their associated uncertainties, by using two measurements of y at the values x_1 and x_2 of x, $0 < x_1 < x_2$, such that

$$y(x_1) = y_1 \pm \sigma_1, \quad y(x_2) = y_2 \pm \sigma_2, \quad \text{cov}[y(x_1), y(x_2)] = \sigma_1 \sigma_2 \rho.$$

Using these two measurements, it follows that

$$y\left(x\right) = \frac{1}{\Delta x}\left[\left(y_2 - y_1\right)x + \left(x_2 y_1 - x_1 y_2\right)\right], \Delta x \triangleq x_2 - x_1.$$

From the above expression, the uncertainties in the coefficients a and b are computed as

$$\text{var}\left(a\right) = \frac{1}{\left(\Delta x\right)^2}\left(\sigma_1^2 + \sigma_2^2 - 2\sigma_1\sigma_2\rho\right),$$

$$\text{var}\left(b\right) = \frac{1}{\left(\Delta x\right)^2}\left(x_2^2\sigma_1^2 + x_1^2\sigma_2^2 - 2x_1 x_2\sigma_1\sigma_2\rho\right),$$

$$\text{cov}\left(a,\, b\right) = \frac{1}{\left(\Delta x\right)^2}\left[\left(x_1 + x_2\right)\sigma_1\sigma_2\rho - x_2\sigma_1^2 - x_1\sigma_2^2\right].$$

The significance of correlations with respect to combinations of correlated data becomes evident: if, for example, the two measurements were fully correlated ($\rho = 1$), then the above expressions would reduce to

$$\sigma_a = \frac{1}{\Delta x}\left|\sigma_1 - \sigma_2\right|, \qquad \sigma_b = \frac{1}{\Delta x}\left|x_2\sigma_1 - x_1\sigma_2\right|$$

and

$$\rho\left(a,\, b\right) = -\frac{\left(\sigma_1 - \sigma_2\right)\left(x_2\sigma_1 - x_1\sigma_2\right)}{\left|\left(\sigma_1 - \sigma_2\right)\left(x_2\sigma_1 - x_1\sigma_2\right)\right|}.$$

Since $x_2 > x_1 > 0$, it follows that, if $\rho = 1$, then

$$\rho\left(a,\, b\right) = \begin{cases} +1, & \sigma_1 < \sigma_2 < x_2\sigma_1/x_1, \\ -1, & otherwise. \end{cases}$$

Following the same procedure, it can be shown that the coefficients a and b are partially anti-correlated when $\rho = 0$; finally, the coefficients a and b are fully anti-correlated if $\rho = -1$.

If the variate $y\left(x\right) = ax + b$ is evaluated at two other points, x' and x'', the corresponding variances and covariances are

$$\text{cov}\left[y\left(x'\right),\, y\left(x''\right)\right] = x'x''\,\text{var}\left(a\right) + \text{var}\left(b\right) + \left(x' + x''\right)\text{cov}\left(a,\, b\right)$$

$$\text{var}\left[y\left(x\right)\right] = x^2\,\text{var}\left(a\right) + \text{var}\left(b\right) + 2x\,\text{cov}\left(a,\, b\right).$$

As a function of x, the quantity $\text{var}\,[y\,(x)]$ takes on the minimal value $\text{var}\,(b)\,[1 - \rho^2\,(a,b)]$ at $x = -\text{cov}\,(a,b)/\text{var}\,(a)$. If the coefficients a and b are either fully correlated or fully anti-correlated (i.e., if $|\rho| = 1$), then $\text{var}\,[y\,(x_0)] = 0$ at $x_0 = (x_2\sigma_1 - x_1\sigma_2\rho)/(\sigma_1 - \sigma_2\rho)$.

When evaluating uncertainties associated with products and quotients of variates, it is more convenient to perform the respective computations using relative rather than absolute variances and covariances. Using *capital letters to denote relative variances and covariances*, and noting that $\delta a/a$ represents the relative variation of a variate a, it follows that the relative variance is related to the absolute variance through the formula

$$Var\,(a) \triangleq \left\langle \left(\frac{\delta a}{a}\right)^2 \right\rangle = \frac{1}{a^2}\left\langle (\delta a)^2 \right\rangle = \frac{\text{var}\,(a)}{a^2}.$$

Similarly, the relative covariance of two variates a and b is related to their absolute covariance through the formula

$$Cov\,(a, b) \triangleq \left\langle \frac{\delta a}{a}\frac{\delta b}{b} \right\rangle = \frac{\text{cov}\,(a, b)}{ab}.$$

Hence, the absolute variance of a product $c = ab$ of two variates a and b is

$$\text{var}\,(c) = b^2\,\text{var}\,(a) + a^2\,\text{var}\,(b) + 2ab\,\text{cov}\,(a, b), \qquad (1.91)$$

while the relative variance of c is readily obtained as

$$Var\,(c) = Var\,(a) + Var\,(b) + 2Cov\,(a, b). \qquad (1.92)$$

Note, as an aside, that the above expression has the same form as the expression of the absolute variance of $c = a+b$. Similarly, for the ratio $c = a/b$ of two variates a and b, the corresponding relative variance is

$$Var\,(c) = Var\,(a) + Var\,(b) - 2Cov\,(a, b).$$

Generalizing the above considerations to a variate of the form

$$y = \frac{x_1 x_2 ... x_m}{x_{m+1} ... x_n}$$

leads to the following expression for the relative variance $Var(y)$:

$$Var(y) = \sum_{i=1}^{n} Var(x_i)$$

$$+ 2\left\{ \sum_{i=1}^{m} \left[\sum_{j=i+1}^{m} Cov(x_i, x_j) - \sum_{j=m+1}^{n} Cov(x_i, x_j) \right] + \sum_{i=m+1}^{n} \sum_{j=i+1}^{n} Cov(x_i, x_j) \right\}.$$

As another example of evaluating a covariance matrix, consider a typical measurement of a neutron activation cross section, for which the uncertainty analysis is performed using representative numerical data. A typical experimental setup for measuring a neutron activation cross involves the irradiation of a sample, which comprises N nuclei of an isotope that can be activated by neutrons, in a given constant and uniform neutron field, characterized by the neutron flux φ. The activated nuclei decay by emitting gamma-ray photons of a definite energy E. After the sample is irradiated to saturation, its activity (i.e., the rate of disintegration) equals the production rate of activated nuclei, which is $N\sigma\varphi$, where σ is the unknown microscopic activation cross section (averaged over the energy distribution of the given flux) to be determined/evaluated. Consider also that the efficiency of the gamma-ray detector for photons of energy E is ε. The unknown activation cross section is thus obtained from the simple relation

$$\sigma = \frac{A}{\varepsilon N \varphi}, \tag{1.93}$$

where A is the actual counting rate registered by the detector. Suppose that the irradiated sample contains three isotopes which are activated simultaneously in one irradiation field; in other words, the same neutron flux applies to all of the three different activities: $\varphi_1 = \varphi_2 = \varphi_3 = \varphi$. Hence, these "three" fluxes (in reality, the same flux φ) are fully correlated, i.e., $\rho(\varphi_i, \varphi_j) = 1$ for all i and j. Consider also that the following additional information is available:

1. The uncertainty in N is negligible by comparison to all other uncertainties;

2. The relative standard deviation of φ is 2%;

3. The measured activities A_1, A_2, and A_3, are all uncorrelated, and their respective standard deviations are 0.5%, 1.0%, and 0.3%, respectively;

4. The uncertainties in the detector efficiencies corresponding to the three photon energies are correlated, as given in Table (1.2).

Since the uncertainty in N is negligible, it follows that the relative variation of the cross-section σ_i, computed to first order using Eq. (1.93), is

$$\frac{\delta\sigma_i}{\sigma_i} = \frac{\delta A_i}{A_i} - \frac{\delta\varepsilon_i}{\varepsilon_i} - \frac{\delta\varphi_i}{\varphi_i}, \ i = 1, 2, 3. \qquad (1.94)$$

TABLE 1.2
Uncertainties in the detector efficiencies corresponding to the three correlated photon energies

	Relative Standard Deviation (in %)	Correlation Matrix		
ε_1	1.6	1.0		
ε_2	2.2	0.8	1.0	
ε_3	1.3	0.5	0.9	1.0

The three sources of uncertainty for A, ε, and φ, respectively, are uncorrelated. Hence, the relative variance of σ, and the relative covariances $Cov(\sigma_i, \sigma_j)$ can be computed to first order from Eq. (1.94) to obtain

$$Var(\sigma_i) = Var(A_i) + Var(\varepsilon_i) + Var(\varphi_i),$$

$$Cov(\sigma_i, \sigma_j) = \left\langle \left(\frac{\delta A_i}{A_i} - \frac{\delta\varepsilon_i}{\varepsilon_i} - \frac{\delta\varphi_i}{\varphi_i} \right) \left(\frac{\delta A_j}{A_j} - \frac{\delta\varepsilon_j}{\varepsilon_j} - \frac{\delta\varphi_j}{\varphi_j} \right) \right\rangle =$$

$$= \left\langle \frac{\delta\varepsilon_i}{\varepsilon_i} \frac{\delta\varepsilon_j}{\varepsilon_j} \right\rangle + \left\langle \frac{\delta\varphi_i}{\varphi_i} \frac{\delta\varphi_j}{\varphi_j} \right\rangle = Cov(\varepsilon_i, \varepsilon_j) + Cov(\varphi_i, \varphi_j).$$

Inserting in the above expressions the numerical information provided in the foregoing yields $Var(\sigma_1) = (0.5^2 + 1.6^2 + 2.0^2)\% = 6.81\%$, so that the relative standard deviation is $\sigma_1 = 2.610\%$. A similar computation yields $Var(\sigma_2) = 9.84\%$, or $\sigma_2 = 3.137\%$. The relative covariance of σ_1 and σ_2 is $Cov(\sigma_1, \sigma_2) = 1.6 \times 2.2 \times 0.8 + 2.0 \times 2.0 \times 1.0 = 6.816$, yielding the following value for corresponding correlation coefficient:

$$\rho(\sigma_1, \sigma_2) = \frac{6.816}{2.610 \times 3.137} = 0.832 \ .$$

The remaining computations are left as an exercise for the reader; the complete uncertainty information pertaining to the three measured cross sections is given in Table (1.3):

TABLE 1.3
Uncertainty information pertaining to the three
measured cross sections

	Relative Standard Deviation (in %)	Correlation Matrix		
ε_1	2.610	1.000		
ε_2	3.137	0.832	1.000	
ε_3	2.404	0.803	0.872	1.000

2

Computation of Means and Variances from Measurements

CONTENTS

The full information gained from a measurement of some physical quantity x is not limited to just a single value. Rather, a measurement yields information about a set of discrete probabilities P_j, when the possible values of x form a discrete set. Similarly, when the possible values of x form a continuum with probability density function $p(x)$, a measurement yields information about a set of infinitesimal probabilities $p(x) \, dx$ of the true value lying between x and $x + dx$. Note that x need not be a bona-fide random quantity for the probability distribution to exist, but could be (and in science and technology it usually is) an imperfectly known constant. In practice, users of measured data seldom require knowledge of the complete posterior distribution, but usually request a "recommended value" for the respective quantity, accompanied by "error bars" or some suitably equivalent summary of the posterior distribution. Decision theory can provide such a summary, since it describes the penalty for bad estimates by a loss function. Since the true value is never known in practice, it is not possible to avoid a loss completely, but it is possible to minimize the expected loss, which is what an optimal estimate must accomplish. As will be shown in this Chapter, in the practically most important case of "quadratic loss" involving a multivariate posterior distribution, the "recommended value" turns out to be the vector of mean values, while the "error bars" are provided by the corresponding covariance matrix. Conversely, an experimental result reported in the form $\langle x \rangle \pm \Delta x$, where Δx represents the standard deviation (i.e., the root-mean-square error), is customarily interpreted as a short-hand notation representing a distribution of possible values x that cannot be recovered in detailed form, but is characterized by the mean $\langle x \rangle$ and standard deviation Δx.

Practical applications require not only mathematical relations between probabilities, but also rules for assigning numerical values to probabilities. Thus, Section 2.1 presents simple examples of computing parameter covariances. Furthermore, Section 2.2 discusses the assignment of prior probability distributions under incomplete information. When no prior information is available, numerical values can be assigned to probabilities by using concepts of group theory, reflecting invariance or symmetries properties of the problem under consideration, as discussed in Section 2.2.1. On the other hand, if repeatable trials are not feasible, but some information could nevertheless be inferred by some other means, information theory can be used in conjunction with the maximum entropy principle (the modern generalization of Bernoulli's principle of insufficient reason) to assign numerical values to probabilities, as will be shown in Section 2.2.2.

Section 2.3 presents methods for evaluating unknown parameters from data that is consistent "within error bars." Thus, the evaluation of a location parameter when the scale parameters are known is presented in subsection 2.3.1. Subsection 2.3.2 analyses the situation when both the scale and location parameters are unknown but need to be evaluated. A problem often encountered in practice is illustrated in subsection 2.3.3, which presents the evaluation of a counting rate (a scale parameter) in the presence of background (noise). Section 2.4 addresses the practical situation of evaluating means and covariances from measurements affected by both random (uncorrelated) and systematic (correlated) errors. As will be mathematically shown in Section 2.4, the random errors can be reduced by repeated measurements of the same quantity, but the systematic errors cannot be reduced this way. They remain as a "residual" uncertainty that could possibly be reduced only by additional measurements, using different techniques and instrumentation, geometry, and so on. Eventually, the effect of additional measurements using different techniques would reduce the correlated (systematic) error just as repetitions of the same measurement using the same technique reduce the uncorrelated statistical uncertainty.

As discussed in Chapter 1, unrecognized or ill-corrected experimental effects, including background, dead time of the counting electronics, instrumental resolution, sample impurities, and calibration errors usually yield inconsistent experimental data. Although genuinely discrepant data may occur with a nonzero probability, it is much more likely that apparently discrepant experiments actually indicate the presence of unrecognized errors. Section 2.5 illustrates the basic principles for evaluating discrepant data fraught by unrecognized, including systematic (common), errors. Section 2.6 concludes this chapter by offering some notes and remarks.

2.1 Statistical Estimation of Means, Covariances, and Confidence Intervals

In some practical cases, information comes in the form of observations, usually associated with a frequency distribution. These observations are, in turn, used to estimate the mathematical form and/or the parameters describing the underlying probability distribution function. The use of observations to estimate the underlying features of probability functions is the objective of a branch of mathematical sciences called *statistics*. Conceptually, the objective of *statistical estimation* is to estimate the parameters $(\theta_1, \ldots \theta_k)$ that describe a particular statistical model, by using observations, x_n, of a frequency function $f(x; \theta_1, \ldots \theta_k)$. Furthermore, this statistical estimation process must provide reasonable assurance that the model based on these estimates will fit the observed data within acceptable limits. It is also important that the *statistical estimates* obtained from observational data be *consistent, unbiased, and efficient*. Therefore, the science of statistics embodies both inductive and deductive reasoning, encompassing procedures for estimating parameters from incomplete knowledge and for refining prior knowledge by consistently incorporating additional information. Hence, the solution to practical problems requires a synergetic use of the various interpretations of probability, including the axiomatic, frequency, and Bayesian interpretations and methodologies. For simplicity, the symbol x will be used in this section to represent both a random quantity and a typical value; a distinction between these two uses will be made only when necessary to avoid confusion. Thus, x is considered to be described by a probability density $p(x)$, with a mean value $E(x) = m_o$ and var $(x) = \sigma^2$. Without loss of generality, x can be considered to be continuous, taking values in an uncountably infinite space; the statistical formalisms to be developed in the following can be similarly developed for finite or countably infinite random variables.

In both classical and Bayesian statistics, the estimation procedures are applied to samples of data, since the complete set of data is never available in practice. A sample, $\mathbf{x}_s$, of size n is defined as a collection of n equally distributed random variables $(x_1, x_2, \ldots, x_n)$; each x_i is associated with the same event space and has the same probability density, $p_i(x_i) = p(x_i)$. The random variable x, which each x_i resembles, is customarily called the *parent random variable*. Each sampling step corresponds to the selection of a sample; thus, the first step selects sample 1, which corresponds to x_1, while the last step (i.e., the nth-step) selects sample n, which corresponds to x_n. The selection of values x_i is called a *sampling process*, and the result of this process is the n-tuple of values $\mathbf{x}_s \equiv (x_1, x_2, \ldots, x_n)$. If the sampling is random (i.e., the selection of each x_i is unaffected by the selection of all other x_j, $j \neq i$), then the collection of random variables can be treated as a random vector $\mathbf{x}_s \equiv (x_1, x_2, \ldots, x_n)$ distributed according to the multivariate probability density $p(x_1, x_2, \ldots, x_n) = p(x_1) p(x_2) \ldots p(x_n)$. For random sampling, the components x_i are uncorrelated (i.e., the covariance matrix is diagonal), and have mean values $E(x_i) = m_o$ and variances var $(x_i) = E\left[(x_i - m_o)^2\right] = \sigma^2$, identical to the mean value and standard deviation of the parent distribution x.

A function $\Theta(x_1,\ldots,x_n)$ that acts only on the sample random variables (and, possibly, on well-defined constants) is called a *statistic*. An *estimator*, $T = \Theta(x_1,\ldots,x_n)$, is a statistic specifically employed to provide estimated values for a particular, true yet unknown, constant value T_o of the underlying probability distribution for the parent variable x. The function Θ is called the *estimation rule*. Since this rule is designed to provide specific values of T that are meant to approximate the constant T_o, the estimator T is called a *point estimator*. The process of selecting estimators is not unique, so the criteria used for particular selections are very important, since they determine the properties of the resulting estimated values for the parameters of the chosen model.

An estimator T of a physical quantity T_o is called *consistent* if it approaches the true value T_o of that quantity (i.e., it converges in probability to T_o) as the number of observations x_n of T_o increases:

$$T(x_1,\ldots,x_n) \xrightarrow{n\to\infty} T_o. \tag{2.1}$$

An estimator T of T_o is called *unbiased* if its expectation is equal to the true value of the estimated quantity:

$$E(T) = T_o. \tag{2.2}$$

The bias $B(T,T_o)$ of an estimator is defined as

$$B(T,T_o) \equiv E(T) - T_o. \tag{2.3}$$

If $B(T,T_o) > 0$, then T tends to overestimate T_o; if $B(T,T_o) < 0$, then T tends to underestimate T_o. The quantity $E\left[(T-T_o)^2\right]$ is called the *mean-squared error*. If the estimator $T = \Theta(x_1,\ldots,x_n)$ utilizes all the information in the sample that pertains to T_o, then the respective estimator is called a *sufficient* estimator. In practice, the choice of estimators is further limited by considering unbiased estimators, which, among all similar estimators, have the smallest variance. A consistent, unbiased, and minimum variance estimator is called an *efficient* estimator.

Intuitively, it would be expected that the smaller the variance of an unbiased estimator, the closer the estimator is to the respective parameter value. This intuitive expectation is indeed correct. For example, $T(x_1, x_2, \ldots, x_n) = \sum_{i=1}^{n} a_i x_i$, where $(\alpha_1, \ldots, \alpha_n)$ are constants, would be a *best linear unbiased estimator (BLUE)* of a parameter θ, if $T(x_1, \ldots, x_n)$ is a linear unbiased estimator such that $var\{T(x_1, \ldots, x_n)\} \leq var\{T'(x_1, \ldots, x_n)\}$, where $T'(x_1, \ldots, x_n)$ is any other linear unbiased estimator of θ.

The sample moment of order k, x_k^S, is the statistic defined as

$$x_k^S \equiv (1/n) \sum_{i=1}^{n} x_i^k, \quad (k = 1, 2, \ldots). \tag{2.4}$$

In the above definition, the superscript "S" denotes "sample." In the special case when, Eq. (2.4) defines the *sample mean value*, which is customarily denoted as $\overline{x}$:

$$\overline{x} \equiv (1/n) \sum_{i=1}^{n} x_i. \tag{2.5}$$

Note that

$$E(\overline{x}) = \frac{1}{n} \sum_{i=1}^{n} E(x_i) = \frac{1}{n} n E(x) = m_0, \tag{2.6}$$

which indicates that the expectation value, $E(\overline{x})$, of the sample mean is an unbiased estimator for the distribution mean value $E(x) = m_0$. It can also be shown that the variance of the sample mean, $var(\overline{x})$, is related to $\sigma^2 = var(x)$ by means of the relation

$$var(\overline{x}) = \sigma^2/n = var(x)/n. \tag{2.7}$$

The sample central moment statistic of order k is defined as

$$\mu_k^S \equiv (1/n) \sum_{i=1}^{n} (x_i - \overline{x})^k, \quad (k = 1, 2, 3, \ldots). \tag{2.8}$$

In particular, the *sample variance* is obtained by setting $k = 2$ in the above definition, to obtain:

$$\mu_2^S \equiv (1/n) \sum_{i=1}^{n} (x_i - \overline{x})^2, \tag{2.9}$$

The sample standard deviation [see also Eq. (2.18), below] is calculated using the formula

$$SD\left(\bar{x}\right) = \sqrt{\left(\frac{1}{n-1}\right)\sum_{i=1}^{n}\left(x_i - \bar{x}\right)^2}. \tag{2.10}$$

The properties of sampling distributions from a normally distributed parent random variable x are of particular practical importance due to the prominent practical and theoretical role played by the central limit theorem. Thus, consider a normally distributed parent random variable x, with mean m_o and variance σ^2. Furthermore, consider a sample $(x_1, x_2, \ldots, x_n)$ with sample mean value $\bar{x}$ and sample variance μ_2^S, as defined in Eqs. (2.5) and (2.9), respectively. Then, the following theorems hold and are often used in practice:

1. The quantities $\bar{x}$ and μ_2^S are independent random variables; note that the converse also holds, namely, if $\bar{x}$ and μ_2^S are independent, then the distribution for x must be normal.

2. The random variable $(n\mu_2^S/\sigma^2)$ is distributed according to a chi-square distribution with $(n-1)$ degrees of freedom.

If y is a random variable distributed according to a χ^2-square distribution with n degrees of freedom, and z is distributed as a standard normal random variable, and y and z are independent, then the ratio random variable $r \equiv z/\sqrt{y/n}$ is distributed according to Student's t-distribution with n degrees of freedom. In particular, this theorem holds when r, y, z are random variables defined as $(n\mu_2^S/\sigma^2) \equiv y$, $(\bar{x} - m_o)/\sigma^2 \equiv z$, and $r \equiv z/\sqrt{y/n}$. The ratio r is frequently used in practice to measure the scatter of the actual data relative to the scatter that would be expected from the parent distribution with standard deviation σ.

If y is distributed according to a χ^2-square distribution with n degrees of freedom, and w is distributed according to a χ^2-square distribution with m degrees of freedom, and y and w are independent, then the ratio variable $R \equiv (w/m)/(y/n)$ is distributed according to an F-distribution with degrees of freedom m and n. It is important to note that the sample mean $\bar{x} \equiv (1/n)\sum_{i=1}^{n} x_i$ defined in Eq. (2.5) is the BLUE for the mean $E\left(x\right) = m_o$ of the parent distribution. This property is demonstrated by first showing that a linear estimator $T\left(x_1, x_2, \ldots, x_n\right) \equiv \sum_{i=1}^{n} a_i x_i$ is unbiased if and only if the constraint $\sum_{i=1}^{n} a_i = 1$ is satisfied. Then, using this constraint, the minimum of

$$\text{var}\left\{T\left(x_1, x_2, \ldots, x_n\right)\right\} = \left[\sum_{i=1}^{n-1} a_i^2 + \left(1 - \sum_{i=1}^{n-1} a_i\right)^2\right] \sigma^2$$

is calculated by setting its first derivative to zero, i.e.,

$$\frac{\partial \text{var}\left\{T\left(x_1, x_2, \ldots, x_n\right)\right\}}{\partial a_i} = \left[2a_i - 2\left(1 - \sum_{i=1}^{n-1} a_i\right)\right] \sigma^2 = (2a_i - 2a_n)\,\sigma^2 = 0\,.$$

The solution of the above equation is $a_i = a_n$, $(i = 1, \ldots, n-1)$. Since $\sum_{i=1}^{n} a_i = 1$, it follows that $a_i = 1/n$, $i = 1, \ldots, n$, which proves that $\bar{x}$ is the BLUE for $E(x) = m_o$.

A concept similar to the BLUE is the *maximum likelihood estimator (MLE)*, which can be introduced, for simplicity, by considering that a single parameter, θ_o, is to be estimated from a random sample $(x_1, x_2, \ldots, x_n)$ of size n. Since each sample is selected independently, the conditional multivariate probability density for the observed sample data set $(x_1, x_2, \ldots, x_n)$ is

$$\prod_{i=1}^{n} p\left(x_i | \theta\right) \equiv L\left(x_1, \ldots, x_n | \theta\right), \tag{2.11}$$

where $p\left(x_i | \theta\right)$ is the conditional probability density that the value x_i will be observed in a single trial. The function $L\left(x_1, \ldots, x_n | \theta\right)$ defined in Eq. (2.11) is called the *likelihood function*. The *maximum-likelihood method* for estimating θ_o consists of finding the particular value, $\tilde{\theta} \equiv \tilde{\theta}\left(x_1, \ldots, x_n\right)$, which maximizes $L\left(x_1, \ldots, x_n | \theta\right)$ for the observed data set $(x_1, x_2, \ldots, x_n)$. Thus, the *MLE*, $\tilde{\theta}$, of θ_o is found as the solution to the equation

$$\left. \frac{d \ln L\left(x_1, \ldots, x_n | \theta\right)}{d\theta} \right|_{\theta = \tilde{\theta}} = 0\,. \tag{2.12}$$

If Eq. (2.12) admits multiple solutions, then the solution that yields the largest likelihood function $L\left(x_1, \ldots, x_n | \theta\right)$ is defined to be the *MLE*. The maximum likelihood method sketched above can, of course, be extended to estimate more than a single parameter θ_o.

For a normally distributed sample $(x_1, \ldots, x_n)$, the sample mean $\bar{x}$ is the

MLE for the parent's distribution mean value, m_o, while the sample variance, μ_2^S, is the *MLE* for the variance σ^2 of the parent's distribution. These results are obtained as follows: since the parent distribution is considered to be normal, the probability distribution for observing x_i is

$$p_i\left(x_i|m_o,\sigma^2\right) = \frac{1}{\sigma\sqrt{2\pi}}e^{-(x_i-m_o)^2/2\sigma^2} \ ,$$

where the *unknown parameters* m_o and σ^2 are to be estimated by finding their respective *MLE's*. First, the likelihood function is obtained according to Eq. (2.11) as

$$L = \left(\frac{1}{\sigma\sqrt{2\pi}}\right)^n \exp\left(-\frac{1}{2\sigma^2}\sum_{i=1}^n (x_i - m_o)^2\right) .$$

The maximum of L can be conveniently calculated by setting to zero the partial derivatives of $\ln L$ with respect to the unknown parameters m_o and σ^2, to obtain:

$$\frac{\partial L}{L\partial(m_o)} = \frac{1}{\sigma^2}\sum_{i=1}^n (x_i - m_o) = 0 \ ,$$

$$\frac{\partial L}{L\partial(\sigma^2)} = -\frac{n}{2\sigma^2} + \frac{1}{2\sigma^4}\sum_{i=1}^n (x_i - m_o)^2 = 0 . \qquad (2.13)$$

The first of the equations in (2.13) yields the estimate

$$\tilde{m}_o = \frac{1}{n}\sum_{i=1}^n x_i \equiv \bar{x}. \qquad (2.14)$$

As has already been discussed in the foregoing, $\bar{x}$ is an unbiased estimate for m_o; furthermore, according to the law of large numbers, it is apparent that $\bar{x} \to m_o$ as $n \to \infty$, which indicates that $\bar{x}$ is also a consistent estimate for m_o.

The second equation in Eq. (2.13) gives the estimate

$$\tilde{\sigma}^2 = (1/n)\sum_{i=1}^n (x_i - m_o)^2 . \qquad (2.15)$$

Since m_o is unknown, it is replaced in Eq. (2.15) by the estimate $\tilde{m}_o = \bar{x}$, as obtained in Eq. (2.14); this replacement leads to the estimate

$$\mu_2^S = \frac{1}{n} \sum_{i=1}^{n} (x_i - \bar{x})^2 . \tag{2.16}$$

The quantity $E\left(\mu_2^S\right)$ is calculated as follows:

$$
\begin{aligned}
E\left(\mu_2^S\right) &= E\left(\frac{1}{n} \sum_{i=1}^{n} (x_i - m_o + m_o - \bar{x})^2\right) \\
&= E\left(\frac{1}{n} \sum_{i=1}^{n} (x_i - m_o)^2 - (\bar{x} - m_o)^2\right) \\
&= E\left(\frac{1}{n} \sum_{i=1}^{n} (x_i - m_o)^2\right) + E\left((\bar{x} - m_o)^2\right) \\
&= \sigma^2 (n-1)/n .
\end{aligned}
\tag{2.17}
$$

As Eq. (2.17) indicates, $E\left(\mu_2^S\right) \to \sigma^2$ as $n \to \infty$, which means that the *MLE* estimate μ_2^S is consistent; however, the *MLE* estimate μ_2^S is biased. This result underscores one of the limitations of *MLEs*, namely that these estimators may be biased.

Multiplying μ_2^S with the correction factor $n/(n-1)$ yields the estimate

$$[n/(n-1)]\,\mu_2^S = [1/(n-1)] \sum_{i=1}^{n} (x_i - \bar{x})^2, \tag{2.18}$$

which deviates from the *MLE* value but, on the other hand, is both consistent and unbiased. In practice, a small deviation from the maximum of the likelihood function is less important than a potential bias in the estimate. In view of Eq. (2.18), the sample standard deviation is computed using the formula given in Eq. (2.10). Similarly, it can be shown that the quantity

$$\hat{V}_{xy} \equiv \frac{1}{n-1} \sum_{i=1}^{n} (x_i - \bar{x})(y_i - \bar{y}). \tag{2.19}$$

is an unbiased estimator of the covariance V_{xy} of two random variables x and y of unknown mean.

The probability distribution function (*PDF*) of an *estimator* can be found by using either the characteristic function or the moment generating function. As an illustrative example, consider n independent observations of a random variable x from an exponential distribution $p(x; \xi) = (1/\xi) \exp(-x/\xi)$. As has been previously shown, the maximum likelihood estimator (*MLE*) $\hat{\xi}$ for ξ is the sample mean of the observed x_i, namely

$$\hat{\xi} = \bar{x} = \frac{1}{n} \sum_{i=1}^{n} x_i . \tag{2.20}$$

If the experiment were repeated many times, one would obtain values of $\hat{\xi}$ distributed according to a *PDF*, $f\left(\hat{\xi}|n, \xi\right)$, that depends on the number, n, of observations per experiment and the true value of the parameter ξ. To determine the expression of $f\left(\hat{\xi}|n, \xi\right)$, note that the characteristic function for the exponentially distributed random variable x is

$$\varphi_x(k) = \int e^{ikx} p(x; \xi) \, dx = \frac{1}{1 - ik\xi} . \tag{2.21}$$

On the other hand, the characteristic function for the random quantity $\varsigma \equiv \sum_{i=1}^{n} x_i = n\hat{\xi}$ is

$$\phi_\varsigma(k) = \frac{1}{(1 - ik\xi)^n} . \tag{2.22}$$

The *PDF*, $f_\varsigma(\varsigma)$, for $\varsigma \equiv n\hat{\xi}$ is the inverse Fourier transform of $\phi_\varsigma(k)$, namely

$$f_\varsigma(\varsigma) = \frac{1}{2\pi} \int_{-\infty}^{\infty} \frac{e^{-ik\varsigma}}{(1 - ik\xi)^n} dk . \tag{2.23}$$

The integrand in Eq. (2.23) has a pole of order n at $-i/\xi$ in the complex k-plane, and can be evaluated using the residue theorem to obtain

$$f_\varsigma(\varsigma) = \frac{1}{(n-1)!} \frac{\varsigma^{n-1}}{\xi^n} e^{-\varsigma/\xi} . \tag{2.24}$$

Transforming variables $\varsigma \to \hat{\xi}$ in Eq. (2.24) yields the *PDF* for the *MLE* $\hat{\xi}$ as

$$f\left(\hat{\xi}|n, \xi\right) = \frac{n^n}{(n-1)!} \frac{\hat{\xi}^{n-1}}{(\xi^n)} e^{-n\hat{\xi}/\xi} . \tag{2.25}$$

Note that Eq. (2.25) is a special case of the *Gamma distribution*. In particular, Eq. (2.25) can be used to compute the expectation values and *PDFs* for mean lifetimes and decay constants in nuclear radioactive decay processes. For example, consider n decay-time measurements $(t_1, \ldots, t_n)$, for which $\hat{\tau} = (1/n) \sum_{i=1}^{n} t_i$ provides an estimate of the mean lifetime τ of the respective particle. Then, the expectation of the estimated mean lifetime $\hat{\tau}$ is obtained, using Eq. (2.25), as

$$E\left(\hat{\tau}\right) = \int_0^\infty \hat{\tau} f\left(\hat{\tau}|n, \tau\right) d\hat{\tau} = \tau. \qquad (2.26)$$

The *PDF*, $p\left(\hat{\lambda}|n, \lambda\right)$, for the *MLE* $\hat{\lambda} = 1/\hat{\tau}$ of the decay constant $\lambda = 1/\tau$ can be computed from Eq. (2.25) by changing the variables $\hat{\tau} \to 1/\hat{\lambda}, \quad \tau \to 1/\lambda$, as follows:

$$p\left(\hat{\lambda}|n, \lambda\right) = f\left(\hat{\tau}|n, \tau\right) \left|d\hat{\tau}/d\hat{\lambda}\right| = \frac{n^n}{(n-1)!} \frac{\lambda^n}{\hat{\lambda}^{n+1}} e^{-n\lambda/\hat{\lambda}}. \qquad (2.27)$$

Using the above expression, the expectation value of the *MLE* $\hat{\lambda}$ is obtained as

$$E\left(\hat{\lambda}\right) = \int_0^\infty \hat{\lambda} p\left(\hat{\lambda}|n, \lambda\right) d\hat{\lambda} = \frac{n}{n-1} \lambda. \qquad (2.28)$$

Note that, even though the *MLE* $\hat{\tau} = (1/n) \sum_{i=1}^{n} t_i$ is an unbiased estimator for τ, the estimator $\hat{\lambda} = 1/\hat{\tau}$ is not an unbiased estimator for $\lambda = 1/\tau$. The bias, however, vanishes in the limit as n goes to infinity, since $E\left(\hat{\lambda}\right) \to \lambda$ as $n \to \infty$. As already mentioned, estimators of the form $T = \Theta(x_1, \ldots, x_n)$ are specifically used to provide estimated values for a particular, true yet unknown, value $T_o = \Theta(x_1^o, \ldots, x_n^o)$ of the underlying probability distribution and, since the estimation rule Θ is designed to provide specific values of T for approximating T_o, the estimator T is called a *point estimator*. Once the estimator T has been obtained, it becomes of interest to determine by how much the estimator can change when measurements are repeatedly performed under the same conditions. This issue is addressed by constructing the so-called confidence interval for the true value $T_o = \Theta(x_1^o, \ldots, x_n^o)$ of the measured quantity. The *confidence interval* is defined to be the interval that contains, with a prescribed probability called the *confidence probability*, the true value of the measured quantity. This concept can be illustrated by considering that $(x_1, \ldots, x_n)$ is a set of random variables defining the sample of data under study, and θ_p is a fundamental parameter of the underlying distribution that produced the data. If it were now possible to introduce two statistics, say $\theta_1 = \Theta_1(x_1, \ldots, x_n)$ and $\theta_2 = \Theta_2(x_1, \ldots, x_n)$ which would guarantee that $P\{\theta_1 < \theta_p < \theta_2\} = \alpha$, then the interval $I(\theta_1, \theta_2)$ would be called the $100\alpha\%$ *confidence interval*. The procedure employed to determine an estimate of the confidence interval is called *interval estimation*. Note that a single experiment involving n samples would produce a single sample $(x_1, \ldots, x_n)$ of data; in turn, this sample would yield a single interval. Additional similar experiments would generate different data and, consequently, different intervals. However, if $I(\theta_1, \theta_2)$ is the $100\alpha\%$ *Confidence interval*, then $100\alpha\%$ of all intervals that might be generated this way would contain the true value, θ_p, of the parameter in question.

There are several methods for constructing confidence intervals. Perhaps the simplest and the most general method to construct confidence intervals is by using Chebyshev's theorem, but the intervals obtained by using Chebyshev's theorem are often too large for practical purposes. When the distribution of the sample observations can be assumed to follow a normal distribution with sample mean x^S and *known* standard deviation σ, then the confidence interval which would contain the true value, θ_p, of the parameter in question is constructed based on the expression

$$P\left\{ \left| x^S - \theta_p \right| \le z_\alpha \frac{\sigma}{\sqrt{n}} \right\} = \alpha, \qquad (2.29)$$

where z_α is the *quantile of the normalized Gaussian distribution, corresponding to the selected confidence probability* α. Standard tables of tabulated values for the Gaussian function are customarily used in conjunction with Eq. (2.29).

In practice, however, the sample standard deviation is rarely known; only its estimate, $SD(x^S)$, as given in Eq. (2.10) can be calculated. In such cases, the confidence intervals are constructed based on Student's t-distribution, which is the distribution of the random quantity

$$t \equiv \frac{\bar{x} - \theta_p}{SD\,(\bar{x})}\,, \tag{2.30}$$

where $SD\,(\bar{x})$ is the sample standard deviation as calculated from Eq. (2.10). The confidence interval $[\bar{x} - t_q SD\,(\bar{x})\,, \bar{x} + t_q SD\,(\bar{x})]$ corresponds to the probability

$$P\{\,|\bar{x} - \theta_p| \le t_q SD\,(\bar{x})\,\} = \alpha\,, \tag{2.31}$$

where t_q is the q-percent point of Student's t-distribution with $(n-1)$ degrees of freedom and significance level $q \equiv (1-\alpha)$. The significance level, q, should be consistent with the significance level adopted for verifying the normality of the sample. Although it is possible to verify the admissibility of the hypothesis that the observations are described by a normal distribution (and therefore verify the hypothesis that Student's t-distribution is admissible), in practice, however, confidence intervals are directly constructed based on Student's t-distribution without verifying its admissibility. The observation that this procedure works in practice indirectly confirms the tacit assumption that the truncated distributions usually obeyed by experimental data are often even narrower than normal distributions. In practice, the confidence probability is usually set equal to 0.95.

Confidence intervals for the standard deviation can also be constructed by using the χ^2-distribution. The confidence interval thus constructed has the limits $(\sqrt{n-1}/\chi_L)SD(\bar{x})$ and $(\sqrt{n-1}/\chi_U)SD(\bar{x})$ for the probability

$$P\left\{ \left(\frac{\sqrt{n-1}}{\chi_L}\right) SD\,(\bar{x}) < \sigma < \left(\frac{\sqrt{n-1}}{\chi_U}\right) SD\,(\bar{x}) \right\} = \alpha\,, \tag{2.32}$$

where χ_U^2 and χ_L^2 are found from tables, with χ_U^2 corresponding to $(1+\alpha)/2$, and χ_L^2 corresponding to $(1-\alpha)/2$. All these considerations make it apparent that the purpose of a confidence interval is to obtain a set of values which, despite sampling variations, yields a reasonable range of values for the estimated parameter, based on the data available. Thus, a confidence interval is an *interval estimator* of the parameter under investigation.

A confidence interval should not be confused with a *statistical tolerance interval*, which is defined as the interval that contains, with prescribed probability α, not less than a prescribed fraction p_o of the entire collection of values of the random quantity. Thus, the statistical tolerance interval is an interval for a random quantity, and this distinguishes it from the confidence interval, which is constructed in order to cover the values of a nonrandom quantity. For example, a group of instruments can be measured to find the interval with limits l_1 and l_2 within which not less than the fraction p_o of the entire batch of instruments will fail, with prescribed probability α. The interval between l_1 and l_2 is the statistical tolerance interval. Care must be exercised to avoid confusing the limits of statistical tolerance and/or confidence intervals with the tolerance range for the size of some parameter. The tolerance (or the limits of the tolerance range) is (are) determined prior to the fabrication of a manufactured object, so that the objects for which the value of the parameter of interest falls outside the tolerance range are unacceptable and are discarded. Thus, the limits of the tolerance range are strict limits, which are not associated with any probabilistic relations.

2.2   Assigning Prior Probability Distributions under Incomplete Information

Recall from Chapter 1 that Bayes' theorem provides the foundation for assimilation of new information, indicating how prior knowledge (e.g., a data file) is to be updated with new evidence (new data). Bayes' theorem can be expressed in the form

$$P\,(theory\,|data\,) \propto P\,(data\,|theory\,) \times P\,(theory\,), \qquad (2.33)$$

where $P\,(theory)$ represents the prior probability that the theory is true, while the *likelihood* $P\,(data|\,theory)$ expresses the probability of observing the data that were actually obtained under the assumption that the theory is true. The *posterior probability*, that the theory is correct after seeing the result of the experiment, is given by $P\,(theory\,|data\,)$. In terms of conditional probabilities for continuous variates, Bayes' theorem takes on the form

$$p(A|BC)dA = \frac{p(B|AC)p(A|C)dA}{\int p(B|AC)p(A|C)dA} \quad , \quad A_{\min} \leq A \leq A_{\max}. \qquad (2.34)$$

Since the denominator in Bayes' theorem is simply a normalization constant, the formal rule for updating observations can be briefly stated as "the posterior is proportional to the likelihood times the prior." It should be understood that the terms "posterior" and "prior" have a logical rather than a temporal connotation, implying "with" and, respectively, "without" having assimilated the new data. Once a prior probability has been assigned, Bayesian statistics indicates how one's degree of belief should change after obtaining additional information (e.g., experimental data).

As is well known, Bayesian statistics provides no fundamental rule for assigning the prior probability to a theory. The choice of the "most appropriate" prior distribution lies at the heart of applying Bayes' theorem to practical problems, and has caused considerable debates over the years. Thus, when prior information related to the problem under consideration is available and can be expressed in the form of a probability distribution, this information should certainly be used. In such cases, the repeated application of Bayes' theorem will serve to refine the knowledge about the respective problem. When only scant information is available, it may be possible to use the maximum entropy principle (as described in statistical mechanics and information theory) for constructing a prior distribution. This situation will be presented in Section 2.2.2. In the extreme case when no specific information is available, it may be possible to construct prior distributions using group theoretical arguments that reflect the possible symmetries in the problem under consideration, as will be discussed in the following subsection.

2.2.1   Assigning Prior Distributions Using Group Theory

Recall that, in the absence of any prior information, the principle of insufficient reason (which was invoked for assigning equal probabilities to the two alternatives: "too high" or "too low") works well for coins, dice, playing cards, and all other cases characterized by discrete alternatives. However, this principle appears to lead to difficulties for continuous alternatives characterized by infinitesimal probabilities, $p(x)\,dx$, because a continuous uniform distribution, taken to describe equal probabilities, becomes non-uniform under change of variables. For example, when estimating the decay constant λ of a radionuclide, if all decay constants are equally probable *a priori*, the prior distribution would be $p(\lambda)\,d\lambda \sim d\lambda$. On the other hand, if all mean lives $\tau = 1/\lambda$ were considered to be equally probable *a priori*, the prior distribution would be $p(\tau)\,d\tau \sim d\tau \sim d\lambda/\lambda^2$. In the past, this apparent arbitrariness regarding priors for continuous random quantities had hampered the use of priors and hence the use of Bayes' theorem. However, while establishing decision theory, Wald [211] proved that the optimal strategy for making decisions (i.e., recommending a value for some uncertain quantity, for instance) is based on Bayes' theorem, which renewed the interest in searching logically well-founded prescriptions for assigning priors. A breakthrough in this regard was achieved by Jaynes [98], who demonstrated that in some very important practical cases, *the invariance properties of the problem under investigation uniquely determine the "least informative" prior that describes initial ignorance about numerical values.* Thus, if a *location parameter* is to be estimated (e.g., the center μ of a Gaussian), the form of the prior must be invariant under an arbitrary shift of location. Otherwise, some locations would be preferred to others, which would be contrary to the assumption of total ignorance. This invariance implies that $p(\mu)\,d\mu = p(\mu + c)\,d(\mu + c)$, which is a functional equation that is satisfied by the uniform distribution, i.e.

$$(\mu)\,d\mu \sim d\mu, \quad -\infty < \mu < \infty. \tag{2.35}$$

For a *scale parameter* (e.g., the standard deviation σ of a Gaussian), the form of the prior must be invariant under an arbitrary rescaling, implying that $p(\sigma)\,d\sigma = p(c\sigma)\,d(c\sigma)$. The solution of this functional equation is

$$(\sigma)\,d\sigma \sim \frac{d\sigma}{\sigma}, \quad 0 < \sigma < \infty. \tag{2.36}$$

Since both the mean life τ and the decay constant λ are actually scale parameters, the prior in Eq. (2.36) must be appropriate for both; this assertion can be verified by changing variables to obtain $p(\lambda)\,d\lambda \sim d\lambda/\lambda$

$\sim d\tau/\tau \sim p(\tau)\,d\tau$. The above prior was introduced for scale parameters by Jeffreys (1939), and is therefore called *Jeffreys' prior*.

For Bernoulli trials with probability of success θ, the form of the prior must be invariant under an arbitrary *invariance under change of evidence*; this requirement leads to the following form for the prior:

$$(\theta)\,d\theta \sim \theta^{-1}(1-\theta)^{-1}d\theta, \quad 0 \le \theta \le 1, \tag{2.37}$$

if $\theta = 0$ (only failures) and $\theta = 1$ (only successes) cannot be excluded *a priori*. A least informative prior that describes invariance under a group of continuous transformations is equivalent to the right invariant Haar measure (i.e., to the weight function that ensures invariance of integrals over the whole group). More specifically, the right Haar measure is related to the prior by changing variables from transformation labels to possible parameters. For example, for the additive group associated with location parameters, the change of variables is from c to μ; similarly, for the multiplicative group associated with scale parameters, the change of variables is from c to σ. Although the least informative priors given in Eqs. (2.36) and (2.37) are not normalizable, they may be considered as limits of very broad, normalizable distributions [on a linear scale, $d\mu$, or on a logarithmic scale, $(d\sigma/\sigma) = d(\ln\sigma)$], in the same way as Dirac's delta function may be considered as the limit of extremely narrow, normalizable distributions. Mathematical difficulties are circumvented by noting that the least informative priors and the most informative delta function are just convenient shorthand notations for distributions that are extremely broad or extremely narrow, respectively, compared to the distributions with which they are convoluted.

2.2.2   Assigning Prior Distributions Using Entropy Maximization

Ideally, knowledge of scientific and technological data would be described in terms of complete probability distributions. In practice, though, such distributions are seldom available; furthermore, users are often interested only in practical "recommended values" and "error bars." Prescriptions for recommending rigorously founded "best values" and "error bars" are provided by *"decision theory,"* founded by Wald [211]. Decision theory demonstrates that a disadvantage or penalty arises for any estimate, x_{est}, of an uncertain (unknown) parameter x. This penalty can be described by a loss function that vanishes at the unknown true value x but grows as the deviation $|x_{est} - x|$ increases. Near the true value x, any reasonably smooth loss function can be taken to be proportional to the squared deviation, $(x_{est} - x)^2$, since the Taylor expansion of such a function about zero-error begins with the quadratic term. For a set of given probabilities, P_n, for the various alternatives n, the expected loss

$$\sum_n P_n (x_{est} - x_n)^2 = \min .$$

is minimized when $x_{est} = \langle x \rangle$; hence $x_{est} = \langle x \rangle$ would be "the recommended value." For a continuous random quantity with probability density $p(x)$, the expected loss is minimized in a similar way, i.e.,

$$\int dx\, p(x) (x_{est} - x)^2 = \min .$$

when $x_{est} = \langle x \rangle$. The above considerations indicate that the expected squared error is minimal when $x_{est} = \langle x \rangle$, and is equal, at the minimum, to the variance of the probability distribution. This argument justifies the choice of "mean" and "variance" as the "best indicators" for a distribution under quadratic loss: the "mean" is the optimal estimate of the (always unknown) true value, and the standard deviation (square root of the variance, also called *dispersion,* standard error, or "$1\,\sigma$" uncertainty) is the best indication of its uncertainty. Hence, parameter estimates should be presented in the form $\langle x \rangle \pm \Delta x$, where $\Delta x = \sqrt{\operatorname{var} x}$. Although other types of loss functions may be preferable for some very specific applications, the quadratic loss is employed as the optimal choice in general, and also in particular, when errors are small and data applications are not specified. The generalization of the foregoing arguments to several uncertain quantities is accomplished by replacing the scalar observable x with a vector $\mathbf{x}$ whose components are the respective observables. The "estimate under quadratic loss" is then described by the vector $\langle \mathbf{x} \rangle$ of mean values, and by the covariance matrix $\left\langle (\mathbf{x} - \langle \mathbf{x} \rangle) (\mathbf{x} - \langle \mathbf{x} \rangle)^T \right\rangle$, with the T indicating "transposition" (since all components of the respective vectors and matrices are real numbers). Note that the expectation values $\langle \mathbf{x} \rangle$ are averages over the joint distribution of all the coordinates of the vector of observables, and the covariance matrix contains both the respective variances and correlations.

While establishing *information theory*, Shannon [189] proved that the lack of information implied by a discrete probability distribution, p_n, with mutually exclusive alternatives can be expressed quantitatively (up to a constant) by its information entropy,

$$S = - \sum_{n=1}^{N} p_n \ln p_n. \tag{2.38}$$

Shannon proved that S is the only measure of indeterminacy that satisfies the following three requirements:

1. S is a smooth function of the p_i;

2. If there are N alternatives, all equally probable, then the indeterminacy and hence S must grow monotonically as N increases; and

3. Grouping of alternatives leaves S unchanged (i.e., adding the entropy quantifying ignorance about the true group, and the suitably weighted entropies quantifying ignorance about the true member within each group, must yield the same overall entropy S as for ungrouped alternatives).

For continuous distributions with probability density $p(x)$, the expression for its information entropy becomes

$$S = - \int dx\, p(x) \ln \frac{p(x)}{m(x)}, \tag{2.39}$$

where $m(x)$ is a prior density that ensures form invariance under change of variable.

When only certain information about the underlying distribution $p(x)$ is available but $p(x)$ itself is unknown and needs to be determined, the *principle of maximum entropy* provides the optimal compatibility with the available information, while simultaneously ensuring minimal spurious information content. As an application of the principle of maximal entropy, suppose that an unknown distribution $p(x)$ needs to be determined when the only available information comprises the (possibly non-informative) prior $m(x)$ and "integral data" in the form of moments of several known functions $F_k(x)$ over the unknown distribution $p(x)$, namely

$$\langle F_k \rangle = \int dx\, p(x)\, F_k(x)\,, \qquad k = 1, 2, \ldots, K. \qquad (2.40)$$

According to the principle of maximum entropy, the probability density $p(x)$ would satisfy the "available information" [i.e., would comply with the constraints Eq. (2.40)] without implying any spurious information or hidden assumptions if $p(x)$ would maximize the information entropy defined by Eq. (2.39) subject to the known constraints given in Eq. (2.40). This is a variational problem that can be solved by using Lagrange multipliers, λ_k, to obtain the following expression:

$$p(x) = \frac{1}{Z} m(x) \exp\left[-\sum_{k=1}^{K} \lambda_k F_k(x) \right]. \qquad (2.41)$$

The normalization constant Z in Eq. (2.41) is defined as

$$Z \equiv \int dx\, m(x) \exp\left[-\sum_{k=1}^{K} \lambda_k F_k(x) \right]. \qquad (2.42)$$

In statistical mechanics, the normalization constant Z is called the *partition function* (or sum over states), and carries the entire information available about the possible states of the system. The expected integral data is obtained by differentiating Z with respect to the respective Lagrange multiplier, i.e.,

$$\langle F_k \rangle = -\frac{\partial}{\partial \lambda_k} \ln Z, \quad k = 1, 2, \ldots, K. \qquad (2.43)$$

In the case of discrete distributions, when the integral data $\langle F_k \rangle$ are not yet known, then $m(x) = 1$, and the maximum entropy algorithm described above yields the uniform distribution, as would be required by the principle of insufficient reason. Therefore, the principle of maximum entropy provides a generalization of the principle of insufficient reason, and can be applied equally to discrete and continuous distributions, ranging from problems in which only information about discrete alternatives is available, to problems in which global or macroscopic information is available.

As already discussed in the foregoing, data reported in the form $\langle x \rangle \pm \Delta x$, $\Delta x \equiv \sqrt{\text{var}\, x}$, customarily implies "best estimates under quadratic loss," which in turn implies the availability of the first and second moments, $\langle x \rangle$ and $\langle x^2 \rangle = \langle x \rangle^2 + (\Delta x)^2$, respectively. In such a case, the maximum entropy algorithm above can be used with $K = 2$, in which case Eq. (2.42) yields $p(x) \sim \exp(-\lambda_1 x - \lambda_2 x^2)$ as the most objective probability density for further inference. In terms of the known moments $\langle x \rangle$ and $\langle x^2 \rangle$, $p(x)$ would be a Gaussian having the following specific form:

$$p(x|\langle x \rangle, \Delta x)\, dx = \frac{\exp\left[-\frac{1}{2}\left(\frac{x - \langle x \rangle}{\Delta x}\right)^2\right]}{\left[2\pi(\Delta x)^2\right]^{1/2}} dx, \quad -\infty < x < \infty. \quad (2.44)$$

When several observables x_i, $i = 1, 2, \ldots, n$, are simultaneously measured, the respective results are customarily reported in the form of "best values" $\langle x_i \rangle$, together with covariances matrix elements $c_{ij} \equiv \langle \varepsilon_i \varepsilon_j \rangle \triangleq \text{cov}(x_j, x_k) = c_{ji}$, $c_{jj} \triangleq \text{var}\, x_j$, where the errors are defined through the relation $\varepsilon_j \equiv (x_j - \langle x_j \rangle)$. In this case, Eq. (2.41) takes on the form

$$p(\varepsilon|C)\, d^n\varepsilon = \frac{1}{Z}\, \exp\left(-\sum_{i,j} \varepsilon_i \lambda_{ij} \varepsilon_j\right) d^n\varepsilon = \frac{1}{Z}\, \exp\left(-\frac{1}{2}\varepsilon^T \Lambda \varepsilon\right) d^n\varepsilon, \quad (2.45)$$

where the Lagrange multiplier λ_{ij} corresponds to c_{ij}; the normalization constant is defined as $Z \equiv \int d^n\varepsilon \, \exp\left(-\frac{1}{2}\varepsilon^T \Lambda \varepsilon\right)$; $\varepsilon = (\varepsilon_1, \ldots, \varepsilon_n)$ denotes the n-component, the vector of errors; $\mathbf{C} \triangleq (c_{ij})_{n \times n}$ is the $n \times n$ covariance matrix for the observables x_i, $i = 1, 2, \ldots, n$; and $\Lambda = (\lambda_{ij})_{n \times n}$ denotes the $n \times n$ matrix of Lagrange multipliers. The explicit form of the normalization constant Z in Eq. (2.45) is obtained by performing the respective integrations to obtain

$$Z = \sqrt{\frac{\pi^n}{\det(\Lambda)}}. \quad (2.46)$$

As previously noted in Section 1.3.2, the relationship between the Lagrange multiplier λ_{ij} and the covariance c_{ij} is obtained by differentiating $\ln Z$ with respect to the respective Lagrange multiplier, cf. Eq. (2.43), to obtain

$$c_{ij} = -\frac{\partial}{\partial \lambda_{ij}} \ln Z = \frac{1}{2} \left(\mathbf{\Lambda}^{-1} \right)_{ij}, \tag{2.47}$$

since differentiation of the determinant $\det (\mathbf{\Lambda})$ with respect to an element of $\mathbf{\Lambda}$ yields the cofactor for this element (which, for a nonsingular matrix, is equal to the corresponding element of the inverse matrix times the determinant). Replacing Eqs. (2.46) and (2.47) in Eq. (2.45) transforms the latter into the form

$$p\left(\varepsilon | \mathbf{C}\right) d^n \varepsilon = \frac{\exp\left(-\frac{1}{2}\varepsilon^{\mathbf{T}}\mathbf{C}^{-1}\varepsilon\right)}{\sqrt{\det\left(2\pi \mathbf{C}\right)}} d^n \varepsilon, \quad -\infty < \varepsilon_i < \infty, \tag{2.48}$$

with $\langle \varepsilon \rangle = \mathbf{0}$, $\langle \varepsilon \varepsilon^T \rangle = \mathbf{C}$, which is a n-variate Gaussian centered at the origin. In terms of the expected values $\langle x_i \rangle$, the above expression becomes

$$p\left(\mathbf{x} | \langle \mathbf{x} \rangle, \mathbf{C}\right) d\mathbf{x} = \frac{\exp\left[-\frac{1}{2}(\mathbf{x} - \langle \mathbf{x} \rangle)^T \mathbf{C}^{-1} (\mathbf{x} - \langle \mathbf{x} \rangle)\right] d\mathbf{x},}{\sqrt{\det\left(2\pi \mathbf{C}\right)}}, \quad -\infty < x_j < \infty. \tag{2.49}$$

Thus, the foregoing considerations show that, when only means and co-variances are known, the *maximum entropy algorithm yields the Gaussian probability distribution* in Eq. (2.49) as the most objective probability distribution, where $\mathbf{x}$ is the data vector with coordinates x_j; $\mathbf{C}$ is the covariance matrix with elements C_{jk}; $d\mathbf{x} \triangleq \prod_j dx_j$ is the volume element in the data space. It often occurs in practice that the variances c_{ii} are known but the covariances c_{ij} are not, in which case the covariance matrix $\mathbf{C}$ would *a priori* be diagonal. In such a case, only the Lagrange parameters λ_{ii} would appear in Eq. (2.45), so that the matrix $\mathbf{\Lambda}$ would also be *a priori* diagonal. In other words, in the absence of information about correlations, the maximum entropy algorithm indicates that *unknown covariances can be taken to be zero*.

Gaussian distributions are often considered appropriate only if many independent random deviations act in concert such that the central limit theorem is applicable. Nevertheless, if only "best values" and their (co)variances are available, the maximum entropy principle indicates that the corresponding Gaussian is the best choice for all further inferences, regardless of the actual form of the unknown true distribution. Furthermore, in contrast to the central limit theorem, the maximum entropy principle is also valid for correlated data.

The maximum entropy principle can also be employed to address systematic errors when their possible magnitudes can be (at least vaguely) inferred, but their signs are not known. The maximum entropy principle indicates that such errors should be described by a Gaussian distribution with zero mean and a width corresponding to the (vaguely) known magnitude, rather than by a rectangular distribution.

The maximum entropy principle is a powerful tool for the assignment of prior (or any other) probabilities, in the presence of incomplete information. Although the above results have been derived for observables x_j that vary in the interval $-\infty < x_j < \infty$, these results can also be used for positive observable $(0 < x_j < \infty)$ by considering a logarithmic scale (or lognormal distributions on the original scale).

2.3   Evaluation of Consistent Data with Independent Random Errors

This section presents three paradigm practical examples of evaluating unknown parameters when consistent data is available. The evaluation of a location parameter when the scale parameters are known is presented in subsection 2.3.1. Subsection 2.3.2 analysis the problem of evaluating both the scale and location parameters when they are not explicitly provided by the experimental data. A problem often encountered in practice is illustrated in subsection 2.3.3, which presents the evaluation of a counting rate (a scale parameter) in the presence of background (noise).

2.3.1   Evaluation of Unknown Location Parameter with Known Scale Parameters

Consider n measurements and/or computations of the same (unknown) quantity μ which yielded data $x_1 \pm \sigma_1$, ..., $x_n \pm \sigma_n$, with known standard deviations σ_j, $j = 1, \ldots, n$. If the distances $|x_j - x_k|$ are smaller or not much larger than $(\sigma_j + \sigma_k)$, *the respective data points are considered to be "consistent" or to agree "within error bars."* If the only additional information available about the distributions of possible errors $x_k - \mu$ is that they can take on any value between $-\infty$ and $+\infty$, then the maximum entropy principle would assign a Gaussian distribution of the form given in Eq. (2.44) to each of these errors, with μ being a *location parameter*. Consequently, the likelihood function is the product of these Gaussians. Furthermore, if no prior information about the location parameter μ is available, the least informative uniform prior given by Eq. (2.35) is appropriate, which yields, after multiplication by the likelihood function, the following Gaussian posterior distribution:

$$p\left(\mu | x_1, \sigma_1 \ldots, x_n, \sigma_n\right) d\mu \propto \exp\left[-\frac{1}{2}\sum_{k=1}^{n}\left(\frac{x_k - \mu}{\sigma_k}\right)^2\right] d\mu, \quad -\infty < \mu < \infty.$$

$$(2.50)$$

Expanding the sum in the exponent of Eq. (2.50) makes it possible to recast the exponent in the form $\left[(\mu - \bar{x})^2 + \overline{x^2} - \bar{x}^2\right]/\left(2\overline{\sigma^2}/n\right)$, where

$$\bar{x} \triangleq \frac{\sum\limits_{k}\sigma_k^{-2}x_k}{\sum\limits_{k}\sigma_k^{-2}}, \qquad \overline{x^2} \triangleq \frac{\sum\limits_{k}\sigma_k^{-2}x_k^2}{\sum\limits_{k}\sigma_k^{-2}},$$

$$\overline{\sigma^2} \triangleq \frac{\sum\limits_{k}\sigma_k^{-2}\sigma_k^2}{\sum\limits_{k}\sigma_k^{-2}} = \frac{n}{\sum\limits_{k}\sigma_k^{-2}}. \qquad (2.51)$$

The rewritten exponent shows that the posterior distribution for μ is a Gaussian with mean $\bar{x}$ and variance $\overline{\sigma^2}/n$, i.e.,

$$p\left(\mu | x_1, \sigma_1 \ldots, x_n, \sigma_n\right) d\mu = \frac{\exp\left[-\frac{(\mu - \bar{x})^2}{2\overline{\sigma^2}/n}\right]}{\left(2\pi\overline{\sigma^2}/n\right)^{1/2}} d\mu, \quad -\infty < \mu < \infty. \qquad (2.52)$$

Based on Eq. (2.52), the estimate for the location parameter μ under quadratic loss can be stated as $\langle\mu\rangle \pm \Delta\mu = \bar{x} \pm (\overline{\sigma^2}/n)^{1/2}$, showing that the best estimate is the (σ^{-2})-weighted average of all x-values. This best estimate can also be called the "least-square estimate" because it minimizes the sum of squares in the exponent of Eq. (2.50). The posterior standard deviation shows the familiar $1/\sqrt{n}$ dependence on sample size [cf. Eq. (2.9)].

2.3.2   Evaluation of Unknown Location and Scale Parameters

Often encountered in data evaluation is the estimation of an unknown quantity μ from data $x_1, x_2, \ldots, x_n$, obtained in n repetitions of some measurement, without any information about the accuracy of the experimental method, i.e., when, in contrast to the situation discussed in the foregoing, the individual standard deviations σ_j are not available for each experimental data x_j. Since the experiments are assumed to be consistent, it follows that common errors can be neglected. In this situation, it is assumed that a *location parameter* μ, as well as a scale parameter (i.e., a standard deviation) σ exist, but they are both unknown and are to be estimated from the data. In such a situation, the sampling distribution constructed for each experimental data x_j by using the maximum-entropy principle, cf. Eq. (2.44), is the Gaussian

$$
p\left(x_j | \mu, \sigma\right) dx_j = \frac{\exp\left[-\frac{1}{2}\left(\frac{x_j - \mu}{\sigma}\right)^2\right]}{\left(2\pi\sigma^2\right)^{1/2}} dx_j, \quad -\infty < x_j < \infty. \tag{2.53}
$$

Furthermore, the likelihood function for the independent data $x_1, x_2, \ldots, x_n$ is the product over all of the respective distributions, i.e., $\prod_j p\left(x_j | \mu, \sigma\right)$.

The prior expressing initial ignorance about the location μ is given by Eq. (2.35), while the prior expressing initial ignorance about the scale σ is given by Eq. (2.36). Since these two priors are independent of each other, it follows that the joint prior expressing initial ignorance about both location and scale is the distribution

$$
p\left(\mu, \sigma\right) d\mu d\sigma \propto \frac{d\mu d\sigma}{\sigma}, \quad -\infty < \mu < \infty, \ 0 < \sigma < \infty. \tag{2.54}
$$

Multiplying the above prior with the likelihood function, $\prod_j p\left(x_j | \mu, \sigma\right)$, yields the posterior distribution

$$
p\left(\mu, \sigma | x_1, x_2, \ldots, x_n\right) d\mu d\sigma \propto \frac{1}{\sigma^n} \exp\left[-\frac{1}{2\sigma^2} \sum_{j=1}^{n} \left(x_j - \mu\right)^2\right] \frac{d\mu d\sigma}{\sigma}. \tag{2.55}
$$

Carrying out the summation in the exponent of Eq. (2.55) leads to the expression

$$\frac{1}{2\sigma^2} \sum_{j=1}^{n} (x_j - \mu)^2 = \left[1 + \left(\frac{\mu - \bar{x}}{\mu_2^S} \right)^2 \right] \frac{n\mu_2^S}{2\sigma^2},$$ (2.56)

where $\bar{x}$ is the sample mean already encountered in Eq. (2.5), and μ_2^S is the sample variance already encountered in Eq. (2.9), namely

$$\bar{x} \triangleq \frac{1}{n} \sum_{j=1}^{n} x_j, \quad \mu_2^S \triangleq \frac{1}{n} \sum_{j=1}^{n} (x_j - \bar{x})^2.$$

Replacing Eq. (2.56) in Eq. (2.55) and normalizing the later distribution to unity leads to the following posterior distribution

$$p(\mu, \sigma | x_1, x_2, \ldots, x_n) \, d\mu d\sigma = p(\mu, \sigma | x_1, x_2, \ldots, x_n) \, d\mu \times p(\sigma | x_1, x_2, \ldots, x_n) \, d\sigma$$
$$= \left[\frac{e^{-zy^2}}{\sqrt{\pi}} \sqrt{z} dy \right] \times \left[\frac{e^{-z} z^{(n-1)/2}}{\Gamma[(n-1)/2]} \frac{dz}{z} \right],$$
$$-\infty < y \equiv \frac{\mu - \bar{x}}{\sqrt{\mu_2^S}} < \infty, \quad 0 < z \equiv \frac{n\mu_2^S}{2\sigma^2} < \infty.$$ (2.57)

The above joint distribution depends on the given data only through the sample mean and the sample variance, which are therefore "jointly sufficient statistics" for this situation. The above expression also shows that the distribution for μ given σ is a Gaussian, while the distribution of σ^{-2} given μ is a Gamma distribution.

If only μ is of interest, regardless of σ, then z becomes a "nuisance parameter" which is marginalized by integrating over all of its possible values, to obtain the marginal distribution

$$p(\mu | \bar{x}, \mu_2^S) \, d\mu = \frac{\Gamma(n/2)}{\sqrt{\pi} \, \Gamma[(n-1)/2]} \frac{dy}{(1 + y^2)^{n/2}}, \quad -\infty < y \equiv \frac{\mu - \bar{x}}{\sqrt{\mu_2^S}} < \infty.$$ (2.58)

The above probability distribution is Student's t-distribution for the variable $t \equiv y/\sqrt{n-1}$. The mean value of this distribution vanishes, i.e., $\langle y \rangle = 0$ for $n > 2$, and its variance is $\langle y^2 \rangle = 1/(n-3)$ for $n > 3$. Hence, the estimates under quadratic loss for the location parameter μ are

$$\langle \mu \rangle = \bar{x}, \quad n > 2, \text{ and } \operatorname{var} \mu = \frac{\mu_2^S}{n-3}; \quad n > 3. \tag{2.59}$$

A finite value for the variance of μ can be obtained only after having performed at least four measurements. Although a bona-fide variance cannot be obtained when $n < 3$, the above distribution reduces to the Cauchy distribution for $n = 2$, with a well-defined half-width that can be used as a measure of its spread.

If only σ is of interest, regardless of μ, then μ can be marginalized by integration over y. The resulting marginal distribution is a χ^2-distribution with $y = n - 1$ degrees of freedom for y, i.e.,

$$p\left(\sigma | \bar{x}, \mu_2^S\right) d\sigma = \frac{e^{-z} z^{(n-3)/2} \, dz}{\Gamma\left[(n-1)/2\right]}, \quad 0 < z \equiv \frac{n \mu_2^S}{2\sigma^2} < \infty. \tag{2.60}$$

For this χ^2-distribution, the mean and variance are equal, namely $\langle z \rangle = \operatorname{var} z = (n-1)/2$. Hence, the recommended values under quadratic loss for the scale parameter σ^{-2} are

$$\langle \sigma^{-2} \rangle = \frac{n-1}{n} \left(\mu_2^S\right)^{-2} \equiv s^{-2} = \frac{n-1}{\sum_i \left(x_i - \bar{x}\right)^2}; \tag{2.61}$$

$$\operatorname{var} \sigma^{-2} = 2 \frac{n-1}{n^2} \left(\mu_2^S\right)^{-4} = \frac{2s^{-4}}{n-1}. \tag{2.62}$$

The above results highlight the origin of the counterintuitive practice to use s^2, instead of the sample variance μ_2^S, as an estimate for σ^2, although the estimate s^2 is biased, i.e., $\langle s^2 \rangle \neq \sigma^2$. Actually, the quantity $\langle s^{-2} \rangle /2$ is an unbiased estimate of the "precision" $\sigma^{-2}/2$.

Finally, the estimate under quadratic loss for the covariance between the location parameter μ and the scale parameter σ is obtained as

$$\operatorname{cov}\left(\mu, \sigma^{-2}\right) = 0. \tag{2.63}$$

In the extreme case of a single experiment, when $n = 1$, the sample variance vanishes, i.e., $\mu_2^S = 0$, and the corresponding particular form of Eq. (2.57) reduces to

$$p\left(\mu, \sigma | x_1\right) d\mu \, d\sigma \; \propto \; \frac{\exp\left[-\frac{1}{2}\left(\frac{\mu - x_1}{\sigma}\right)^2\right] d\mu \, d\sigma}{\sqrt{2\pi\sigma^2}} \, \frac{}{\sigma}. \tag{2.64}$$

The marginal distributions of μ and σ, respectively, become

$$p\left(\mu | x_1\right) d\mu \equiv \frac{d\mu}{|\mu - x_1|}, p\left(\sigma | x_1\right) d\sigma \; \propto \; \frac{d\sigma}{\sigma}. \tag{2.65}$$

The above marginal distribution of μ has a sharp maximum at the observed value x_1. On the other hand, the marginal distribution of σ has remained unchanged by the experiment x_1, being still equal to the least informative prior. Together, these two results logically indicate that a sample of size $n = 1$ can provide some information about the location but not about the spread of a distribution. This example underscores that diverging expectation values in Bayesian parameter estimation indicate that the given information is not sufficient for definite conclusions, and that a more informative prior or better data are needed.

The joint posterior distribution of the mean μ and the variance σ^2 is the complete information about the parameters of a Gaussian that can be extracted from the data in this example. This posterior can be used for recommending values and their uncertainties for quadratic or other loss functions. However, predictions of the outcome of further measurements are also of interest. Thus, for predicting "under quadratic loss" the outcome of one further measurement, $x \equiv x_{n+1}$, the posterior distribution in Eq. (2.57) is averaged over the posterior distribution of its parameters. This averaging ensures that the expected squared error (i.e., the error in the value of the chosen Gaussian caused by incorrect parameters) is smallest for any x. Hence, for the sampling distribution in Eq. (2.57), the "best" probability for the outcome of one further measurement to be found in dx at x is given by

$$p\left(x | \bar{x}, \mu_2^S\right) dx \; \propto \; dx \int d\sigma \int d\mu \, p\left(x | \mu, \sigma\right) p\left(\mu, \sigma | \bar{x}, \mu_2^S\right)$$

$$\propto \; dx \int_0^\infty dz \int_{-\infty}^\infty e^{-(y-w)^2 z/n} \, e^{-z} z^{(n-1)/2} \, dy, \tag{2.66}$$

where $w \triangleq (x - \bar{x})/\sqrt{\mu_2^S}$. Integrating first over the Gaussian (over all y), then over the remaining Gamma distribution (over all z) leads to the following form for the "predictive" distribution

$$p\left(x|\bar{x}, \mu_2^S\right) dx = \frac{\Gamma\left(n/2\right)}{\sqrt{\pi}\,\Gamma\left(\frac{n-1}{2}\right)} \frac{dt}{\left(1+t^2\right)^{n/2}},$$

$$-\infty < t \equiv \frac{x - \bar{x}}{\sqrt{\mu_2^S}\sqrt{n+1}} < \infty, \quad n > 1. \tag{2.67}$$

The above results shows that, although the sampling distribution is Gaussian, the best predictive distribution for the outcome of an additional measurement is not a Gaussian but a Student's t-distribution, which is always broader than the Gaussian for finite n (although it approaches a Gaussian as $n \to \infty$). Furthermore, the best estimate for any function $f(x)$ of the next datum is given by its expectation value $\langle f \rangle$ with respect to the predictive distribution.

2.3.3   Scale Parameter (Count Rate) Evaluation in the Presence of Background Noise

Consider that n events are counted during a recording time t, with a well-defined mean-time interval $\langle t \rangle$ between counts. The average count rate, λ, is defined as the reciprocal of the mean interval, $\lambda \triangleq 1/\langle t \rangle$. In an ideal experimental setting, these count rates would correspond to the actual physical events ("signal") under consideration. In practice, however, the measured counts include not only the signal but also "noise" caused by background events. Thus, λ is the sum of a "signal count rate," λ_s, and the background (noise) count rate λ_b, i.e., $\lambda = \lambda_s + \lambda_b$. A separate measurement needs to be performed in order to evaluate the impact of the background noise. Consider, therefore, that the background noise does not vary in time, so that n_b counts are measured within a time interval t_b. Since only the first moment $\langle t \rangle$ of the unknown interval distribution is known, the maximum entropy algorithm presented in Section. 2.2 provides an exponential form for the unknown interval distribution, namely

$$p\left(t|\lambda\right) dt = e^{-\lambda t}\lambda dt, \quad 0 \le t \le \infty. \tag{2.68}$$

Using the interval distribution above to construct the joint probability that counts are registered in infinitesimal time intervals dt_1, dt_2, ..., dt_n within some time span t, and integrating this joint probability over all possible locations of the intervals dt_1, dt_2, ..., dt_n yields the probability (as a Poisson distribution) of n counts being registered at arbitrary times within the interval t, given the count rate λ

$$P(n|\lambda, t) = \frac{e^{-\lambda t}(\lambda t)^n}{n!}, \quad n = 0, 1, 2, \dots . \tag{2.69}$$

The above Poisson distribution provides the likelihood function for the estimation of the count rate λ from given data n and t, and it highlights the role of λ as a scale parameter since all times are multiplied (scaled) by it. Therefore, as discussed in Section 2.2, Jeffreys' prior, $p(\lambda) \sim d\lambda/\lambda$, is appropriate if we are completely ignorant initially, even about the order of magnitude. The posterior, properly normalized with $\Gamma(n) = (n-1)!$, is a Gamma (or χ^2) distribution,

$$p(\lambda|n, t)\, d\lambda = \frac{e^{-\lambda t}(\lambda t)^n}{\Gamma(n)} \frac{d\lambda}{\lambda}, \quad 0 < \lambda < \infty. \tag{2.70}$$

Changing variable in the above distribution from λ to $\lambda_s = \lambda - \lambda_b$, with fixed background count rate λ_b yields the following conditional probability for λ_s,

$$p(\lambda_s|\lambda_b, n, t)\, d\lambda_s = \frac{e^{-(\lambda_b+\lambda_s)t}[(\lambda_b + \lambda_s)t]^n}{\Gamma(n)} \frac{d\lambda_s}{\lambda_b + \lambda_s}, \quad 0 < \lambda_s < \infty. \tag{2.71}$$

For the auxiliary background measurement, the same arguments as those leading to Eq. (2.70) can be used to obtain the following Gamma distribution for λ_b:

$$p(\lambda_b|n_b, t_b)\, d\lambda_b = \frac{e^{-\lambda_b t_b}(\lambda_b t_b)^{n_b}}{\Gamma(n_b)} \frac{d\lambda_b}{\lambda_b}, \quad 0 < \lambda_b < \infty. \tag{2.72}$$

The joint probability for the two unknown count rate parameters, λ_b and λ_s is obtained by multiplying Eqs. (2.71) and (2.72). Finally, the posterior distribution for λ_s is obtained by using the binomial expansion for $(\lambda_b + \lambda_s)^{n-1}$ in the joint distribution of λ_b and λ_s, and subsequently integrating the latter over λ_b. This sequence of operations yields, after normalization, the posterior marginal distribution for λ_s as a superposition of Gamma distributions of the form:

$$p\left(\lambda_s|n_b, t_b, n, t\right) d\lambda_s = \sum_{k=1}^{n} w_k \frac{e^{-\lambda_s t}(\lambda_s t)^k}{\Gamma(k)} \frac{d\lambda_s}{\lambda_s}, \qquad 0 < \lambda_s < \infty, \qquad (2.73)$$

where the weights w_k are defined as

$$w_k \triangleq \frac{Nb\left(n-k|\left(\frac{t}{t+t_b}\right), n_b\right)}{\sum\limits_{j=1}^{n} Nb\left(n-j|\left(\frac{t}{t+t_b}\right), n_b\right)}, \qquad k = 1, \ldots, n, \qquad (2.74)$$

and where $Nb\left(N|\frac{t}{t+t_b}, n_b\right)$ denotes the following *negative-binomial distribution* with integer index n_b and parameter $t/(t+t_b)$:

$$Nb\left(N|\frac{t}{t+t_b}, n_b\right) \triangleq \frac{(N+n_b-1)!}{N!(n_b-1)} \frac{t^n}{(t+t_b)^{n+1}}$$

$$= \binom{N+n_b-1}{N}\left(\frac{t}{t+t_b}\right)^n\left(1-\frac{t}{t+t_b}\right)^N, \qquad N = 0,1,2,\ldots .$$

$$(2.75)$$

Recall that the standard form of the negative binomial distribution is

$$Nb\left(x|\theta, r\right) = \binom{r+x-1}{r-1}\theta^r(1-\theta)^x, \qquad x = 0,1,2,\ldots . \qquad (2.76)$$

The mean and variance are $\langle x \rangle = r\theta$ and $\text{var}(x) = r(1-\theta)/\theta^2$; it is also important to note that the sum of n independent negative binomial quantities with parameters $\{\theta, r_i\}$, $i = 1, \ldots, n$, is also a negative binomial random quantity with parameters θ and $r = \sum\limits_{i=1}^{n} r_i$.

The normalizing sum in the denominator of Eq. (2.74) arises because the number of signal counts cannot exceed the total number of counts, $k \leq n$, thus yielding a truncated negative binomial distribution. The posterior marginal distribution for λ_s in Eq. (2.73) comprises the sum of individual Gamma distributions corresponding to each nonvanishing number of signal counts, weighted by the probability of this number conditional on the results of the background measurement. The mean and variance of λ_s can be computed by using Eq. (2.73) in conjunction with the relation

$$\int_0^\infty x^{\nu-1} \exp\left(-\mu x\right) dx = \mu^{-\nu}\Gamma\left(\nu\right), \quad \text{Re}\left(\mu > 0\right), \quad \text{Re}\left(\nu > 0\right). \quad (2.77)$$

After using the above relation to compute the first and second moments of λ_s in Eq. (2.73), the mean $\langle\lambda_s\rangle$ and relative standard deviation $\Delta\lambda_s / \langle\lambda_s\rangle$ are obtained as

$$\langle\lambda_s\rangle = \frac{\sum\limits_{k=1}^{n} w_k k}{t} \simeq \frac{n}{t} - \frac{n_b}{t_b}; \quad (2.78)$$

$$\frac{\Delta\lambda_s}{\langle\lambda_s\rangle} = \left\{\left(\sum_{k=1}^{n} k w_k\right)^{-1} + \left(\sum_{k=1}^{n} k^2 w_k\right)\left(\sum_{k=1}^{n} k w_k\right)^{-2} - 1\right\}^{1/2}$$

$$\simeq \frac{\left(n/t^2 + n_b/t_b^2\right)^{1/2}}{\left|n/t - n_b/t_b\right|}. \quad (2.79)$$

When the counts n and n_b are large, the posterior distribution is sharply peaked, the upper bound n on the sums over k can be extended to replace the mean and variance of the truncated negative binomial distribution by those corresponding to the full one, and Stirling's approximation for factorials can be used to obtain the approximate expressions shown in Eqs. (2.78) and (2.79). In this case, Eq. (2.78) indicates that the average count rate is the difference between the observed total rate and the background rate, while Eq. (2.79) indicates that the corresponding variance is given by the sum of the two empirical variances, $\text{var}\,\lambda_s \simeq n/t^2 + n_b/t_b^2$. When the signal-to-noise ratio is unfavorable, the two count rates are comparable, $n/t \simeq n_b/t_b$, and the relative standard error becomes large. At the other extreme, in the absence of noise, $n_b \cong 0$, so that Eqs. (2.78) and (2.79) reduce to:

$$\langle\lambda_s\rangle \simeq \frac{n}{t}; \quad \frac{\Delta\lambda_s}{\langle\lambda_s\rangle} \simeq \frac{1}{\sqrt{n}}. \quad (2.80)$$

In particular, Eq. (2.80) underscores the inverse-square-root dependence of the relative standard deviations on the number of counts encountered; this relationship is important for statistical tallies, in particular for Monte Carlo results.

The Gamma distribution given in Eq. (2.70) can also be used for predicting the number of counts N in some other time interval T. The joint probability that, given n counts registered during the time t, the propositions "N counts within T" and "λ within $d\lambda$" are both true, is

$$p\left(N, \lambda | T, n, t\right) d\lambda = \frac{e^{-\lambda T}(\lambda T)^N}{\Gamma\left(N\right)} \frac{e^{-\lambda t}(\lambda t)^n}{\Gamma\left(n\right)} \frac{d\lambda}{\lambda},$$

$$0 < \lambda < \infty, \quad N = 0, 1, 2, ... \tag{2.81}$$

Marginalizing (integrating over) the "nuisance parameter" λ yields the probability of registering "N counts within T, given n counts registered during the time t, regardless of λ"; this probability is a negative binomial distribution of the form

$$Nb\left(N | \frac{t}{t+T}, n\right) = \binom{N+n-1}{N} \left(\frac{t}{t+T}\right)^n \left(1 - \frac{t}{t+T}\right)^N, \quad N = 0, 1, 2, ...$$

$$\tag{2.82}$$

The mean and variance of the above distribution are $\langle N \rangle = nt/\left(t+T\right)$ and, respectively, $\mathrm{var}(N) = nt/T[t/\left(t+T\right)]^2$.

2.4   Evaluation of Consistent Data with Random and Systematic Errors

In the previous section, only random or uncorrelated errors were considered. In practice, however, some of the errors affecting measurements are random (uncorrelated) while the others are systematic (correlated). As will be shown in this section, the independent errors can be reduced by repeated measurements of the same quantity, but the correlated errors cannot be reduced this way. They remain as a "residual" uncertainty that could be reduced only by additional measurements, using different techniques and instrumentation, geometry, etc. Eventually, the effect of additional measurements using different techniques would reduce the correlated (systematic) error just as repetitions of the same measurement using the same technique reduce the uncorrelated statistical uncertainty. Both subsections 2.4.1 and 2.4.2 underscore these facts: subsection 2.4.1 refers to the discrete variates, while subsection 2.4.2 refers to continuous variates.

2.4.1 Discrete Outcomes: Correlated and Uncorrelated Relative Frequencies

Consider n measurements of the same quantity, x, with discrete possible outcomes (observables), x_ν, as are typically obtained in the channels of an analogue-to-digital converter. Each measurement yields one of the N possible values for the observable outcomes x_ν, $n = (1, 2, \ldots, N)$, and each of the values x_ν differs from the true value x because of errors. The outcomes x_ν are often considered to follow a Poisson distribution or, in certain cases, a multinomial distribution. It is convenient to label the measurements (trials) using the index $k = (1, 2, \ldots, n)$; the result of the kth measurement is denoted by r_k, $k = (1, 2, \ldots, n)$. Each of the N^n possible outcomes of the n trials occurs with a finite probability $P(r_1, \ldots, r_n)$, normalized such that

$$\sum_{r_1} \cdots \sum_{r_n} P(r_1, \ldots, r_n) = 1. \tag{2.83}$$

The result r occurs within the sequence $\{r_1, \ldots, r_n\}$ with the relative frequency f_r, which is not known *a priori*, but its mean $\langle f_r \rangle$ and variance var(f_r) can be evaluated using the given probabilities $P(r_1, \ldots, r_n)$. Using the Kronecker-delta symbol δ_{rr_k}, the relative frequency f_r can be expressed as

$$f_r = \frac{1}{n} \sum_{k=1}^{n} \delta_{rr_k}. \tag{2.84}$$

The mean, $\langle f_r \rangle$, of f_r, is computed by using Eq. (2.84) to obtain

$$\langle f_r \rangle = \frac{1}{n} \sum_{k=1}^{n} \left[\sum_{r_1} \cdots \sum_{r_n} P(r_1, \ldots, r_n) \, \delta_{rr_k} \right] \triangleq \frac{1}{n} \sum_{k=1}^{n} P(k|r), \tag{2.85}$$

where $P(k|r)$ denotes the probability that the kth-trial yields the result r. Considering next the relative frequencies f_r and f_s of two results, r and s, respectively, in conjunction with Eq. (2.84) yields the following expression for the mixed second moments $\langle f_r f_s \rangle$:

$$\langle f_r f_s \rangle = \frac{1}{n^2} \sum_{j=1}^{n} \sum_{k=1}^{n} \left[\sum_{r_1} \cdots \sum_{r_n} P(r_1, \ldots, r_n) \, \delta_{rr_j} \, \delta_{ssk} \right]$$

$$\triangleq \frac{1}{n^2} \sum_{j=1}^{n} \sum_{k=1}^{n} P(j|r; k|s), \tag{2.86}$$

where $P(j|r; k|s)$ denotes the joint probability that the jth trial yields the result r and the kth trial yields the result s. The particular case when $j = k$ indicates consideration of the same measurement (i.e., r must be identical to s), implying that $P(k|r; k|s) = \delta_{rs}P(k|s)$. The covariances, cov (f_r, f_s), of the estimated frequencies can be computed from their definition, as follows:

$$\text{cov}(f_r, f_s) \triangleq \langle f_r f_s \rangle - \langle f_r \rangle \langle f_s \rangle$$

$$= \frac{1}{n} \left(\delta_{rs} P_r - P_r P_s \right) + \left(1 - \frac{1}{n} \right) \left(P_{rs} - P_r P_s \right), \tag{2.87}$$

where P_s denotes the probability of observing the value x_s in some unspecified single trial, as defined and normalized below

$$P_s \triangleq \frac{1}{n} \sum_{j=1}^{n} P(j|s), \quad \sum_s P_s = 1, \tag{2.88}$$

while P_{rs} denotes the probability of observing the values x_r and x_s in some unspecified pair of distinct trials, as defined and normalized below

$$P_{rs} \triangleq \frac{1}{n(n-1)} \sum_j \sum_{k \neq j} P(j|r; k|s), \quad \sum_{r,s} P_{rs} = 1. \tag{2.89}$$

The variance $var(f_r)$ is obtained as the particular case of Eq. (2.87), namely:

$$var(f_r) \triangleq \langle f_r^2 \rangle - \langle f_r \rangle^2 = \frac{1}{n} \left(P_r - P_r^2 \right) + \left(1 - \frac{1}{n} \right) \left(P_{rr} - P_r^2 \right). \tag{2.90}$$

When the equivalent trials are independent of each other, the relation $P_{rs} = P_r P_s$ holds, including the particular case $P_{rr} = P_r^2$. It follows from Eq. (2.87) that cov $(f_r, f_s) = -P_r P_s/n$, implying that the relative frequencies are *anti-correlated* (rather than truly uncorrelated), but in the limit of many mea-surements ($n \to \infty$), these anti-correlations vanish. These anti-correlations stem from the normalization condition which imposes the constraint that if one frequency varies in a certain sense, the others must vary in the opposite sense in order to keep their normalized sum, $\sum_r f_r = 1$, unaffected. Further-more, Eq. (2.90) shows that the variances also vanish as $n \to \infty$. In summary, when the *measurements are independent*, then

$$\text{cov}\,(f_r, f_s) = -P_r P_s/n \ \sim \frac{1}{n} \ \xrightarrow{n \to \infty} 0,$$

$$\text{var}\,(f_r) \sim \frac{1}{n} \ \xrightarrow{n \to \infty} 0. \tag{2.91}$$

When the *trials* are *correlated*, however, the relation $P_{\mu\nu} \neq P_\mu P_\nu$ holds. Hence, as Eq. (2.87) indicates, the variances cannot become smaller than $P_{\nu\nu} - P_\nu^2$ even for arbitrarily good statistics, since

$$\text{cov}\,(f_r, f_s) \ \xrightarrow{n \to \infty} \ P_{rs} - P_r P_s. \tag{2.92}$$

Similarly, the variances var (f_ν) tend to the value $(P_{\nu\nu} - P_\nu^2)$ in the limit of very many trials ("good statistics") as $n \to \infty$. These results indicate that any correlation imposes an irreducible limit on the accuracy with which frequencies can be predicted from probabilities.

The relationships between probabilities and frequencies derived in the fore-going can be applied to the observable experimental outcomes x_ν in order to estimate the sample mean, $\bar{x}$, of the observed experimental values, as well as the corresponding variance, var $(\bar{x})$. In terms of the actual frequencies f_r, the *sample mean*, $\bar{x}$, is expressed as

$$\bar{x} = \sum_\nu f_\nu x_\nu. \tag{2.93}$$

Since the actual frequencies f_ν are unknown, $\bar{x}$ cannot be computed di-rectly from Eq. (2.93). Nevertheless, the expectation value, $\langle \bar{x} \rangle$, of the sample mean $\bar{x}$, and the corresponding variance, var $(\bar{x})$, can be obtained from Eqs. (2.85), (2.86), and (2.93), as follows:

$$\langle \bar{x} \rangle = \sum_r \langle f_r \rangle x_r = \sum_r P_r x_r = \langle x \rangle , \tag{2.94}$$

$$\text{var}\left(\bar{x}\right) \triangleq \left\langle \bar{x}^2 \right\rangle - \left\langle \bar{x} \right\rangle^2 = \frac{1}{n}\left(\left\langle x^2 \right\rangle - \left\langle x \right\rangle^2\right) + \left(1 - \frac{1}{n}\right)\left(\left\langle x,\, x \right\rangle - \left\langle x \right\rangle^2\right), \quad (2.95)$$

respectively. In Eqs. (2.94) and (2.95), the quantities

$$\left\langle x \right\rangle \triangleq \sum_r P_r x_r \quad \text{and} \quad \left\langle x^2 \right\rangle \triangleq \sum_r P_r x_r^2, \quad (2.96)$$

are expectation values for a single unspecified measurement, while the quantity

$$\left\langle x,\, x \right\rangle \triangleq \sum_{r,\, s} P_{rs} x_r x_s. \quad (2.97)$$

refers to an unspecified pair of distinct and possibly correlated measurements. As indicated by Eq. (2.95), the variance of the sample average, $\text{var}\left(\bar{x}\right)$, comprises the sum of two contributions, namely: (i) a contribution from uncorrelated (statistical) errors of single measurements; and (ii) a contribution from correlated (systematic) errors of pairs of measurements, which can arise from common errors such as errors in standards, detector calibration, and electronic settings.

When all n measurements are equivalent, the probabilities $P\left(k|s\right)$ must be the same for all trials, i.e., $P\left(k|s\right) = P_s$. Similarly, the probabilities $P\left(j|r;\ k|s\right)$ must be the same for all pairs of trials, i.e., $P\left(j|r;\ k|s\right) = P_{rs}$ when $j \neq k$. When the measurements are uncorrelated (which implies that common or systematic errors are absent), the relation $P_{rs} = P_r P_s$ holds, hence $\left\langle x,\, x \right\rangle = \left\langle x \right\rangle^2$. Thus, when the measurements are uncorrelated, Eq. (2.95) implies that

$$\text{var}\left(\bar{x}\right) = \frac{1}{n}\left(\left\langle x^2 \right\rangle - \left\langle x \right\rangle^2\right) = \frac{1}{n}\,\text{var}\,x; \quad \Rightarrow \quad \text{var}\left(\bar{x}\right) \xrightarrow{n \to \infty} 0. \quad (2.98)$$

The above result indicates that additional experiments would indeed help to increase measurement accuracy, but only if the measurements are uncorrelated. On the other hand when the errors are common or systematic, the measurements are fully correlated, so the result x_r of the first trial implies the same result for all following trials, which means that $P_{rs} = \delta_{rs} P_r$ or, equivalently, $\left\langle x,\, x \right\rangle = \left\langle x^2 \right\rangle$. Replacing this result in Eq. (2.98) shows that, when the *measurements are correlated*, the following relation holds

$$\text{var}\left(\overline{x}\right) = \langle x^2 \rangle - \langle x \rangle^2 \triangleq \text{var}\left(x\right), \tag{2.99}$$

The above result indicates that repetition of correlated measurements does not help to reduce the uncertainty, which remains the same as it was after the first trial, as long as the unknown systematic errors do not change. As already mentioned, correlated errors can be reduced only by measuring the quantity of interest anew by means of different techniques, with different instrumentation, geometry, etc., in order to obtain systematic errors as different and as independent of each other as possible. Eventually, several measurements of the same quantity with different techniques could reduce the overall systematic uncertainty in the same way in which a comparable number of repetitions with the same technique would reduce the uncorrelated statistical uncertainty.

2.4.2 Continuous Outcomes: Consistent Data with Random and Systematic Errors

Consider n measurements and/or computations of the same quantity μ. The *measurements are considered to be consistent "within error bars," but are affected by both random and systematic errors.* Usually the systematic error would be common to all data (e.g., because the experiments were all performed using the same calibrated detector) while the random components (such as arrising from counting statistics) would differ from each other. Thus, each measurement i will yield a distinct result (data) x_i, $i = 1, \ldots, n$, of the form

$$x_i - \mu = \varepsilon_i + \varepsilon_0; \quad i = 1, \ldots, n, \tag{2.100}$$

where ε_0 denotes the common systematic error, while the random errors ε_i can be considered as uncorrelated, without loss of generality. In the absence of information to the contrary, the errors ε_i, $i = 0, 1, \ldots, n$, can be considered to take any real value from $-\infty$ to $+\infty$, without preference for negative or positive values (or, equivalently, positive values are considered to be as equally probable as negative ones). In this case, their unknown probability distributions would have vanishing first moments, i.e., $\langle \varepsilon_i \rangle = 0$, $i = 0, 1, \ldots n$. Estimates τ_i^2, $i = 0, 1, \ldots, n$, of the variances of the statistical errors ε_i

are considered to be available; hence, the second moments $\langle \varepsilon_i^2 \rangle$ of the unknown probability distributions for the errors ε_i would take on the values $\langle \varepsilon_i^2 \rangle = \tau_i^2$, $i = 0, 1, \ldots, n$. When the available information is restricted to only the first and second moments of the otherwise unknown distributions of possible errors, the maximum entropy principle presented in Section 2.2 implies that Gaussian distributions should be used for all further inferences, since all other distributions with the same first and second moments would have lower information entropy. Since the errors considered in Eq. (2.100) do not depend on each other (i.e., knowing one of the errors indicates nothing about the others), their joint distribution is the product of n independent Gaussians, i.e.,

$$p\left(\varepsilon_0, \ldots, \varepsilon_n | \tau_0, \ldots, \tau_n\right) d\varepsilon_0 \ldots d\varepsilon_n \propto \exp\left(-\frac{1}{2} \sum_{i=0}^{n} \frac{\varepsilon_i^2}{\tau_i^2}\right) d\varepsilon_0 \ldots d\varepsilon_n,$$

$$-\infty < \varepsilon_i < \infty, \quad i = 0, \ldots, n.$$

$$(2.101)$$

The Gaussian distribution above encodes all of the available (incomplete) information and serves as the starting point for further inferences. The next step is to find the "likelihood," i.e., the probability of finding the sample $\mathbf{x} = (x_1, \ldots, x_n)$ in intervals $(dx_1, \ldots, dx_n)$, given the parameters $(\mu, \tau_0, \tau_1, \ldots, \tau_n)$. This requires a change of variables from ε_i to x_i, which transforms Eq. (2.101) into the likelihood distribution

$$p\left(\varepsilon_0, \mathbf{x} | \mu, \tau_0, \tau_1, \ldots, \tau_n\right) d\varepsilon_0 \, dx_1 \ldots dx_n$$

$$\propto \exp\left[-\frac{1}{2}\left(\frac{\varepsilon_0^2}{\tau_0^2} + \sum_{i=1}^{n} \frac{(x_i - \mu - \varepsilon_0)^2}{\tau_i^2}\right)\right] d\varepsilon_0 \, dx_1 \ldots dx_n,$$

$$-\infty < \varepsilon_0, x_1, \ldots, x_n < \infty. \qquad (2.102)$$

The uninteresting variate ("nuisance parameter") ε_0 is "marginalized" by integrating over all of its possible values to obtain a weighted average over all possible values of ε_0 with their various probabilities as weights. This integration (as well as several other integrations to follow in this section), is performed using the relationship

$$\int_{-\infty}^{\infty} x^m \exp\left(-px^2 + 2qx\right) dx = \frac{1}{2^{m-1}p} \sqrt{\frac{\pi}{p}} \frac{d^{m-1}}{dq^{m-1}} \left[q \exp\left(\frac{q^2}{p}\right)\right]$$

$$= m! \sqrt{\frac{\pi}{p}} \left(\frac{q}{p}\right)^m e^{\frac{q^2}{p}} \sum_{k=0}^{E\left(\frac{m}{2}\right)} \frac{1}{(n-2k)!\,(k)!} \left(\frac{p}{4q^2}\right)^k ; \quad p > 0.$$

$$(2.103)$$

Using the above formula with $m = 0$ for integration over ε_0, yields

$$p\left(\mathbf{x}|\mu, \tau_0, \tau_1, \ldots, \tau_n\right) dx_1 \ldots dx_n =$$

$$= \frac{\exp\left[-\frac{1}{2}Q\left(\mu, \mathbf{x}, \tau_0, \tau_1, \ldots, \tau_n\right)\right]}{\sqrt{(2\pi)^n \det(\mathbf{C})}} dx_1 \ldots dx_n, \quad -\infty < x_1, \ldots, x_n < \infty,$$

$$(2.104)$$

where

$$Q\left(\mu, \mathbf{x}, \tau_0, \tau_1, \ldots, \tau_n\right) \triangleq \sum_{i=1}^{n} (x_i - \mu)^2 \tau_i^{-2} - \frac{\left(\sum_{i=1}^{n} (x_i - \mu)\tau_i^{-2}\right)^2}{\left(\sum_{i=0}^{n} \tau_i^{-2}\right)}$$

$$= (\mathbf{x} - \mu)^T \mathbf{C}^{-1} (\mathbf{x} - \mu), \qquad (2.105)$$

where the vector $\mu \triangleq (\mu, \ldots, \mu)$ has all of its n-components equal to the location parameter μ, and $\mathbf{C}^{-1}$ is the inverse of the covariance matrix $\mathbf{C}$ for the n measurements, defined in this case as

$$
\mathbf{C}^{-1} =
\begin{pmatrix}
\tau_1^{-2} & 0 & 0 & \cdots & 0 \\
0 & \tau_2^{-2} & 0 & \cdots & 0 \\
0 & 0 & \tau_3^{-2} & \ddots & \vdots \\
\vdots & \vdots & \ddots & \ddots & 0 \\
0 & 0 & \cdots & 0 & \tau_n^{-2}
\end{pmatrix}
$$

$$
-
\begin{pmatrix}
\tau_1^{-4} & \tau_1^{-2}\tau_2^{-2} & \tau_1^{-2}\tau_3^{-2} & \cdots & \tau_1^{-2}\tau_n^{-2} \\
\tau_2^{-2}\tau_1^{-2} & \tau_2^{-4} & \tau_2^{-2}\tau_3^{-2} & \cdots & \tau_2^{-2}\tau_n^{-2} \\
\tau_3^{-2}\tau_1^{-2} & \tau_3^{-2}\tau_2^{-2} & \tau_3^{-4} & \cdots & \vdots \\
\vdots & \vdots & \vdots & \ddots & \vdots \\
\tau_n^{-2}\tau_1^{-2} & \tau_n^{-2}\tau_2^{-2} & \tau_n^{-2}\tau_3^{-2} & \cdots & \tau_n^{-4}
\end{pmatrix}
\left(\sum_{i=0}^{n} \tau_i^{-2} \right)^{-1}.
$$

$$(2.106)$$

Having marginalized ε_0, the next step is to find a prior distribution for μ which encodes the available knowledge about μ before any measurements and/or computations are performed. If no information whatsoever is available regarding the location parameter μ, all real values must appear equally probable to an observer, so the least informative prior is the uniform distribution $p(\mu)\, d\mu \propto d\mu$, $-\infty < \mu < \infty$, as indicated in Eq. (2.83). Next, Bayes' theorem is employed to obtain the posterior distribution for the location parameter $\mu, -\infty < \mu < \infty$, as the product of this prior and the likelihood given in Eq. (2.104), thus giving the following Gaussian as the posterior distribution

$$
p(\mu|\mathbf{x}; \tau_0; \tau_1, \ldots, \tau_n)\, d\mu \propto \exp\left[-\frac{1}{2} Q(\mu, \mathbf{x}, \tau_0, \tau_1, \ldots, \tau_n) \right] d\mu,
$$

$$
-\infty < \mu < \infty, \tag{2.107}
$$

with $Q(\mu, \mathbf{x}, \tau_0, \tau_1, \ldots, \tau_n)$ rearranged as a quadratic function of μ, i.e.,

$$Q\left(\mu, \mathbf{x}, \tau_0, \tau_1, \ldots, \tau_n\right) \triangleq A\mu^2 - 2B\mu + C;$$

$$A \triangleq \tau_0^{-2} \left(\sum_{i=1}^{n} \tau_i^{-2}\right) \left(\sum_{i=0}^{n} \tau_i^{-2}\right)^{-1}; \quad B \triangleq 5\tau_0^{-2} \left(\sum_{i=1}^{n} x_i \tau_i^{-2}\right) \left(\sum_{i=0}^{n} \tau_i^{-2}\right)^{-1};$$

$$C \triangleq \sum_{i=1}^{n} x_i^2 \tau_i^{-2} - \left(\sum_{i=1}^{n} x_i \tau_i^{-2}\right)^2 \left(\sum_{i=0}^{n} \tau_i^{-2}\right)^{-1}. \tag{2.108}$$

When estimates of the values of the systematic error τ_0 and, respectively, random errors $\tau_1, \ldots, \tau_n$ are available, the explicit expressions for the mean and variance of μ can be determined by computing the first and second moments of the above Gaussian with respect μ, by using Eq. (2.103) with $m = 1$, and $m = 2$, respectively. Alternatively, the quadratic form $Q\left(\mu, \mathbf{x}, \tau_0, \tau_1, \ldots, \tau_n\right)$ can be rearranged in the standard form for a univariate normal distribution:

$$Q\left(\mu, \mathbf{x}, \tau_0, \tau_1, \ldots, \tau_n\right) = \frac{\left(\mu - \langle\mu\rangle\right)^2}{\text{var}\left(\mu\right)}. \tag{2.109}$$

Either way, the results for the mean and, respectively, variance of μ are

$$\langle\mu\rangle = \left(\sum_{i=1}^{n} x_i \tau_i^{-2}\right) \left(\sum_{i=1}^{n} \tau_i^{-2}\right)^{-1}, \quad \text{var}(\mu) = \tau_0^2 + \left(\sum_{i=1}^{n} \tau_i^{-2}\right)^{-1}. \tag{2.110}$$

In the particular case when all of the experiments can be considered as "exchangeable trials," i.e., when they are all affected by the same common error and have equally good (or bad) statistical errors $\tau_1^2 = \tau_2^2 = \ldots = \tau_n^2$, Eq. (2.110) reduces to

$$\langle\mu\rangle = \frac{1}{n} \sum_{i=1}^{n} x_i \triangleq \bar{x}, \quad \text{var}(\mu) = \tau_0^2 + \frac{\tau_1^2}{n}. \tag{2.111}$$

The expressions in Eq. (2.111) indicate that the mean $\langle\mu\rangle$ is unaffected by the systematic error, but the variance var(μ) is affected by both systematic and statistical errors. It is furthermore apparent that only the standard deviation of the statistical component $\left(\sum_{i=1}^{n} \tau_i^{-2}\right)^{-1}$ of the total uncertainty

$\text{var}(\mu)$ decreases, as $1/\sqrt{n}$, when measurements and/or computations are repeated n times. In contradistinction, the systematic component, τ_0^2, of the total error $\text{var}(\mu)$ remains as a residual uncertainty that cannot be reduced by mere repetition. Recall that this is the same conclusion which was reached in Section 2.4.1, based on relative frequency considerations. Only new measurements and/or computations with nonequivalent techniques, within a generalized Bayesian simultaneous combination of all data, can help reduce the systematic uncertainties. Note that the results shown in Eq. (2.110), including the simplified expressions given in Eq. (2.111) in the case of repeated exchangeable experiments, can be used only if the statistical error τ_0^2 and the systematic mean-square errors τ_i^2 are actually known.

In cases when the errors τ_0^2 and τ_i^2 are not actually known, they must be estimated simultaneously with the location parameter μ. The procedure for such cases can be illustrated by considering, for simplicity, that all experiments are "exchangeable trials," i.e., they are all affected by the same common error and have equally good (or bad) statistical errors $\tau_1^2 = \tau_2^2 = \ldots = \tau_n^2$. However, in contradistinction to the situation leading to Eq. (2.107), the actual values of τ_1^2 and τ_0^2 are now considered unknown and need to be determined along the location parameter μ. The likelihood function for this extended problem is still provided by the distribution in Eq. (2.104), particularized to perfectly repeatable experiments, i.e., $\tau_1^2 = \tau_2^2 = \ldots = \tau_n^2$. In this case, the matrix $\mathbf{C}^{-1}$ in Eq. (2.106) reduces to:

$$\mathbf{C}^{-1} = \frac{1}{\tau_1^2} \left[\begin{pmatrix} 1 & 0 & 0 & \cdots & 0 \\ 0 & 1 & 0 & \cdots & 0 \\ 0 & 0 & 1 & \ddots & \vdots \\ \vdots & \vdots & \ddots & \ddots & 0 \\ 0 & 0 & \cdots & 0 & 1 \end{pmatrix} - \frac{\tau_0^2}{\tau_1^2 + n\tau_0^2} \begin{pmatrix} 1 & 1 & 1 & \cdots & 1 \\ 1 & 1 & 1 & \cdots & 1 \\ 1 & 1 & 1 & \ddots & \vdots \\ \vdots & \vdots & \ddots & \ddots & 1 \\ 1 & 1 & \cdots & 1 & 1 \end{pmatrix} \right]. $$

$$(2.112)$$

The determinant of the matrix inside the brackets of Eq. (2.112) can be computed by subtracting the first column from all the others, and subsequently adding all resulting rows to the first one. Denoting, for convenience, $\frac{\tau_0^2}{\tau_1^2 + n\tau_0^2} \triangleq \beta$, and performing these operations yields

$$
\begin{vmatrix}
1-\beta & -\beta & -\beta & \cdots & -\beta \\
-\beta & 1-\beta & -\beta & \cdots & -\beta \\
-\beta & -\beta & 1-\beta & \ddots & \vdots \\
\vdots & \vdots & \ddots & \ddots & -\beta \\
-\beta & -\beta & -\beta & \cdots & 1-\beta
\end{vmatrix}
\begin{vmatrix}
1-\beta & -1 & -1 & \cdots & -1 \\
-\beta & 1 & 0 & \cdots & 0 \\
-\beta & 0 & 1 & \ddots & \vdots \\
\vdots & \vdots & \ddots & \ddots & 0 \\
-\beta & 0 & 0 & \cdots & 1
\end{vmatrix}
$$

$$
= \begin{vmatrix}
1-n\beta & 0 & 0 & \cdots & 0 \\
-\beta & 1 & 0 & \cdots & 0 \\
-\beta & 0 & 1 & \ddots & \vdots \\
\vdots & \vdots & \ddots & \ddots & 0 \\
-\beta & 0 & 0 & \cdots & 1
\end{vmatrix}
= 1 - n\beta = 1 - n\frac{\tau_0^2}{\tau_1^2 + n\tau_0^2} = \frac{\tau_1^2}{\tau_1^2 + n\tau_0^2}.
$$

$$
\tag{2.113}
$$

Using the above result together with Eq. (2.112) yields

$$
\det\left(\mathbf{C}^{-1}\right) = \frac{\tau_1^2}{\tau_1^2 + n\tau_0^2}\left(\frac{1}{\tau_1^2}\right)^n = \left(\frac{1}{\tau_1^2}\right)^{n-1}\frac{1}{\tau_1^2 + n\tau_0^2} = (\det \mathbf{C})^{-1}. \tag{2.114}
$$

Taking into account the specific form of $\mathbf{C}^{-1}$ in Eq. (2.112), and introducing the sample mean $\bar{x}$ and sample variance μ_2^S for this particular situation, makes it possible to rearrange the quadratic form $Q\left(\mu, \mathbf{x}, \tau_0, \tau_1, \ldots, \tau_n\right)$ in Eq. (2.105) to read

$$
Q = \frac{n\mu_2^S}{\tau_1^2} + \frac{n(\bar{x}-\mu)^2}{\tau_1^2 + n\tau_0^2}, \tag{2.115}
$$

where

$$
\bar{x} \triangleq (1/n)\sum_j x_j; \quad \mu_2^S \triangleq (1/n)\sum_j (x_j - \bar{x})^2 = \overline{x^2} - (\bar{x})^2. \tag{2.116}
$$

Using the expression for $Q\left(\mu, \mathbf{x}, \tau_0, \tau_1, \ldots, \tau_n\right)$ given in Eq. (2.115) in the likelihood distribution, cf. Eq. (2.104), properly normalized using Eq. (2.114), leads to the following expression:

$$p\left(\mathbf{x}|\mu, \tau_0, \tau_1\right) dx_1 \ldots dx_n = \frac{\exp\left[-\frac{n\mu_2^S}{2\tau_1^2}\right]}{\sqrt{\left(2\pi\tau_1^2\right)^{n-1}}} \times \frac{\exp\left[-\frac{n}{2}\frac{(\bar{x}-\mu)^2}{\tau_1^2 + n\tau_0^2}\right]}{\sqrt{2\pi\left(\tau_1^2 + n\tau_0^2\right)}} dx_1 \ldots dx_n.$$

(2.117)

The least informative prior for the location parameter μ is again the uniform distribution on the linear scale. However, the least informative priors for each of the positive scale parameters τ_0 and τ_1, respectively, is Jeffreys' prior, which is uniform on the logarithmic scale. All three parameters are statistically independent, hence their joint prior is just the product of the three priors, i.e.,

$$p\left(\mu, \tau_0, \tau_1\right) d\mu\, d\tau_0\, d\tau_1 \propto d\mu \frac{d\tau_0}{\tau_0} \frac{d\tau_1}{\tau_1},$$

$$-\infty < \mu < \infty, \quad 0 \le \tau_0, \tau_1, \sigma < \infty.$$

(2.118)

Multiplying the likelihood function in Eq. (2.117) and the least informative prior in Eq. (2.118) yields the posterior as a product of three distributions,

$$p\left(\mu, \tau_0, \tau_1 | x_1, \ldots, x_n\right) d\mu\, d\tau_0\, d\tau_1 = p\left(\mu|\bar{x}, \tau_0, \tau_1\right) d\mu$$

$$\times p\left(\tau_1|\mu_2^S\right) d\tau_1 \times p\left(\tau_0\right) d\tau_0, \quad -\infty < \mu < \infty, \quad 0 \le \tau_0, \tau_1 < \infty,$$

(2.119)

where the respective distributions, properly normalized to unity are

$$p\left(\mu|\bar{x}, \tau_0, \tau_1\right) d\mu = \frac{1}{\sqrt{2\pi\left(\tau_1^2 + n\tau_0^2\right)/n}} \exp\left[-\frac{n}{2}\frac{(\mu - \bar{x})^2}{\tau_1^2 + n\tau_0^2}\right] d\mu,$$

(2.120)

$$p\left(\tau_1|\mu_2^S\right) d\tau_1 = \frac{2\left(\frac{n\mu_2^S}{2\tau_1^2}\right)^{(n-1)/2}}{\Gamma\left(\frac{n-1}{2}\right)} \exp\left[-\frac{n\mu_2^S}{2\tau_1^2}\right] \frac{d\tau_1}{\tau_1}.$$

(2.121)

Note that the Gaussian distribution $p\left(\mu|\bar{x}, \tau_0, \tau_1\right) d\mu$ is normalized using Eq. (2.103) with $m = 0$, while the Gamma distribution $p\left(\tau_1|\mu_2^S\right) d\tau_1$, with argument $\frac{n\mu_2^S}{2\tau_1^2}$, is normalized using the formula

$$\int_0^\infty x^{\nu-1} \exp(-\mu x)\, dx = \mu^{-\nu} \Gamma(\nu), \quad \mathrm{Re}(\mu > 0), \quad \mathrm{Re}(\nu > 0). \quad (2.122)$$

Since the quantities τ_0 and τ_1 are unknown, we must still marginalize (i.e., average) the posterior distribution given in Eq. (2.119) over the distribution of statistical errors, τ_1, and also over the distribution for systematic ones, τ_0. Thus, the mean $\langle \mu \rangle$ is obtained by integrating (over μ) the three distributions given in Eqs. (2.119), (2.120), and (2.121), to obtain

$$\langle \mu \rangle \triangleq \int_0^\infty p(\tau_0)\, d\tau_0 \int_0^\infty p\left(\tau_1 | \mu_2^S\right) d\tau_1 \int_{-\infty}^\infty \mu p(\mu | \bar{x}, \tau_0, \tau_1)\, d\mu$$

$$= \int_0^\infty p(\tau_0)\, d\tau_0 \int_0^\infty p\left(\tau_1 | \mu_2^S\right) d\tau_1 \, (\bar{x}) = (\bar{x}) \int_0^\infty p(\tau_0)\, d\tau_0.$$

If Jeffreys' prior $p(\tau_0)\, d\tau_0 = d\tau_0/\tau_0$ were used in the above integral to complete the averaging over the systematic error τ_0, the result for $\langle \mu \rangle$ would be unbounded (logarithmically divergent), which implies that the available information about the systematic error τ_0 (which is at this point severely limited to knowing just that τ_0 is a scale parameter) is insufficient for a definite estimate.

Therefore, a finite result for $\langle \mu \rangle$ can only be obtained if Jeffreys' prior is replaced with a somewhat more informative prior distribution. Such a slightly more informative distribution is provided by the exponential distribution, which, as will be shown shortly, suffices to obtain finite estimates for the various moments (mean and variance) of μ. In particular, considering a value for the size of the systematic errors (e.g., their order of magnitude according to the present state-of-the-art), and taking this value as the mean (first moment), $\langle \tau_0 \rangle$, of a normalizable prior distribution, the maximum entropy principle yields the following exponential distribution to be used as the prior,

$$p(\tau_0 | \langle \tau_0 \rangle)\, d\tau_0 = \exp\left(-\frac{\tau_0}{\langle \tau_0 \rangle}\right) \frac{d\tau_0}{\langle \tau_0 \rangle}, \quad 0 \le \tau_0 \le \infty. \quad (2.123)$$

The above exponential distribution resembles closely Jeffreys' prior in the vicinity of the specified mean $\langle \tau_0 \rangle$, but very small and very large values of τ_0 are sufficiently less probable to enable the existence of a finite normalization.

Also, the standard deviation of the exponential distribution is equal to the mean $\langle \tau_0 \rangle$, indicating that the prior information encoded by the exponential distribution is still rather vague.

Using the exponential distribution shown in Eq. (2.123) as the prior for the systematic error τ_0 makes it possible to obtain finite estimated values (under quadratic) for various means and variances. Thus, the mean $\langle \mu \rangle$ is obtained by integrating (over μ) the three distributions given in Eqs. (2.120), (2.121), and (2.123), to obtain

$$\langle \mu \rangle \triangleq \int_0^\infty p\left(\tau_0|\langle\tau_0\rangle\right) d\tau_0 \int_0^\infty p\left(\tau_1|\mu_2^S\right) d\tau_1 \int_{-\infty}^\infty \mu p\left(\mu|\bar{x},\tau_0,\tau_1\right) d\mu$$

$$= \int_0^\infty p\left(\tau_0|\langle\tau_0\rangle\right) d\tau_0 \int_0^\infty p\left(\tau_1|\mu_2^S\right) d\tau_1 \left(\bar{x}\right) = \bar{x}. \tag{2.124}$$

The above result shows that $\langle \mu \rangle$ is just the sample average, regardless of whether the measurements errors are statistical or systematic. The second moment, $\langle \mu^2 \rangle$, is needed in order to compute the variance, $\mathrm{var}(\mu)$; this second moment is computed from its definition, i.e.,

$$\langle \mu^2 \rangle \triangleq \int_0^\infty p\left(\tau_0|\langle\tau_0\rangle\right) d\tau_0 \int_0^\infty p\left(\tau_1|\mu_2^S\right) d\tau_1 \int_{-\infty}^\infty \mu p\left(\mu|\bar{x},\tau_0,\tau_1\right) d\mu$$

$$= \int_0^\infty p\left(\tau_0|\langle\tau_0\rangle\right) d\tau_0 \int_0^\infty p\left(\tau_1|\mu_2^S\right) d\tau_1 \left[\left(\bar{x}\right)^2 + \tau_0^2 + \tau_1^2/n\right]$$

$$= \left(\bar{x}\right)^2 + 2\langle\tau_0\rangle^2 + \frac{\mu_2^S}{n-3}. \tag{2.125}$$

Using the above result in conjunction with the result obtained in Eq. (2.124) yields

$$\mathrm{var}\,\mu = \langle \mu^2 \rangle - \langle \mu \rangle^2 = \left(\bar{x}\right)^2 + 2\langle\tau_0\rangle^2 + \frac{\mu_2^S}{n-3} - \left(\bar{x}\right)^2$$

$$= 2\langle\tau_0\rangle^2 + \frac{\mu_2^S}{n-3}. \tag{2.126}$$

The first term in Eq. (2.126) represents the vague estimate of possible systematic errors. The second term in Eq. (2.126) is the average of τ_1^2/n over the Gamma distribution $p\left(\tau_1|\mu_2^S\right) d\tau_1$, and it reflects the random scatter (uncertainties) of the experimental data. In the limit of no systematic errors, as $\langle \tau_0 \rangle \to 0$, the first term in Eq. (2.126) vanishes and the resulting estimate coincides with the estimate for purely random errors, as already obtained in Eq. (2.59).

Expectation values for the unknown statistical variance component τ_1^2 and for its reciprocal (the precision) τ_1^{-2}, can also obtained by using the posterior distribution. Thus, the mean $\langle \tau_1^2 \rangle$ of τ_1^2 is obtained by performing the integrations

$$\langle \tau_1^2 \rangle \triangleq \int_{-\infty}^{\infty} \tau_1^2 p\left(\tau_1 | \mu_2^S\right) d\tau_1 \int_0^{\infty} p\left(\tau_0 | \langle \tau_0 \rangle\right) d\tau_0 \int_0^{\infty} p\left(\mu | \bar{x}, \tau_0, \tau_1\right) d\mu$$

$$= \frac{n\mu_2^S}{n-3}. \tag{2.127}$$

The second moment $\langle \tau_1^4 \rangle$ of τ_1^2 is obtained similarly:

$$\langle \tau_1^4 \rangle \triangleq \int_{-\infty}^{\infty} \tau_1^4 p\left(\tau_1 | \mu_2^S\right) d\tau_1 \int_0^{\infty} p\left(\tau_0 | \langle \tau_0 \rangle\right) d\tau_0 \int_0^{\infty} p\left(\mu | \bar{x}, \tau_0, \tau_1\right) d\mu$$

$$= \frac{\left(n\mu_2^S\right)^2}{(n-3)(n-5)}. \tag{2.128}$$

Using the results in Eqs. (2.127) and (2.128) leads to

$$\operatorname{var}\left(\tau_1^2\right) = \langle \tau_1^4 \rangle - \langle \tau_1^2 \rangle^2 = \frac{\left(n\mu_2^S\right)^2}{(n-3)(n-5)} - \left(\frac{n\mu_2^S}{n-3}\right)^2 = \frac{2}{n-5}\left(\frac{n\mu_2^S}{n-3}\right)^2. \tag{2.129}$$

Hence, the relative standard deviation (also referred to as the "uncertainty of the uncertainty"), which is usually reported under "quadratic loss" in the form $\langle \tau_1^2 \rangle \pm \sqrt{\operatorname{var}\left(\tau_1^2\right)}$, can be obtained from the results in Eqs. (2.128) and (2.129) as

$$\frac{\sqrt{\operatorname{var}\left(\tau_1^2\right)}}{\langle \tau_1^2 \rangle} = \sqrt{\frac{2}{n-5}}. \tag{2.130}$$

Similar computations for the precision τ_1^{-2} lead to the following results:

$$\langle \tau_1^{-2} \rangle = \frac{n-1}{n}\left(\mu_2^S\right)^{-1} \triangleq s^{-2}; \qquad \frac{\sqrt{\operatorname{var}\left(\tau_1^{-2}\right)}}{\langle \tau_1^{-2} \rangle} = \sqrt{\frac{2}{n-1}}. \tag{2.131}$$

The quantity $s^2 = \mu_2^S n/(n-1)$, which is the reciprocal of the first result shown in Eq. (2.131) is sometimes called the *"experimental variance,"* and is often recommended as an unbiased estimate for an unknown true variance σ^2. This recommendation stems from the fact that, for a given parent distribution, the variance is equal to the average of s^2 over all possible samples, i.e., $\sigma^2 = \langle s^2 \rangle$, and the hope is that performing many repeated experiments might bring the experimental variance close to the true variance. However, the practical interest is not in an average over all possible samples for a given parent distribution, but in the parameters of a (Gaussian) parent distribution for the one specific sample under consideration. For such practical situations, it is correct to use the (reciprocal) relationship given in Eq. (2.131) for the "precision," rather than the relation $s^2 = \mu_2^S n/(n-1)$.

To summarize, it is important to note *that the sample variance* μ_2^S, *i.e., the scatter of the data, provides information only about the uncorrelated (statistical) error component but not about the correlated (systematic) errors. Therefore, the experimental data (considered in this section to comprise both random and systematic errors)* only allow the estimation of the statistical uncertainty of a location parameter μ, and this only if at least four equivalent measurements have been made, to ensure that $n \geq 4$. For large n, the statistical root-mean-square random error has the familiar $1/\sqrt{n}$ dependence. On the other hand, systematic errors do not essentially alter the scatter; they just shift or scale the cluster of data as a whole. Common errors remain undiminished regardless of the number of identical repetitions of the measurement, providing thus a residual uncertainty, as has been repeatedly illustrated in this and in the previous chapter. Since systematic errors cannot be inferred from the scatter of the data, they must be estimated from additional prior information. The only way to reduce systematic errors is by performing measurements with several nonequivalent methods, with common errors as diverse as possible.

2.5   Evaluation of Discrepant Data with Unrecognized Random Errors

Consider n measurements and/or computations of the same (unknown) quantity μ, which yielded data $x_1 \pm \sigma_1$, ..., $x_n \pm \sigma_n$; with *known standard deviations* σ_j, $j = 1, \ldots, n$. Previously, in Sections 2.2 and 2.3, the distance $|x_i - x_j|$ between any two measured values was considered to be smaller, or at least not much larger, than the sum of the corresponding uncertainties, $\sigma_i + \sigma_j$, and the data was said to be consistent or to agree "within error bars." However, *if the distances $|x_j - x_k|$ are larger than $(\sigma_j + \sigma_k)$, the data are considered to be inconsistent (discrepant).* The evaluation of such data is the subject of this section, which follows closely the recent work by Cacuci and Ionescu-Bujor [28].

As discussed in Chapter 1, inconsistencies can be caused by unrecognized or ill-corrected experimental effects, including background, dead time of the counting electronics, instrumental resolution, sample impurities, and calibration errors. Note that the probability that two equally precise measurements yield a separation greater than $\sigma_i + \sigma_j = 2\sigma$ is very small, namely $erfc\,(1) \simeq 0.157$ for Gaussian sampling distributions with standard deviation σ. Thus, although there is a nonzero probability that legitimately discrepant data do occur, it is much more likely *that apparently discrepant experiments actually indicate the presence of unrecognized errors ε_j*. Consequently, such apparent inconsistencies indicate that the total errors must actually be of the form $(x_j - \mu + \varepsilon_j)$ rather than just of the form $(x_j - \mu)$.

With the known data $\mathbf{x} \equiv (x_1, \ldots, x_n)$ and $\boldsymbol{\sigma} \equiv (\sigma_1, \ldots, \sigma_n)$, and both μ and the vector of unrecognized errors $\boldsymbol{\varepsilon} \equiv (\varepsilon_1, \ldots, \varepsilon_n)$ unknown, the starting point for determining the posterior distribution for μ and $\boldsymbol{\varepsilon}$ is

$$p\,(\mu, \boldsymbol{\varepsilon} | \boldsymbol{\sigma}, \mathbf{x}, \boldsymbol{\tau})\, d\mu\, d\,(\boldsymbol{\varepsilon}) \quad \propto \quad p\,(\mathbf{x} | \mu, \boldsymbol{\varepsilon}, \boldsymbol{\sigma}) \times p\,(\mu)\, d\mu \times p\,(\boldsymbol{\varepsilon})\, d\,(\boldsymbol{\varepsilon})\,, \qquad (2.132)$$

where $p\,(\mathbf{x} | \mu, \boldsymbol{\varepsilon}, \boldsymbol{\sigma})$ denotes the likelihood probability distribution of errors $(x_j - \mu)$ given the quantities $\boldsymbol{\sigma}$ and $\boldsymbol{\varepsilon}$, while $p\,(\mu)\, d\mu$ and $p\,(\boldsymbol{\varepsilon})\, d\,(\boldsymbol{\varepsilon})$ denote the prior probabilities reflecting what is already known about the respective quantities. Since only the data $x_1 \pm \sigma_1$, ..., $x_n \pm \sigma_n$ are known, the maximum entropy principle for constructing the respective prior distributions would yield independent Gaussians having μ as the common location parameter. In turn, the appropriate uninformative prior for μ would be $p(\mu) \, d\mu \propto d\mu$.

When the unrecognized errors ε_j are uncorrelated (as would be the case when knowing one of them would not yield information about the other unknowns ones) their joint prior distribution would be the product of the individual distribution of each of the unrecognized error, i.e., $p(\varepsilon) \, d(\varepsilon) = \prod_j p(\varepsilon_j) \, d\varepsilon_j$. Furthermore, in the absence of any information regarding the way in which the data was measured, positive and negative errors would be equally probable. Hence, the probability distribution for the unrecognized errors, ε_i, of the i-th experiment would be symmetric about zero, and the same distribution would apply to all measurements. This consideration implies that the first moments would vanish, i.e., $\langle \varepsilon_j \rangle = 0$. Recall, though, that the *actual value of the variance*, $\text{var}(\varepsilon_j) = \langle \varepsilon_j^2 \rangle$, *is unknown*, since otherwise the unrecognized errors would no longer be "unrecognized." Perhaps the simplest nontrivial way to model the unknown true variance is to consider it to be of the form

$$\text{var}(\varepsilon_j) = \langle \varepsilon_j^2 \rangle = \tau_i^2/c, \tag{2.133}$$

where τ_i^2 denotes an *estimate of the actual variance* of ε_i (estimated, for example, from the accuracy of the techniques employed), and c is *an adjustable common scale parameter with* prior $p(c) \, dc$. Values $\tau_i = 0$ would be assigned to experiments that are free of unrecognized errors. Based solely on the information $\langle \varepsilon_j \rangle = 0$ and $\langle \varepsilon_j^2 \rangle = \tau_i^2/c$, with τ_i^2/c considered to be known, application of the maximum entropy principle yields the following Gaussian form for the prior distributions for each ε_i:

$$p(\varepsilon_i | \tau_i) \, d\varepsilon_i = \frac{\exp\left(-\frac{1}{2} \frac{c\varepsilon_i^2}{\tau_i^2}\right)}{\sqrt{2\pi\tau_i^2/c}} d\varepsilon_i , \quad -\infty < \varepsilon_i < \infty. \tag{2.134}$$

On the other hand, the probability to measure the value x_i, given the true value μ, the unrecognized error ε_i, and the recognized uncertainty σ_i, is given by

$$
p\left(x_i|\mu, \sigma_i, \varepsilon_i\right) dx_i = \frac{\exp\left[-\frac{1}{2}\left(\frac{x_i-\mu-\varepsilon_i}{\sigma_i}\right)^2\right]}{\sqrt{2\pi\sigma_i^2}} dx_i, \quad -\infty < x_i < \infty. \quad (2.135)
$$

Aggregating all of the above considerations leads to the following joint posterior for μ and ε, given $\boldsymbol{\sigma} \triangleq (\sigma_1, \ldots, \sigma_n)$, $\mathbf{x} \triangleq (x_1, \ldots, x_n)$, and $\boldsymbol{\tau} \triangleq (\tau_1, \ldots, \tau_n)$

$$
p\left(\mu, \boldsymbol{\varepsilon}|\boldsymbol{\sigma}, \mathbf{x}, \boldsymbol{\tau}\right) d\mu\, d\left(\boldsymbol{\varepsilon}\right) \propto
$$

$$
d\mu \int_0^\infty dc\, p\left(c\right) \prod_{j=1}^n d\varepsilon_j \frac{\exp\left[-\frac{(x_j-\mu-\varepsilon_j)^2}{2\sigma_j^2} - \frac{c\varepsilon_j^2}{2\tau_j^2}\right]}{\sqrt{2\pi\sigma_j^2}\sqrt{2\pi\tau_j^2/c}}.
$$

$$(2.136)$$

The marginal posterior distribution for the errors ε is obtained by marginalizing (integrating over) μ and c in Eq. (2.136). On the other hand, the marginal posterior distribution for the location parameter μ is obtained by marginalizing each ε_i and c in Eq. (2.136). Both of these marginal distributions will be examined, in turn.

Consider first the marginal posterior distribution for the location parameter μ. The integration over each ε_i in Eq. (2.136) can be performed by using the formula given in Eq. (2.103), which is reproduced below for convenience:

$$
\int_{-\infty}^\infty x^m \exp\left(-px^2 + 2qx\right) dx = \frac{1}{2^{m-1}p} \sqrt{\frac{\pi}{p}} \frac{d^{m-1}}{dq^{m-1}} \left[q \exp\left(\frac{q^2}{p}\right)\right]
$$

$$
= m! \sqrt{\frac{\pi}{p}} \left(\frac{q}{p}\right)^m e^{\frac{q^2}{p}} \sum_{k=0}^{E\left(\frac{m}{2}\right)} \frac{1}{(n-2k)!\,(k)!} \left(\frac{p}{4q^2}\right)^k; \quad p > 0.
$$

$$(2.137)$$

Expanding the exponent in Eq. (2.136) and using Eq. (2.137) leads to the result

$$p(\mu|\boldsymbol{\sigma},\mathbf{x},\boldsymbol{\tau})\,d\mu \; \propto$$

$$d\mu \int_0^\infty dc\, p(c) \prod_{j=1}^n \left(\sigma_j^2 + \tau_j^2/c\right)^{-1/2} \exp\left[-\frac{1}{2}\frac{(\mu - x_j)^2}{\sigma_j^2 + \tau_j^2/c}\right].$$

$$(2.138)$$

2.5.1 Using Jeffreys' Prior for the Scale Factor c

Since c is a scale factor, the (Jeffreys' prior) least informative prior (Jeffreys' prior) $p(c)\,dc \propto dc/c$ would be appropriate to use in Eq. (2.138). It is apparent that the integration over c cannot be carried out exactly in terms of elementary functions. To gain insight into the nature of the distribution $p(\mu|\boldsymbol{\sigma},\mathbf{x},\boldsymbol{\tau})\,d\mu$, it is useful to consider the two limiting cases, namely the case when the recognized uncertainties are unimportant, i.e., $\sigma_j = 0$, and the opposite limiting case, when all errors are known (i.e., no unrecognized errors, so that $\tau_j = 0$).

When the *recognized uncertainties are unimportant*, it is permissible to set $\sigma_j = 0$ in Eq. (2.138), which consequently describes a situation in which all unrecognized errors are known only up to the scale factor c. Setting $\sigma_j = 0$ in Eq. (2.138) and integrating the resulting expression over c using Jeffreys' prior dc/c yields a marginal Student's t-distribution of the form

$$p(\mu|\boldsymbol{\sigma}\to 0,\mathbf{x},\boldsymbol{\tau})\,d\mu \; \propto \; d\mu \int_0^\infty dc\,\left(c^{\frac{n}{2}-1}\right)\exp\left[-\frac{c}{2}\sum_{j=1}^n \frac{(\mu - x_j)^2}{\tau_j^2}\right]$$

$$\propto \; d\mu \left[\sum_{j=1}^n \frac{(\mu - x_j)^2}{\tau_j^2}\right]^{-\frac{n}{2}}. \qquad (2.139)$$

The last expression on the right side of Eq. (2.139) can be rearranged by expanding the sum, to obtain

$$\left[\sum_{j=1}^{n}\frac{(\mu-x_j)^2}{\tau_j^2}\right]^{-\frac{n}{2}} \propto \left[1+\frac{(\mu-\langle x\rangle)^2}{\langle x^2\rangle-\langle x\rangle^2}\right]^{-\frac{n}{2}}$$

$$\propto St\left(\mu\Big|\langle x\rangle,(n-1)\left[\langle x^2\rangle-\langle x\rangle^2\right]^{-1},n-1\right),$$

$$(2.140)$$

where

$$\langle x\rangle\triangleq\left(\sum_{i=1}^{n}x_i\tau_i^{-2}\right)\left(\sum_{i=1}^{n}\tau_i^{-2}\right)^{-1}; \quad \langle x^2\rangle\triangleq\left(\sum_{i=1}^{n}x_i^2\tau_i^{-2}\right)\left(\sum_{i=1}^{n}\tau_i^{-2}\right)^{-1},$$

$$(2.141)$$

and where

$$St\left(\mu\mid y,\lambda,\alpha\right)\triangleq C\left[1+\frac{\lambda}{\alpha}(\mu-y)^2\right]; \quad C\triangleq\frac{\Gamma\left(\alpha+1/2\right)}{\Gamma\left(1/2\right)\Gamma\left(\alpha/2\right)}\left(\frac{\lambda}{\alpha}\right)^{1/2} \quad (2.142)$$

denotes the Student's t-distribution with the mean and variance given by

$$\langle\mu\rangle=y,\text{ for }\alpha>1;\quad var\left(\mu\right)=\frac{\alpha}{\lambda}\frac{1}{\alpha-2},\text{ for }\alpha>2. \quad (2.143)$$

In view of Eqs. (2.138) through (2.143), it follows that the marginal distribution $p\left(\mu|\boldsymbol{\sigma},\mathbf{x},\boldsymbol{\tau}\right)d\mu$ for the location parameter μ has the mean and variance given by

$$\langle\mu\rangle=\langle x\rangle,\ n>2;\quad var\left(\mu\right)=\frac{\mu_2^S}{n-3},\ n>3;\quad \mu_2^S\triangleq\langle x^2\rangle-\langle x\rangle^2, \quad (2.144)$$

with weights for the sample averages which are proportional to τ_j^{-2}. For $n=2$, Eq. (2.139) is a Cauchy distribution centred at $\langle x\rangle$, with infinite standard deviation but finite half–width at half–maximum $\sqrt{\mu_2^S}$. For $n\geq 4$, Eq. (2.144) indicates that the mean $\langle\mu\rangle$ of the location parameter μ is given by the sample average, regardless of the nature (i.e., recognized or unrecognized) errors. Furthermore, the variance of the location parameter μ is determined by the scatter in the data x_k, as reflected by the sample variance μ_2^S. This result is the same as was obtained in Eq. (2.59), *when the location parameter μ was estimated from a sample drawn from a Gaussian distribution with unknown* *variance.* This case (i.e., when the recognized uncertainties are unimportant) is thus the same as the case considered in Section 2.3.2. Recall that Student's t-distribution approaches a Gaussian as $n \to \infty$, but is always broader than a Gaussian for finite n.

The opposite limiting case is when all errors are known, i.e., the case of zero *unrecognized errors*, $\tau_j = 0$. In this case, the distribution in Eq. (2.138) reduces to the Gaussian already encountered in Section 2.3.1, cf. Eq. (2.52), *when the location parameter μ was estimated from a sample drawn from a Gaussian distribution with known variance, namely*

$$p\left(\mu|\boldsymbol{\sigma}, \mathbf{x}, \boldsymbol{\tau} \to \mathbf{0}\right) d\mu = \frac{\exp\left[-\frac{(\mu-\bar{x})^2}{2\sigma^2/n}\right]}{\left(2\pi\sigma^2/n\right)^{1/2}} d\mu, \quad -\infty < \mu < \infty, \qquad (2.145)$$

with

$$\bar{x} \triangleq \frac{\sum\limits_k \sigma_k^{-2} x_k}{\sum\limits_k \sigma_k^{-2}}, \quad \overline{\sigma^2} \triangleq \frac{\sum\limits_k \sigma_k^{-2} \sigma_k^2}{\sum\limits_k \sigma_k^{-2}} = \frac{n}{\sum\limits_k \sigma_k^{-2}}. \qquad (2.146)$$

In the general case, $\sigma_i > 0$, $\tau_i > 0$, the main contribution to the distribution $p\left(\mu|\boldsymbol{\sigma}, \mathbf{x}, \boldsymbol{\tau}\right) d\mu$ can be obtained by using the saddle-point approximation to determine the leading term of the integral over c in Eq. (2.138). For an integral of this form, the leading term in the saddle-point (Laplace's) approximation is given by

$$\int_0^\infty \exp\left[-nh\left(c\right)\right] dc \sim \sqrt{2\pi}\, n^{-1/2} \left|h''\left(c^*\right)\right|^{-1/2} \exp\left[-nh\left(c^*\right)\right], \quad n \to \infty, \qquad (2.147)$$

with c^* being defined as the point where $[-h\left(c\right)]$ attains its supremum over c, i.e., $-h\left(c_*\right) = \sup\,[-h\left(c\right)]$, $0 < c < \infty$. In this case, c^* is the point where the first derivative of $[-h\left(c\right)]$ with respect to c vanishes. The distribution $p\left(\mu|\boldsymbol{\sigma}, \mathbf{x}, \boldsymbol{\tau}\right) d\mu$ in Eq. (2.138) can be cast in the form of Eq. (2.147), i.e., $p\left(\mu|\boldsymbol{\sigma}, \mathbf{x}, \boldsymbol{\tau}\right) d\mu \sim d\mu \int_0^\infty \exp\left[-nh\left(c\right)\right] dc$ by defining

$$h\left(c\right) \triangleq -\frac{1}{2}\log c - \frac{1}{n}\log p\left(c\right) + \frac{1}{2n}\sum_{j=1}^n \log\left(c\sigma_j^2 + \tau_j^2\right) + \frac{c}{2n}\sum_{j=1}^n \frac{(\mu - x_j)^2}{c\sigma_j^2 + \tau_j^2}. \qquad (2.148)$$

Setting $p(c) = 1/c$ in the right-side of Eq. (2.148) and setting the first derivative (with respect to c) of the resulting expression to zero at c^* yields

$$h'(c^*) = 0 = -\frac{1}{2c^*} + \frac{1}{nc^*} + \frac{1}{2n} \sum_{j=1}^{n} \frac{\sigma_j^2}{c^* \sigma_j^2 + \tau_j^2} + \frac{1}{2n} \sum_{j=1}^{n} \frac{\tau_j^2 (\mu - x_j)^2}{(c^* \sigma_j^2 + \tau_j^2)^2},$$

and, consequently,

$$c^* = (n-2) \left[\sum_{j=1}^{n} \frac{\sigma_j^2}{c^* \sigma_j^2 + \tau_j^2} + \sum_{j=1}^{n} \frac{\tau_j^2 (\mu - x_j)^2}{(c^* \sigma_j^2 + \tau_j^2)^2} \right]^{-1}. \tag{2.149}$$

Furthermore, the expression of the second derivative of $h(c)$ is

$$h''(c) = \frac{n-2}{2nc^2} - \frac{1}{2n} \sum_{j=1}^{n} \frac{\sigma_j^4}{(c\sigma_j^2 + \tau_j^2)^2} - \frac{1}{n} \sum_{j=1}^{n} \frac{\tau_j^2 \sigma_j^2 (\mu - x_j)^2}{(c\sigma_j^2 + \tau_j^2)^3} \tag{2.150}$$

Thus, Eqs. (2.148) through (2.150) provide all the information needed to compute the quantity on the right-side of Eq. (2.147), which in turn yields

$$p(\mu|\boldsymbol{\sigma}, \mathbf{x}, \boldsymbol{\tau}) \, d\mu \propto d\mu \frac{(c^*)^{\frac{n}{2}-1}}{\prod\limits_{j=1}^{n} (c^* \sigma_j^2 + \tau_j^2)^{1/2}} \exp\left[-\frac{c^*}{2} \sum_{j=1}^{n} \frac{(\mu - x_j)^2}{c^* \sigma_j^2 + \tau_j^2} \right]. \tag{2.151}$$

When $\tau_j = 0$, Eq. (2.151) reduces to the Gaussian given by Eq. (2.145) and Eq. (2.52). In the opposite limit, as $\sigma_j \to 0$, the expression in Eq. (2.151) reduces uniformly to Eq. (2.139), since

$$c^* \xrightarrow{\sigma_j \to 0} (n-2) \left[\sum_{j=1}^{n} \tau_j^{-2} (\mu - x_j)^2 \right]^{-1}, \tag{2.152}$$

$$h''(c^*) \xrightarrow{\sigma_j \to 0} \frac{1}{2n(n-2)} \left[\sum_{j=1}^{n} \tau_j^{-2} (\mu - x_j)^2 \right]^2, \tag{2.153}$$

$$p(\mu|\boldsymbol{\sigma} \to 0, \mathbf{x}, \boldsymbol{\tau}) \, d\mu \to d\mu \left[\sum_{j=1}^{n} \frac{(\mu - x_j)^2}{\tau_j^2} \right]^{-\frac{n}{2}}. \tag{2.154}$$

In the general case, when $\sigma_i > 0$, $\tau_i > 0$, the mean $\langle \mu \rangle$ and variance var (μ) of the location parameter μ can be obtained by using once again the saddle-point integration technique to determine the respective leading terms of the corresponding integrals over $d\mu$, of the first and second moments of the distribution $p(\mu|\sigma, \mathbf{x}, \tau) \, d\mu$ in Eq. (2.151). Prior to using the saddle-point technique, though, the value of c^* would need to be determined by solving Eq. (2.149) iteratively, using Picard iterations or some variant of Newton's method. For example, the Picard iterations would be

$$c_{k+1} = (n-2) \left[\sum_{j=1}^{n} \frac{\sigma_j^2}{c_k \sigma_j^2 + \tau_j^2} + \sum_{j=1}^{n} \frac{\tau_j^2(\mu - x_j)^2}{\left(c_k \sigma_j^2 + \tau_j^2\right)^2} \right]^{-1} ; \quad \lim_{k \to \infty} c_k \to c^*.$$

(2.155)

Alternatively, Newton's (or a quasi-Newton) method could be applied directly to Eq. (2.148), to compute the supremum of $[-h(c)]$, which would involve the iterative process

$$c_{k+1} = c_k - \frac{h'(c_k)}{h''(c_k)}; \quad \lim_{k \to \infty} c_k \to c^*. \tag{2.156}$$

Any iterative procedure would need a starting value, c_0, on the respective right sides of Eqs. (2.155) or (2.156). Since the values τ_i are the best available estimates of the uncertainties caused by unrecognized errors, the scale factor c would be expected to be close to unity, so that a reasonable starting value would be $c_0 = 1$. The corresponding starting form, $p(\mu|\sigma, \mathbf{x}, \tau, c_0 = 1) \, d\mu$, of the marginal distribution for μ would be a Gaussian of the same form as in the case when all errors are known, cf. Eq. (2.52), except that the weightings would be $(\sigma_k^2 + \tau_k^2)^{-2}$, as can be seen by setting $c^* = 1$ in Eq. (2.151) to obtain

$$p(\mu|\sigma, \mathbf{x}, \tau, c_0 = 1) \, d\mu = \frac{\exp\left[-\frac{(\mu - \bar{x})^2}{2(\sigma^2 + \tau^2)/n}\right]}{\left[2\pi(\sigma^2 + \tau^2)/n\right]^{1/2}} d\mu, \quad -\infty < \mu < \infty, \quad (2.157)$$

with

$$\bar{x} \triangleq \frac{\sum\limits_{k} \left(\sigma_k^2 + \tau_k^2\right)^{-2} x_k}{\sum\limits_{k} \left(\sigma_k^2 + \tau_k^2\right)^{-2}}; \quad \overline{\left(\sigma^2 + \tau^2\right)} \triangleq \frac{n}{\sum\limits_{k} \left(\sigma_k^2 + \tau_k^2\right)^{-2}}. \tag{2.158}$$

2.5.2 Using an Exponential Prior for the Scale Factor c

If the scale factor c is expected to be close to unity, then its mean value would also be expected to be close to unity, i.e., $\langle c \rangle = 1$. In such a case, the maximum-entropy construction presented in Section 2.2.2 could be used with the constraint $\langle c \rangle = 1$ to obtain the following prior for the scale factor c:

$$p(c)\, dc = e^{-c} dc, \quad 0 < c < \infty. \tag{2.159}$$

The above prior is almost as noncommittal as Jeffreys' prior dc/c, decreasing also monotonically as c increases, but gives less weight to extreme values approaching the two ends of the positive real axis, leading thus to finite integrals over c. Using the above exponential prior in Eq. (2.139) gives

$$p(\mu|\boldsymbol{\sigma}, \mathbf{x}, \boldsymbol{\tau})\, d\mu \; \propto$$

$$d\mu \int_0^\infty dc \, \frac{(c)^{\frac{n}{2}-1}}{\prod\limits_{j=1}^{n} \left(c\sigma_j^2 + \tau_j^2\right)^{1/2}} \exp\left\{(-c)\left[1 + \frac{1}{2}\sum_{j=1}^{n} \frac{(\mu - x_j)^2}{c\sigma_j^2 + \tau_j^2}\right]\right\}. \tag{2.160}$$

The form of the above distribution for the limiting case of negligible known errors can be readily derived by setting $\sigma_j \to 0$ in Eq. (2.160) and performing the integration over c, to obtain:

$$p\left(\mu|\boldsymbol{\sigma}\rightarrow\mathbf{0},\mathbf{x},\boldsymbol{\tau}\right)d\mu \;\rightarrow\; d\mu\int_{0}^{\infty}dc\,(c)^{\frac{n}{2}}\exp\left\{(-c)\left[1+\frac{1}{2}\sum_{j=1}^{n}\tau_{j}^{-2}(\mu-x_{j})^{2}\right]\right\}$$

$$\propto\,d\mu\left[1+\frac{1}{2}\sum_{j=1}^{n}\tau_{j}^{-2}(\mu-x_{j})^{2}\right]^{-\left(\frac{n}{2}+1\right)}\qquad\propto\,d\mu\left[1+A(\mu-\langle\mu\rangle)^{2}\right]^{-\left(\frac{n}{2}+1\right)}$$

$$\propto\,St\left(\mu\,|\langle\mu\rangle,(n+1)\,A,n+1\right)d\mu.$$

$$(2.161)$$

Hence, the mean $\langle\mu\rangle$ and variance $\mathrm{var}\,(\mu)$ are given in this case by

$$\langle\mu\rangle=\langle x\rangle,\;\; n>0;\quad \mathrm{var}\,(\mu)=\frac{n+1}{\lambda\,(n-1)}=\frac{A}{n-1},\;\; n>1, \qquad (2.162)$$

where

$$\langle x\rangle\triangleq\left(\sum_{j=1}^{n}x_{j}\tau_{j}^{-2}\right)\left(\sum_{j=1}^{n}\tau_{j}^{-2}\right)^{-1};\quad \langle x^{2}\rangle\triangleq\left(\sum_{j=1}^{n}x_{j}^{2}\tau_{j}^{-2}\right)\left(\sum_{j=1}^{n}\tau_{j}^{-2}\right)^{-1},$$

$$A\triangleq\left[2\left(\sum_{j=1}^{n}\tau_{j}^{-2}\right)^{-1}+\langle x^{2}\rangle-\langle x\rangle^{2}\right]^{-1}. \qquad (2.163)$$

Comparing the mean obtained in Eq. (2.162) for μ to the corresponding expression obtained for Jeffreys' prior in Eq. (2.143) indicates that the mean $\langle\mu\rangle$ is still given by the sample average with weights proportional to τ_{j}^{-2}, but the mean is valid already for two (as opposed to three) experiments. Comparing now the corresponding expressions for $\mathrm{var}\,(\mu)$ in Eq. (2.162) and Eq. (2.144), respectively, indicates that the use of the exponential distribution for c brings in the additional term $\left(\sum_{j=1}^{n}\tau_{j}^{-2}\right)^{-1}$ in Eq. (2.162), extending the validity of the latter to already two experiments (as opposed to four, as required when Jeffreys' prior is used).

The other limiting case, that of no unrecognized errors, is obtained by setting $\tau_{j}=0$ in Eq. (2.160) and performing the integration over c. This procedure leads to the same Gaussian already encountered in Section 2.3.1, cf. Eq. (2.52), when the location parameter μ was estimated from a sample drawn from a Gaussian distribution with known variance.

In the general case, when $\sigma_i > 0$, $\tau_i > 0$, the mean $\langle \mu \rangle$ and variance var (μ) of the location parameter μ can be obtained by using once again the saddle-point integration technique to determine the respective leading terms of the corresponding integrals over $d\mu$, of the first and second moments of the distribution $p(\mu | \sigma, \mathbf{x}, \tau) \, d\mu$ in Eq. (2.160). In this case, the distribution $p(\mu | \sigma, \mathbf{x}, \tau) \, d\mu$ in Eq. (2.160) can be cast in the form $p(\mu | \sigma, \mathbf{x}, \tau) \, d\mu \sim d\mu \int_0^\infty \exp[-nh(c)] \, dc$ by defining

$$-nh(c) = (-n) \left\{ -\frac{1}{2} \log c + \frac{1}{2n} \sum_{j=1}^n \log \left(c\sigma_j^2 + \tau_j^2 \right) + \frac{c}{n} \left[1 + \frac{1}{2} \sum_{j=1}^n \frac{(\mu - x_j)^2}{c\sigma_j^2 + \tau_j^2} \right] \right\}. \tag{2.164}$$

The point c^*, where $[-h(c)]$ defined by Eq. (2.164) attains its supremum over c, is obtained by solving the equation

$$h'(c^*) = 0 = -\frac{1}{2c^*} + \frac{1}{n} + \frac{1}{2n} \sum_{j=1}^n \frac{\sigma_j^2}{c^*\sigma_j^2 + \tau_j^2} + \frac{1}{2n} \sum_{j=1}^n \frac{\tau_j^2 (\mu - x_j)^2}{\left(c^*\sigma_j^2 + \tau_j^2 \right)^2}, \tag{2.165}$$

which yields the root

$$c^* = \frac{n}{2} \left[1 + \frac{1}{2} \sum_{j=1}^n \frac{\sigma_j^2}{c^*\sigma_j^2 + \tau_j^2} + \frac{1}{2} \sum_{j=1}^n \frac{\tau_j^2 (\mu - x_j)^2}{\left(c^*\sigma_j^2 + \tau_j^2 \right)^2} \right]^{-1}. \tag{2.166}$$

The expression of the second derivative of $h(c)$ is

$$h''(c) = \frac{1}{2c^2} - \frac{1}{2n} \sum_{j=1}^n \frac{\sigma_j^4}{\left(c\sigma_j^2 + \tau_j^2 \right)^2} - \frac{1}{n} \sum_{j=1}^n \frac{\tau_j^2 \sigma_j^2 (\mu - x_j)^2}{\left(c\sigma_j^2 + \tau_j^2 \right)^3}. \tag{2.167}$$

The expressions in Eqs. (2.164) through (2.167) provide all the information needed to compute the quantity on the right side of Eq. (2.147), which in turn yields

$$p\left(\mu|\boldsymbol{\sigma},\mathbf{x},\boldsymbol{\tau}\right)d\mu \sim d\mu\sqrt{2\pi}\,n^{-1/2}\left|h''\left(c^*\right)\right|^{-1/2}\exp\left[-nh\left(c^*\right)\right]$$

$$\propto \frac{d\mu\left(c^*\right)^{\frac{n}{2}}}{\sqrt{2\pi n|h''(c^*)|}\prod_{j=1}^{n}\left(c^*\sigma_j^2+\tau_j^2\right)^{1/2}}\exp\left\{\left(-c^*\right)\left[1+\frac{1}{2}\sum_{j=1}^{n}\frac{(\mu-x_j)^2}{c^*\sigma_j^2+\tau_j^2}\right]\right\}.$$

$$(2.168)$$

When $\sigma_j \to 0$, the above expression reduces uniformly to Eq. (2.161), since

$$c^* \xrightarrow{\sigma_j\to 0} \frac{n}{2}\left[1+\frac{1}{2}\sum_{j=1}^{n}\tau_j^{-2}(\mu-x_j)^2\right]^{-1}, \qquad (2.169)$$

$$h''\left(c^*\right) \xrightarrow{\sigma_j\to 0} \frac{2}{n^2}\left[1+\frac{1}{2}\sum_{j=1}^{n}\tau_j^{-2}(\mu-x_j)^2\right]^2, \qquad (2.170)$$

$$p\left(\mu|\boldsymbol{\sigma}\to\mathbf{0},\mathbf{x},\boldsymbol{\tau}\right)d\mu \to d\mu\left[1+\frac{1}{2}\sum_{j=1}^{n}\tau_j^{-2}(\mu-x_j)^2\right]^{-\left(\frac{n}{2}+1\right)}. \qquad (2.171)$$

In the opposite limit, as $\tau_j = 0$, Eq. (2.168) reduces again to the Gaussian given by Eq. (2.52). In the general case, $\sigma_i > 0$, $\tau_i > 0$, the mean $\langle\mu\rangle$ and variance var (μ) of the location parameter μ can be obtained by using once again the saddle-point integration technique to determine the respective leading terms of the corresponding integrals over $d\mu$, of the first and second moments of the distribution $p\left(\mu|\boldsymbol{\sigma},\mathbf{x},\boldsymbol{\tau}\right)d\mu$ in Eq. (2.151). Prior to using the saddle-point technique, though, the value of c^* would need to be determined by solving Eq. (2.166) iteratively, using Picard iterations or some variant of Newton's method.

2.5.3 Marginal Posterior Distribution for the Unrecognized Errors

The marginal posterior distribution for the unrecognized errors $\boldsymbol{\varepsilon}$ is obtained by marginalizing the distribution $p\left(\mu,\boldsymbol{\varepsilon}|\boldsymbol{\sigma},\mathbf{x},\boldsymbol{\tau}\right)d\mu\,d\left(\boldsymbol{\varepsilon}\right)$ given in Eq. (2.136) over μ and c. Marginalizing this distribution over μ gives

$$p\left(\mu, \varepsilon \mid \boldsymbol{\sigma}, \mathbf{x}, \boldsymbol{\tau}\right) d\left(\varepsilon\right)$$

$$\propto d\left(\varepsilon\right) \int_{-\infty}^{\infty} dc\, p\left(c\right) \frac{\exp\left(-\frac{c\varepsilon_1^2}{2\tau_1^2}\right)}{\sqrt{2\pi\sigma_1^2}\sqrt{2\pi\tau_1^2/c}} \cdots \frac{\exp\left(-\frac{c\varepsilon_n^2}{2\tau_n^2}\right)}{\sqrt{2\pi\sigma_n^2}\sqrt{2\pi\tau_n^2/c}}$$

$$\times \int_{-\infty}^{\infty} d\mu \exp\left[-\frac{1}{2}\sum_{j=1}^{n}\frac{(x_j - \mu - \varepsilon_j)^2}{\sigma_j^2}\right]. \tag{2.172}$$

Performing the integration over μ in the last integral on the right side of Eq. (2.172) gives

$$\int_{-\infty}^{\infty} d\mu \exp\left[-\frac{1}{2}\sum_{j=1}^{n}\frac{(x_j - \mu - \varepsilon_j)^2}{\sigma_j^2}\right]$$

$$= \sqrt{\frac{2\pi}{\sum_{j=1}^{n}\sigma_j^{-2}}} \exp\left[\frac{1}{2}\left(\sum_{j=1}^{n}\frac{x_j - \varepsilon_j}{\sigma_j^2}\right)^2\left(\sum_{j=1}^{n}\sigma_j^{-2}\right)^{-1} - \frac{1}{2}\sum_{j=1}^{n}\frac{(x_j - \varepsilon_j)^2}{\sigma_j^2}\right]$$

$$\propto \exp\left[-\frac{1}{2}(\boldsymbol{\varepsilon} - \mathbf{x})^T \mathbf{C}_\sigma^{-1}\left(\mathbf{I} - \mathbf{W}_\sigma\right)(\boldsymbol{\varepsilon} - \mathbf{x})\right], \tag{2.173}$$

where $\boldsymbol{\varepsilon} \triangleq (\varepsilon_1, \ldots, \varepsilon_n)$, $\mathbf{x} \triangleq (x_1, \ldots, x_n)$, $\mathbf{I} \triangleq (\delta_{ij})_{n \times n}$, and

$$\mathbf{W}_\sigma \triangleq \begin{pmatrix} \sigma_1^{-2} & \sigma_2^{-2} & \sigma_3^{-2} & \cdots & \sigma_n^{-2} \\ \sigma_1^{-2} & \sigma_2^{-2} & \sigma_3^{-2} & \cdots & \sigma_n^{-2} \\ \sigma_1^{-2} & \sigma_2^{-2} & \sigma_3^{-2} & \cdots & \vdots \\ \vdots & \vdots & \vdots & \ddots & \vdots \\ \sigma_1^{-2} & \sigma_2^{-2} & \sigma_3^{-2} & \cdots & \sigma_n^{-2} \end{pmatrix} \left(\sum_{i=0}^{n}\sigma_i^{-2}\right)^{-1}; \quad \mathbf{C}_\sigma^{-1} \triangleq \left(\sigma_j^{-2}\delta_{ij}\right)_{n \times n}.$$

$$\tag{2.174}$$

using *Jeffreys' prior*, $p(c) dc \propto dc/c$, the integration over c in Eq. (2.172) yields

$$\int_{-\infty}^{\infty} dc \, c^{\frac{n}{2}-1} \exp\left(-\frac{c}{2}\sum_{j=1}^{n}\frac{\varepsilon_j^2}{\tau_j^2}\right) \propto \left[\frac{1}{2}\sum_{j=1}^{n}\frac{\varepsilon_j^2}{\tau_j^2}\right]^{-\frac{n}{2}}$$

$$\propto \left[\varepsilon^T \mathbf{C}_\tau^{-1}\varepsilon\right]^{-\frac{n}{2}}; \quad \mathbf{C}_\tau^{-1} \triangleq \left(\tau_j^{-2}\delta_{ij}\right)_{n \times n}. \tag{2.175}$$

Using now Eqs. (2.173) and (2.175) in Eq. (2.172) yields

$$p(\mu, \varepsilon | \boldsymbol{\sigma}, \mathbf{x}, \boldsymbol{\tau}) \, d\varepsilon$$

$$= \frac{d\varepsilon \left[\varepsilon^T \mathbf{C}_\tau^{-1}\varepsilon\right]^{-\frac{n}{2}} \exp\left[-\frac{1}{2}(\varepsilon - \mathbf{x})^T \mathbf{C}_\sigma^{-1}(\mathbf{I} - \mathbf{W}_\sigma)(\varepsilon - \mathbf{x})\right]}{\int_{-\infty}^{\infty} d\varepsilon \left[\varepsilon^T \mathbf{C}_\tau^{-1}\varepsilon\right]^{-\frac{n}{2}} \exp\left[-\frac{1}{2}(\varepsilon - \mathbf{x})^T \mathbf{C}_\sigma^{-1}(\mathbf{I} - \mathbf{W}_\sigma)(\varepsilon - \mathbf{x})\right]}. \tag{2.176}$$

The *maximum likelihood estimator (MLE)*, cf. Eq. (2.12), can be used on the numerator of Eq. (2.176) to obtain a first approximation of the vector of mean values, $\langle \varepsilon \rangle \triangleq (\langle \varepsilon_1 \rangle, \ldots, \langle \varepsilon_n \rangle)$ and covariance matrix, $\mathbf{C}_\varepsilon \triangleq [\text{cov}(\varepsilon_i \varepsilon_j)]_{n \times n}$. In this approximation, Eq. (2.176) takes the form of the following Gaussian

$$p(\mu, \varepsilon | \boldsymbol{\sigma}, \mathbf{x}, \boldsymbol{\tau}) \, d\varepsilon \propto \exp\left\{-\frac{1}{2}(\hat{\varepsilon} - \mathbf{x})^T \left[\nabla^2 h(\hat{\varepsilon})\right](\hat{\varepsilon} - \mathbf{x})\right\}, \tag{2.177}$$

where the scalar-valued function $h(\varepsilon)$ is defined as

$$h(\varepsilon) \triangleq \frac{1}{2}\log\left(\varepsilon^T \mathbf{C}_\tau^{-1}\varepsilon\right) + \frac{1}{2n}(\varepsilon - \mathbf{x})^T \mathbf{C}_\sigma^{-1}(\mathbf{I} - \mathbf{W}_\sigma)(\varepsilon - \mathbf{x}), \tag{2.178}$$

and where $\hat{\varepsilon}$ denotes the vector-valued point where the gradient $\nabla h(\hat{\varepsilon})$ of $h(\varepsilon)$ vanishes, and $\left[\nabla^2 h(\hat{\varepsilon})\right]$ denotes the Hessian matrix of $h(\varepsilon)$ evaluated at $\hat{\varepsilon}$. The gradient $\nabla h(\hat{\varepsilon})$ and, respectively, Hessian of $h(\varepsilon)$ can be readily computed from Eq. (2.178), to obtain

$$\nabla h(\hat{\varepsilon}) = \left\{\frac{\mathbf{C}_\tau^{-1}\varepsilon}{\varepsilon^T \mathbf{C}_\tau^{-1}\varepsilon} + \frac{1}{n}\mathbf{C}_\sigma^{-1}(\mathbf{I} - \mathbf{W}_\sigma)(\varepsilon - \mathbf{x})\right\}_{\varepsilon = \hat{\varepsilon}} = 0, \tag{2.179}$$

$$\nabla^2 h\left(\varepsilon\right) = \frac{\mathbf{C}_\tau^{-1}}{\varepsilon^T \mathbf{C}_\tau^{-1} \varepsilon} - \frac{2\mathbf{C}_\tau^{-1}\varepsilon\varepsilon^T \mathbf{C}_\tau^{-1}}{\left(\varepsilon^T \mathbf{C}_\tau^{-1}\varepsilon\right)^2} + \frac{1}{n}\mathbf{C}_\sigma^{-1}\left(\mathbf{I} - \mathbf{W}_\sigma\right). \tag{2.180}$$

Solving Eq. (2.179) yields the following fixed-point equation for $\hat{\varepsilon}$:

$$\hat{\varepsilon} = \mathbf{x} + \mathbf{W}_\sigma\left(\hat{\varepsilon} - \mathbf{x}\right) - \frac{n\mathbf{C}_\sigma\mathbf{C}_\tau^{-1}\hat{\varepsilon}}{\hat{\varepsilon}^T \mathbf{C}_\tau^{-1}\hat{\varepsilon}} = \mathbf{x} - \langle \mathbf{x} \rangle + \langle \hat{\varepsilon} \rangle - \frac{n\mathbf{C}_\sigma\mathbf{C}_\tau^{-1}\hat{\varepsilon}}{\hat{\varepsilon}^T \mathbf{C}_\tau^{-1}\hat{\varepsilon}}, \tag{2.181}$$

where

$$\langle x \rangle \triangleq \left(\sum_{j=1}^n \frac{x_j}{\sigma_j^2}\right)\left(\sum_{j=1}^n \sigma_j^{-2}\right)^{-1} ; \quad \mathbf{W}_\sigma\mathbf{x} = \langle x \rangle \begin{pmatrix} 1 \\ \vdots \\ 1 \end{pmatrix} \triangleq \langle \mathbf{x} \rangle ;$$

$$\langle \hat{\varepsilon} \rangle \triangleq \left(\sum_{j=1}^n \frac{\hat{\varepsilon}_j}{\sigma_j^2}\right)\left(\sum_{j=1}^n \sigma_j^{-2}\right)^{-1} ; \quad \mathbf{W}_\sigma\hat{\varepsilon} = \langle \hat{\varepsilon} \rangle \begin{pmatrix} 1 \\ \vdots \\ 1 \end{pmatrix} \triangleq \langle \hat{\varepsilon} \rangle .$$

$$\tag{2.182}$$

The above fixed-point equation can be solved for $\hat{\varepsilon}$ using some variant of Newton's method or, more simply, by Picard iterations of the form

$$\hat{\varepsilon}^{(k+1)} = \mathbf{x} - \langle \mathbf{x} \rangle + \left\langle \hat{\varepsilon}^{(k)} \right\rangle - \frac{n\mathbf{C}_\sigma\mathbf{C}_\tau^{-1}\hat{\varepsilon}^{(k)}}{\left(\hat{\varepsilon}^{(k)}\right)^T \mathbf{C}_\tau^{-1}\hat{\varepsilon}^{(k)}}, \tag{2.183}$$

$$\hat{\varepsilon}^{(0)} = \mathbf{x} - \langle \mathbf{x} \rangle , \quad \hat{\varepsilon}^{(1)} = \left[1 - \frac{n\mathbf{C}_\sigma\mathbf{C}_\tau^{-1}}{(\mathbf{x} - \langle \mathbf{x} \rangle)^T \mathbf{C}_\tau^{-1}(\mathbf{x} - \langle \mathbf{x} \rangle)}\right](\mathbf{x} - \langle \mathbf{x} \rangle), \tag{2.184}$$

or, in component form,

$$\hat{\varepsilon}_i^{(k+1)} = x_i - \langle x \rangle + \left\langle \hat{\varepsilon}_i^{(k)} \right\rangle - n\sigma_i^2\tau_i^{-2}\hat{\varepsilon}_i^{(k)}\left(\sum_{j=1}^n \tau_j^{-2}\left(\hat{\varepsilon}_i^{(k)}\right)^2\right)^{-1}, \tag{2.185}$$

$$\hat{\varepsilon}_i^{(0)} = (x_i - \langle x \rangle), \quad \hat{\varepsilon}_i^{(1)} = \left[1 - \frac{n\sigma_i^2\tau_i^{-2}}{\sum\limits_{j=1}^{n} \tau_j^{-2}(x_j - \langle x \rangle)^2} \right] (x_i - \langle x \rangle), \quad (2.186)$$

$$\frac{\partial^2 h\,(\varepsilon)}{\partial \varepsilon_i^2} = \tau_i^{-2} \left(\sum_{k=1}^{n} \tau_k^{-2}\varepsilon_k^2 \right)^{-1} - 2\left(\tau_i^{-4}\varepsilon_i^2 \right) \left(\sum_{k=1}^{n} \tau_k^{-2}\varepsilon_k^2 \right)^{-2}$$
$$+ \frac{\sigma_i^{-2}}{n} \left[1 - \sigma_i^{-2} \left(\sum_{k=1}^{n} \sigma_k^{-2} \right)^{-1} \right], \quad (2.187)$$

$$\frac{\partial^2 h\,(\varepsilon)}{\partial \varepsilon_i \partial \varepsilon_j} = -2\left(\tau_i^{-2}\tau_j^{-2}\varepsilon_i\varepsilon_j \right) \left(\sum_{k=1}^{n} \tau_k^{-2}\varepsilon_k^2 \right)^{-2} + \frac{1}{n} \left[\sigma_i^{-2}\sigma_j^{-2} \left(\sum_{k=1}^{n} \sigma_k^{-2} \right)^{-1} \right].$$
$$(2.188)$$

It follows from Eq. (2.177) that the mean value, $\langle \varepsilon \rangle^{MLE}$, and covariance matrix, $\mathbf{C}_\varepsilon^{MLE}$, of ε in the *MLE*-approximation are:

$$\langle \varepsilon \rangle^{MLE} = \hat{\varepsilon}; \qquad \mathbf{C}_\varepsilon^{MLE} = \left[\nabla^2 h\,(\hat{\varepsilon}) \right]^{-1}. \quad (2.189)$$

A more accurate expression for $\langle \varepsilon \rangle$ and $\mathbf{C}_\varepsilon$ can be obtained by evaluating these quantities using the saddle-point (Laplace) approximation. In general, the Laplace approximation applied to a posterior feature of interest, $E\left[f\,(\varepsilon)\right]$, leads to the following result:

$$E\left[f\,(\varepsilon)\right] \triangleq \frac{\int f\,(\varepsilon)\,p\,(\mathbf{y}|\varepsilon)\,\pi\,(\varepsilon)\,d\varepsilon}{\int p\,(\mathbf{y}|\varepsilon)\,\pi\,(\varepsilon)\,d\varepsilon} \triangleq \frac{\int_{-\infty}^{\infty} \exp\left[-nh^*\,(\varepsilon)\right]d\varepsilon}{\int_{-\infty}^{\infty} \exp\left[-nh\,(\varepsilon)\right]d\varepsilon}$$

$$\sim e^{\{-n[h^*(\varepsilon^*)-h(\hat{\varepsilon})]\}} \sqrt{\frac{Det\left[\nabla^2 h\,(\hat{\varepsilon})\right]}{Det\left[\nabla^2 h^*\,(\varepsilon^*)\right]}} \left[1 + O\left(n^{-2}\right)\right]$$

$$\sim \frac{f\,(\varepsilon^*)\,p\,(\mathbf{y}|\varepsilon^*)\,\pi\,(\varepsilon^*)}{p\,(\mathbf{y}|\hat{\varepsilon})\,\pi\,(\hat{\varepsilon})} \sqrt{\frac{Det\left[\nabla^2 h\,(\hat{\varepsilon})\right]}{Det\left[\nabla^2 h^*\,(\varepsilon^*)\right]}} \left[1 + O\left(n^{-2}\right)\right], \quad (2.190)$$

where ε^* and $\hat{\varepsilon}$ denote parameter values that maximize $h^*\,(\varepsilon)$ and $h\,(\varepsilon)$, respectively. For the expectation value $E\left[f\,(\varepsilon)\right]$ of a function $f\,(\varepsilon)$ taken over the probability distribution $p\,(\mu, \varepsilon|\sigma, \mathbf{x}, \tau)\,d\varepsilon$ given in Eq. (2.176), the application of Laplace's method involves the function $h\,(\varepsilon)$ as defined in Eq. (2.178), so that $\hat{\varepsilon}$ is the solution of Eq. (2.181), where the gradient $\nabla h^* (\varepsilon^*)$ vanishes. On the other hand, the function $h^* (\varepsilon)$ is defined as:

$$h^* (\varepsilon) \triangleq \frac{1}{2} \log \left(\varepsilon^T \mathbf{C}_\tau^{-1} \varepsilon \right)$$
$$+ \frac{1}{2n} (\varepsilon - \mathbf{x})^T \mathbf{C}_\sigma^{-1} (\mathbf{I} - \mathbf{W}_\sigma) (\varepsilon - \mathbf{x}) - \frac{1}{n} \log f (\varepsilon) , \qquad (2.191)$$

and the supremum of $h^* (\varepsilon)$ is attained at the point ε^* where the gradient $\nabla h^* (\varepsilon^*)$ vanishes. Hence, ε^* is obtained by solving the following system of nonlinear equations

$$\nabla h^* (\varepsilon) = \left\{ \frac{\mathbf{C}_\tau^{-1} \varepsilon}{\varepsilon^T \mathbf{C}_\tau^{-1} \varepsilon} + \frac{1}{n} \left[\mathbf{C}_\sigma^{-1} (\mathbf{I} - \mathbf{W}_\sigma) (\varepsilon - \mathbf{x}) - \frac{\nabla f (\varepsilon)}{f (\varepsilon)} \right] \right\}_{\varepsilon = \varepsilon^*} = \mathbf{0}. \qquad (2.192)$$

In particular, the mean values $\langle \varepsilon_i \rangle^{SP}$, $(i = 1, \dots, n)$ in the saddle-point approximation are obtained by setting $f (\varepsilon) = \varepsilon_i$ in Eqs. (2.190) through (2.192), thus obtaining

$$\langle \varepsilon \rangle^{SP} = (\varepsilon^*) \frac{p (\mathbf{y}|\varepsilon^*) \pi (\varepsilon^*)}{p (\mathbf{y}|\hat{\varepsilon}) \pi (\hat{\varepsilon})} \sqrt{\frac{Det \left[\nabla^2 h (\hat{\varepsilon}) \right]}{Det \left[\nabla^2 h^* (\varepsilon^*) \right]}} \left[1 + O \left(n^{-2} \right) \right]. \qquad (2.193)$$

The solution ε_i^* is obtained by solving iteratively the following particular form of Eq. (2.192):

$$\left(\varepsilon_i^* \right)^{(k+1)} = x_i - \langle x \rangle + \left\langle \left(\{ \varepsilon_i^* \}^{(k)} \right\rangle + \sigma_i^2 \left[\left(\varepsilon_i^* \right)^{(k)} \right]^{-1} \right.$$
$$- n \sigma_i^2 \tau_i^{-2} \left(\varepsilon_i^* \right)^{(k)} \left(\sum_{j=1}^{n} \tau_j^{-2} \left(\left(\varepsilon_i^* \right)^{(k)} \right)^2 \right)^{-1} , \qquad (2.194)$$

with the zeroth and first-order approximations given, respectively, by

$$\left(\varepsilon_i^* \right)^{(0)} = (x_i - \langle x \rangle) , \qquad (2.195)$$

$$\left(\varepsilon_i^* \right)^{(1)} = \left[1 + \frac{\sigma_i^2}{(x_j - \langle x \rangle)^2} - \frac{n \sigma_i^2 \tau_i^{-2}}{\sum\limits_{j=1}^{n} \tau_j^{-2} (x_j - \langle x \rangle)^2} \right] (x_i - \langle x \rangle) . \qquad (2.196)$$

Comparing the above results with those in Eq. (2.186) indicates that

$$(\varepsilon_i^*)^{(0)} = \hat{\varepsilon}_i^{(0)}; \quad (\varepsilon_i^*)^{(1)} = \hat{\varepsilon}_i^{(1)} + \frac{\sigma_i^2}{x_j - \langle x \rangle}. \tag{2.197}$$

The above results already indicate that even if the zeroth-order approximation $(\varepsilon_i^*)^{(0)}$ were used to evaluate the functions $p(\mathbf{y}|\boldsymbol{\varepsilon}^*), \pi(\boldsymbol{\varepsilon}^*)$, and $Det\left[\nabla^2 h^*(\boldsymbol{\varepsilon}^*)\right]$ on the right side of Eq. (2.193), the saddle-point result

$$\left\langle \varepsilon_i^{(0+)} \right\rangle^{SP} \simeq \left[1 + \frac{\sigma_i^2}{(x_j - \langle x \rangle)^2} - \frac{n\sigma_i^2 \tau_i^{-2}}{\sum_{j=1}^{n} \tau_j^{-2}(x_j - \langle x \rangle)^2} \right] (x_i - \langle x \rangle)$$

$$= \left\langle \varepsilon_i^{(1)} \right\rangle^{MLE} + \frac{\sigma_i^2}{(x_j - \langle x \rangle)^2}, \tag{2.198}$$

already contains the additional (correction) term $\sigma_i^2(x_j - \langle x \rangle)^{-2}$ when compared to the expression of the full first-order approximation $\left\langle \varepsilon_i^{(1)} \right\rangle^{MLE}$. The full first-order approximation would of course include additional correction terms stemming from the functions $p(\mathbf{y}|\boldsymbol{\varepsilon}^*), \pi(\boldsymbol{\varepsilon}^*)$, and $Det\left[\nabla^2 h^*(\boldsymbol{\varepsilon}^*)\right]$ evaluated at $(\varepsilon_i^*)^{(1)}$.

1. If a *uniform prior* $p(c)\,dc \sim dc$ were used for Eq. (2.172), then the above results would hold but with n being replaced by $(n+2)$, as can be readily noticed from Eq. (2.175).

2. If the *exponential prior* given in Eq. (2.159) were used in Eq. (2.172), performing the respective integration over c would give

$$\int_{-\infty}^{\infty} dc\, c^{\frac{n}{2}} \exp\left[(-c) \left(1 + \frac{1}{2} \sum_{j=1}^{n} \frac{\varepsilon_j^2}{\tau_j^2} \right) \right] \propto \left[1 + \frac{1}{2} \boldsymbol{\varepsilon}^T \mathbf{C}_\tau^{-1} \boldsymbol{\varepsilon} \right]^{-\frac{n}{2}-1}. \tag{2.199}$$

Inserting the above result together with Eq. (2.173) in Eq. (2.172) yields

$$p\left(\mu, \varepsilon \mid \sigma, \mathbf{x}, \tau\right) d\varepsilon =$$

$$\frac{d\varepsilon \left[1 + \frac{1}{2}\varepsilon^T \mathbf{C}_\tau^{-1}\varepsilon\right]^{-\frac{n}{2}-1} \exp\left[-\frac{1}{2}(\varepsilon - \mathbf{x})^T \mathbf{C}_\sigma^{-1}(\mathbf{I} - \mathbf{W}_\sigma)(\varepsilon - \mathbf{x})\right]}{\int_{-\infty}^{\infty} d\varepsilon \left[1 + \frac{1}{2}\varepsilon^T \mathbf{C}_\tau^{-1}\varepsilon\right]^{-\frac{n}{2}-1} \exp\left[-\frac{1}{2}(\varepsilon - \mathbf{x})^T \mathbf{C}_\sigma^{-1}(\mathbf{I} - \mathbf{W}_\sigma)(\varepsilon - \mathbf{x})\right]}.$$

$$(2.200)$$

In this case, the scalar-valued function $h\left(\varepsilon\right)$ is defined as

$$h\left(\varepsilon\right) \triangleq \frac{n+2}{2n}\log\left(1 + \frac{1}{2}\varepsilon^T \mathbf{C}_\tau^{-1}\varepsilon\right) + \frac{1}{2n}(\varepsilon - \mathbf{x})^T \mathbf{C}_\sigma^{-1}(\mathbf{I} - \mathbf{W}_\sigma)(\varepsilon - \mathbf{x}).$$

$$(2.201)$$

The gradient $\nabla h\left(\varepsilon\right)$ and, respectively, Hessian of $h\left(\varepsilon\right)$ can be readily computed from Eq. (2.201), to obtain

$$\nabla h\left(\varepsilon\right) = \frac{n+2}{2n}\frac{\mathbf{C}_\tau^{-1}\varepsilon}{1 + \varepsilon^T \mathbf{C}_\tau^{-1}\varepsilon/2} + \frac{1}{n}\mathbf{C}_\sigma^{-1}(\mathbf{I} - \mathbf{W}_\sigma)(\varepsilon - \mathbf{x}), \qquad (2.202)$$

$$\nabla^2 h\left(\varepsilon\right) = \frac{n+2}{2n}\left[\frac{\mathbf{C}_\tau^{-1}}{1 + \varepsilon^T \mathbf{C}_\tau^{-1}\varepsilon/2} - \frac{\mathbf{C}_\tau^{-1}\varepsilon\varepsilon^T \mathbf{C}_\tau^{-1}}{\left(1 + \varepsilon^T \mathbf{C}_\tau^{-1}\varepsilon/2\right)^2}\right] + \frac{1}{n}\mathbf{C}_\sigma^{-1}(\mathbf{I} - \mathbf{W}_\sigma).$$

$$(2.203)$$

Setting $\nabla h\left(\hat{\varepsilon}\right) = \mathbf{0}$ at $\varepsilon = \hat{\varepsilon}$, leads to the following fixed-point equation for $\hat{\varepsilon}$:

$$\hat{\varepsilon} = \mathbf{x} - \langle \mathbf{x} \rangle + \langle \hat{\varepsilon} \rangle - \frac{n+2}{2}\frac{\mathbf{C}_\sigma \mathbf{C}_\tau^{-1}\hat{\varepsilon}}{1 + \hat{\varepsilon}^T \mathbf{C}_\tau^{-1}\hat{\varepsilon}/2}, \qquad (2.204)$$

which can again be solved for $\hat{\varepsilon}$ by using some variant of Newton's method or, more simply, by Picard iterations of the form

$$\hat{\varepsilon}^{(k+1)} = \mathbf{x} - \langle \mathbf{x} \rangle + \left\langle \hat{\varepsilon}^{(k)} \right\rangle - \frac{n+2}{2}\frac{\mathbf{C}_\sigma \mathbf{C}_\tau^{-1}\hat{\varepsilon}^{(k)}}{1 + \left(\hat{\varepsilon}^{(k)}\right)^T \mathbf{C}_\tau^{-1}\hat{\varepsilon}^{(k)}/2}, \qquad (2.205)$$

$$\hat{\varepsilon}^{(0)} = \mathbf{x} - \langle \mathbf{x} \rangle, \quad \hat{\varepsilon}^{(1)} = \left[1 - \frac{n+2}{2}\frac{\mathbf{C}_\sigma \mathbf{C}_\tau^{-1}}{1 + (\mathbf{x} - \langle \mathbf{x} \rangle)^T \mathbf{C}_\tau^{-1}(\mathbf{x} - \langle \mathbf{x} \rangle)/2}\right](\mathbf{x} - \langle \mathbf{x} \rangle),$$

$$(2.206)$$

or, in component form,

$$\hat{\varepsilon}_i^{(k+1)} = x_i - \langle x \rangle + \left\langle \hat{\varepsilon}^{(k)} \right\rangle - \frac{n+2}{2} \sigma_i^2 \tau_i^{-2} \hat{\varepsilon}_i^{(k)} \left(1 + \tfrac{1}{2} \sum_{j=1}^{n} \tau_j^{-2} \left(\hat{\varepsilon}_i^{(k)} \right)^2 \right)^{-1},$$

$$(2.207)$$

$$\hat{\varepsilon}_i^{(0)} = x_i - \langle x \rangle \,, \hat{\varepsilon}_i^{(1)} = \left[1 - \frac{n+2}{2} \frac{\sigma_i^2 \tau_i^{-2}}{1 + \tfrac{1}{2} \sum\limits_{j=1}^{n} \tau_j^{-2} (x_j - \langle x \rangle)^2} \right] (x_i - \langle x \rangle) .$$

$$(2.208)$$

The components of the Hessian matrix $\nabla^2 h\left(\varepsilon\right)$ are as follows:

$$\frac{\partial^2 h\left(\varepsilon\right)}{\partial \varepsilon_i^2} = \frac{n+2}{2n} \left[\tau_i^{-2} \left(1 + \frac{1}{2} \sum_{k=1}^{n} \tau_k^{-2} \varepsilon_k^2 \right)^{-1} - 2 \left(\tau_i^{-4} \varepsilon_i^2 \right) \left(1 + \frac{1}{2} \sum_{k=1}^{n} \tau_k^{-2} \varepsilon_k^2 \right)^{-2} \right]$$

$$+ \frac{\sigma_i^{-2}}{n} \left[1 - \sigma_i^{-2} \left(\sum_{k=1}^{n} \sigma_k^{-2} \right)^{-1} \right] ,$$

$$(2.209)$$

$$\frac{\partial^2 h\left(\varepsilon\right)}{\partial \varepsilon_i \partial \varepsilon_j} = -\frac{n+2}{2n} \left(\tau_i^{-2} \tau_j^{-2} \varepsilon_i \varepsilon_j \right) \left(1 + \frac{1}{2} \sum_{k=1}^{n} \tau_k^{-2} \varepsilon_k^2 \right)^{-2}$$

$$- \frac{1}{n} \left[\sigma_i^{-2} \sigma_j^{-2} \left(\sum_{k=1}^{n} \sigma_k^{-2} \right)^{-1} \right] .$$

$$(2.210)$$

In the saddle-point approximation, cf. Eq. (2.190), the scalar-valued function $h^*\left(\varepsilon\right)$ with the *exponential prior* takes on the form

$$h^*\left(\varepsilon\right) \triangleq \frac{n+2}{2n} \log\left(1 + \frac{1}{2} \varepsilon^T \mathbf{C}_\tau^{-1} \varepsilon \right)$$

$$+ \frac{1}{2n} (\varepsilon - \mathbf{x})^T \mathbf{C}_\sigma^{-1} \left(\mathbf{I} - \mathbf{W}_\sigma \right) (\varepsilon - \mathbf{x}) - \frac{1}{n} \log f\left(\varepsilon\right) . \qquad (2.211)$$

In particular, the mean values $\langle \varepsilon_i \rangle^{SP}$, $(i = 1, \ldots, n)$ in the saddle-point approximation given by Eq. (2.193) would be computed by using the solution

ε_i^* of the following fixed-point equation obtained by setting to zero the gradient $\nabla h^* (\varepsilon^*)$ of Eq. (2.211):

$$\varepsilon_i^* = x_i - \langle x \rangle + \langle \varepsilon^* \rangle + \sigma_i^2 [\varepsilon_i^*]^{-1}$$

$$- \frac{n+2}{2} \sigma_i^2 \tau_i^{-2} \varepsilon_i^* \left(1 + \frac{1}{2} \sum_{j=1}^{n} \tau_j^{-2} (\varepsilon_i^*)^2 \right)^{-1} . \tag{2.212}$$

The solution ε_i^* would be obtained by solving the above fixed-point equation iteratively. Picard iterations, for example, would yield

$$(\varepsilon_i^*)^{(k+1)} = x_i - \langle x \rangle + \left\langle (\varepsilon^*)^{(k)} \right\rangle + \sigma_i^2 \left[(\varepsilon_i^*)^{(k)} \right]^{-1} \tag{2.213}$$

$$- \frac{n+2}{2} \sigma_i^2 \tau_i^{-2} (\varepsilon_i^*)^{(k)} \left(1 + \frac{1}{2} \sum_{j=1}^{n} \tau_j^{-2} \left((\varepsilon_i^*)^{(k)} \right)^2 \right)^{-1} , \tag{2.214}$$

with the zero- and first-order approximations given, respectively, by

$$(\varepsilon_i^*)^{(0)} = x_i - \langle x \rangle , \tag{2.215}$$

$$(\varepsilon_i^*)^{(1)} = \left[1 + \frac{\sigma_i^2}{(x_j - \langle x \rangle)^2} - \frac{n+2}{2} \frac{\sigma_i^2 \tau_i^{-2}}{1 + \frac{1}{2} \sum_{j=1}^{n} \tau_j^{-2} (x_j - \langle x \rangle)^2} \right] (x_i - \langle x \rangle) . \tag{2.216}$$

Comparing Eq. (2.216) with Eq. (2.208) indicates that

$$(\varepsilon_i^*)^{(0)} = \hat{\varepsilon}_i^{(0)}, \quad (\varepsilon_i^*)^{(1)} = \hat{\varepsilon}_i^{(1)} + \frac{\sigma_i^2}{x_j - \langle x \rangle}, \tag{2.217}$$

a result similar to that indicated by Eq, (2.197). Furthermore, just as Eq. (2.198) indicated, if the zero-order approximation $(\varepsilon_i^*)^{(0)}$ is used to evaluate the functions $p(\mathbf{y}|\varepsilon^*), \pi(\varepsilon^*)$, and $Det \left[\nabla^2 h^* (\varepsilon^*) \right]$, then the saddle-point result for $\left\langle \varepsilon_i^{(0+)} \right\rangle^{SP}$ using the exponential prior is

$$\left\langle \varepsilon_i^{(0+)} \right\rangle^{SP} \simeq \left[1 + \frac{\sigma_i^2}{(x_j - \langle x \rangle)^2} - \frac{n+2}{2} \frac{\sigma_i^2 \tau_i^{-2}}{1 + \frac{1}{2} \sum\limits_{j=1}^{n} \tau_j^{-2}(x_j - \langle x \rangle)^2} \right] (x_i - \langle x \rangle)$$

$$= \left\langle \varepsilon_i^{(1)} \right\rangle^{MLE} + \frac{\sigma_i^2}{(x_j - \langle x \rangle)^2}, \tag{2.218}$$

which contains the additional (correction) term $\sigma_i^2(x_j - \langle x \rangle)^{-2}$ as compared to the expression of the first-order approximation $\left\langle \varepsilon_i^{(1)} \right\rangle^{MLE}$. The complete first-order approximation would of course include additional correction terms stemming from the functions $p\left(\mathbf{y}|\boldsymbol{\varepsilon}^*\right), \pi\left(\boldsymbol{\varepsilon}^*\right)$, and $Det\left[\nabla^2 h^*\left(\boldsymbol{\varepsilon}^*\right)\right]$ evaluated at $(\varepsilon_i^*)^{(1)}$.

The above treatment of unrecognized systematic errors is an example of a two-stage "hierarchical" Bayesian method, involving a twofold application of Bayes' theorem to the sampling distribution that depends on parameters ε_i having a Gaussian prior, namely Eq. (2.134), which in turn depended on the so-called "hyper-parameter" c, which had itself a "hyper-prior" distribution. Furthermore, the first-order approximations $\left\langle \varepsilon_i^{(1)} \right\rangle^{MLE}$ are similar to the James-Stein estimators, which were shown (see, e.g., ref. [15]) to have sometimes lower risk than the estimates resulting from Bayesian estimation under quadratic loss (that minimize the square error averaged over all possible parameters, for the sample at hand). It is important to note, though, that the two-stage method Bayesian used above yields results that are superior to the James-Stein estimators, especially for small samples. Moreover, the iterations shown in the foregoing yield further improvements in a systematic and unambiguous way, without the discontinuities, questions of interpretation, and restrictions (e.g., $n \geq 2$) associated with James-Stein estimators. This fact is particularly valuable and relevant for scientific data evaluation, where best values must often be inferred (for quadratic or any other loss) from just a single available sample.

2.6 Notes and Remarks

Section 2.2 is not exhaustive regarding the use of group theory and symmetries for assigning priors, but has been limited to presenting only the most commonly encountered priors in practice, which are also encountered throughout this book. The topic of assigning priors is an area of active research, and the recent book by Berger [15] provides considerably more information on this topic. The reference books by Tribus [202] and Jaynes [99], contain detailed presentations on the use of probabilities for addressing practical situations encountered by experimentalists. The case studies presented in Section 2.3 are extensions of the work of Fröehner ([68], [69]).

It is probably not too unfair to say that, although measurements without systematic errors are the exception rather than the rule, conventional (frequentist) sampling theory has not much to offer to practitioners in science and technology who are confronted with systematic errors and correlations. This is in marked contrast to the wealth of material on statistical errors, for which satisfactory techniques are available, based on counting (Poisson) statistics or on Gaussian models for the scatter of repeatedly measured data. The material in Chapter 2 attempts to clarify some of these issues with the help of modern probabilistic tools such as Bayesian parameter estimation under quadratic loss, group-theoretical least informative priors, and probability assignment by entropy maximization. We have also noted how common errors, the most frequent type of systematic error, invariably induce correlations, and how correlations are described by nondiagonal covariance matrices. The least informative prior for the nonnegative correlation coefficients of repeated equivalent measurements turns out to be Haldane's prior. Concerning repeated measurements, probability theory asserts, in agreement with experience, that only the statistical uncertainty component decreases as $1/\sqrt{n}$ if a measurement is repeated n times, whereas the systematic component constitutes a residual uncertainty that cannot be reduced by mere repetition. Only re-measurement with several nonequivalent techniques, and generalized Bayesian least-squares fitting to all data simultaneously, also can reduce the systematic uncertainties. An ideal least-squares fit should use, apart from prior information, raw data rather than reduced data as input, and include the data reduction step. In this way both the parameters of interest (for instance resonance parameters or optical-model parameters) and the auxiliary quantities (e.g., normalization constants, backgrounds, flight paths, etc.) can be adjusted simultaneously while all prior uncertainties, including the systematic ones, are properly propagated to the final parameter estimates. After having performed the adjustment, the auxiliary quantities ("nuisance parameters") can be removed by marginalization.

3

Optimization Methods For Large-Scale Data Assimilation

CONTENTS

3.1　Introduction

This chapter presents minimization algorithms that are best suited for unconstrained and constrained minimization of large-scale systems such as *four-dimensional variational* data assimilation, referred to as "4–D VAR," in weather prediction and similar applications in the geophysical sciences. The operational implementation of the 4–D VAR method hinges crucially upon fast convergence of efficient gradient-based large-scale unconstrained minimization algorithms. These algorithms generally minimize a cost function which attempts to quantify the discrepancies between forecast and observations in a window of assimilation, where the model is used as a strong constraint. Data assimilation problems in oceanography and meteorology contain many degrees of freedom ($\approx 10^7$). Consequently, conjugate-gradient methods and limited-memory quasi-Newton (LMQN) methods come into consideration since they require storage of only a few vectors, containing information from a few iterations. Studies (e.g., refs. [74] and [126]) indicate that "Limited Memory Broyden Fletcher–Goldfarb and Shanno" (L-BFGS) and its French equivalent M1QN3 are among the best LMQN methods available to date.

Global minimization algorithms are required to compute the unique lowest minimum, in data assimilation problems. Consequently, this chapter also presents a short review of several global minimization algorithms. The user-friendly minimization software available within the International Mathematical Subroutine Library (IMSL), Numerical Algorithms Group (NAG), and Harwell libraries will be mentioned along with the newly developed Large and Nonlinear Constrained Extended Lagrangian Optimization Techniques (LANCELOT) library for large-scale nonlinear constrained optimization.

Optimization aims at finding a global minimizer $\mathbf{x}^*$ of function $F(\mathbf{x}^*)$ where the function attains its least value. By definition, a point $\mathbf{x}^*$ is a global minimizer of a function $F(\mathbf{x})$ if

$$F(\mathbf{x}^*) \leq F(\mathbf{x}), \quad for\ all\ \mathbf{x} \in \mathbb{R}^n. \tag{3.1}$$

In general, only local knowledge about $F(\mathbf{x})$ is available. Therefore, it is very difficult to find the global optimum except in particular instances, e.g., when $F(\mathbf{x})$ is a convex set. The basic definitions needed for local optimization are presented next. By definition, a set $S \in \mathbb{R}^n$ is called a *convex set* if the straight line segment connecting any two points in S lies entirely inside S. Formally, any two points $\mathbf{x} \in S$ and $\mathbf{y} \in S$, satisfy the relation, $a\mathbf{x} + (1-a)\mathbf{y} \in S$ for all $a \in [0, 1]$.

In particular, if the objective function and the feasible region in the optimization problem are both convex, then any local minimizer is also a global minimizer.

$$\min f(\mathbf{x}) \ such \ that \ \begin{cases} c_i(\mathbf{x}) = 0, & i \in E \\ c_i(\mathbf{x}) \geq 0, & i \in I \end{cases}, \quad \mathbf{x} \in \mathbb{R}^n, \quad (3.2)$$

In Eq. (3.2), the symbols E and I denotes the equality and inequality constraints index sets, respectively while $c_i(\mathbf{x})$ denote constraints imposed on the function f.

By definition, a point $\mathbf{x}^*$ is a local minimizer of a function f if there is a neighborhood of $\mathbf{x}^*$ such that $f(\mathbf{x}^*) \leq f(\mathbf{x})$ for all $\mathbf{x} \in \mathbf{N}$. Nonlinear problems may admit more than a single local minimum. For example Daescu et al. [42] have shown that the potential energy function that determines the 3-D molecular configuration of certain substances admits many millions of local minima. Taylor's theorem plays a fundamental role in identifying local minima. For a functional $f : \mathbb{R}^n \rightarrow \mathbb{R}$ that is continuously differentiable, Taylor's theorem states that

$$f(\mathbf{x} + \mathbf{p}) = f(\mathbf{x}) + \nabla f(\mathbf{x} + t\mathbf{p})^T \mathbf{p}, \ for \ p \in \mathbb{R}^n. \quad (3.3)$$

and when the functional is twice continuously differentiable the following relations hold:

$$\nabla f(\mathbf{x} + \mathbf{p}) = \nabla f(\mathbf{x}) + \int_0^1 \nabla^2 f(\mathbf{x} + t\mathbf{p})\mathbf{p}, \quad (3.4)$$

and

$$f(\mathbf{x}+\mathbf{p}) = f(\mathbf{x}) + \nabla f(\mathbf{x} + t\mathbf{p})^T \mathbf{p} + \frac{1}{2}\nabla^2 f(\mathbf{x}+t\mathbf{p})\mathbf{p}, \quad for \ some \ t \in (0,1). \quad (3.5)$$

Recall that $\mathbf{x}^*$ is called a *stationary point* if $\nabla f(\mathbf{x}^*) = 0$. Recall also that a matrix $\mathbf{B}$ is called *positive definite* if $\mathbf{p}^T\mathbf{B}\mathbf{p} > 0$, for all $\mathbf{p} \neq \mathbf{0}$. A matrix $\mathbf{B}$ is called positive semidefinite if $\mathbf{p}^T\mathbf{B}\mathbf{p} \geq 0$ for all $\mathbf{p}$.

Theorem: If $\mathbf{x}^*$ is a *local minimizer* of a functional $f:R^n \rightarrow R$ and if $\nabla^2 f$ exists and is continuous in an open neighborhood $\mathbf{x}^*$ then $\nabla f(\mathbf{x}^*) = 0$ and $\nabla^2 f(\mathbf{x}^*)$ is positive semidefinite.

Any descent direction that makes an angle of strictly less than $\pi/2$ with $-\nabla f_k$ will produce decrease in f for a small enough step. In other words writing Taylor's theorem in the form

$$f(\mathbf{x}_k + \varepsilon\mathbf{p}_k) = f(\mathbf{x}_k) + \varepsilon\mathbf{p}_k^T\nabla f(\mathbf{x}_k) + O(\varepsilon^2), \tag{3.6}$$

indicates that when $\mathbf{p}_k$ is any *descent direction* which implies that, the angle θ between $\mathbf{p}_k$ and ∇f_k is such that $\cos\theta_k < 0$

$$\mathbf{p}_k^T\nabla f(\mathbf{x}_k) = \|\mathbf{p}_k\| \, \|\nabla f(\mathbf{x}_k)\| \cos\theta_k < 0, \tag{3.7}$$

From this relation and Taylor's theorem it follows that $f(\mathbf{x}_k + \varepsilon\mathbf{p}_k) < f(\mathbf{x}_k)$ for all positive but sufficiently small values of ε. In particular, the *steepest descent direction* $-\nabla f_k$, is defined as the direction along which f decreases most rapidly. Taylor's Theorem indicates that for any search direction $\mathbf{p}$ and step-length parameter α satisfy the relation

$$f(\mathbf{x}_k + \alpha\mathbf{p}) = f(\mathbf{x}_k) + \alpha\mathbf{p}^T\nabla^2 f(\mathbf{x}_k + t\mathbf{p})\mathbf{p}, \text{ for some } t \in (0, \alpha). \tag{3.8}$$

The coefficient of α is the rate of change of f in the direction $\mathbf{p}$ at $\mathbf{x}_k$ is obtained by minimizing $\mathbf{p}^T\nabla f_k$ subject to the condition that $\|\mathbf{p}\| = 1$. $\mathbf{p}^T\nabla f_k = \|\mathbf{p}\| \, \|\nabla f_k\| \cos\theta = \|\nabla f_k\| \cos\theta$ where θ is the angle between $\mathbf{p}$ and ∇f_k. It follows that the minimum of $\mathbf{p}^T\nabla f_k$ is attained when $\cos\theta = -1$, which yields $\mathbf{p} = -\nabla f_k/\|\nabla f_k\|$. This direction is orthogonal to the contours of the objective function f. In summary the steepest descent method is a line-search method that moves along $\mathbf{p}_k = -\nabla f_k$ at every iterative step. The advantage of the steepest descent method is that it only requires knowledge of gradient ∇f_k. However, this method can be very slow on difficult problems.

The *Newton search direction* is the vector $\mathbf{p}$ which minimizes the quantity

$m_\alpha(\mathbf{p})$ defined by the second-order Taylor approximation

$$f(\mathbf{x}_k + \mathbf{p}) = f(\mathbf{x}_k) + \mathbf{p}^T \nabla f(\mathbf{x}_k) + \frac{1}{2} \mathbf{p}^T \nabla^2 f(\mathbf{x}_k) = m_k(\mathbf{p}), \qquad (3.9)$$

with $\nabla^2 f_k$ positive definite.

Setting the derivative of $m_k(\mathbf{p})$ with respect to $\mathbf{p}$ to zero yields the explicit formula for the Newton direction $\mathbf{p}_k^N$

$$\mathbf{p}_k^N = -(\nabla^2 f(\mathbf{x}_k))^{-1} \nabla f(\mathbf{x}_k). \qquad (3.10)$$

Comparing Eq. (3.9) with Eq. (3.3) indicates that $\nabla^2 f(\mathbf{x}_k + t\mathbf{p})$ was replaced by $\nabla^2 f_k$. If $\nabla^2 f_k$ is smooth, the difference introduces a perturbation of $O(\|\mathbf{p}^3\|)$; Therefore, for small $\|\mathbf{p}\|$ the approximation $f(\mathbf{x}_k + \mathbf{p}) \approx m_k(\mathbf{p})$ is acceptable. The Newton direction can be used in a line-search method when $\nabla^2 f_k$ is positive definite. Unlike the steepest descent direction, there is a step-length of 1 associated with the Newton direction. Most line-search implementations of Newton's method use the unit step a $= 1$ where possible, and adjust α only when it does not produce a satisfactory reduction in the value of f.

When $\nabla^2 f_k$ is not positive definite, the Newton direction may not exist. Even when it is defined, it may not satisfy the descent property $\nabla f_k^T \mathbf{p}_k^n < 0$, in which case it is unsuitable as a search direction. In such situations, line-search methods modify the definition of $\mathbf{p}_k$ to ensure that the descent condition is satisfied while retaining the benefit of the second-order information contained in $\nabla^2 f_k$. Methods that use the Newton direction have a fast rate of local convergence, typically quadratic. After a neighborhood of the solution is reached, convergence to high accuracy often occurs in just a few iterations. However, the main shortcoming of the Newton direction is the requirement for the Hessian $\nabla^2 f_k$.

The typical minimization iteration reads

$$\mathbf{x}_{k+1} = \mathbf{x}_k + \alpha_k \mathbf{p}_k,$$

where the positive scalar α_k is called *step-length*, while $\mathbf{p}_k$ is called the *search direction*; $\mathbf{p}_k$ should be a *descent direction* as we search for the minimum of the function f. The search direction $\mathbf{p}_k$ is computed from an expression of the general form

$$\mathbf{p}_k = -\mathbf{B}_k^{-1}\nabla f_k, \tag{3.11}$$

where $\mathbf{B}_k$ is a symmetric nonsingular matrix.

The steepest descent methods use $\mathbf{B}_k = \mathbf{I}$ while the Newton's method uses $\mathbf{B}_k = \nabla^2 f(\mathbf{x}_k)$. As will be shown in subsequent sections, quasi-Newton methods use matrices $\mathbf{B}_k$ which are computed by updating an approximation of the Hessian matrix, at every iteration, by a low rank (rank one or rank two) matrix. Whenever $\mathbf{B}_k$ is a positive-definite matrix, Eq. (3.11) ensures that

$$\mathbf{p}_k\nabla f_k = -\nabla f_k^T \mathbf{B}_k^{-1}\nabla f_k < 0, \tag{3.12}$$

thereby ensuring that $\mathbf{p}_k$ is a descent direction.

The essence of minimization consists in choosing suitable directions $\mathbf{p}_k$ and step-lengths α_k such that we obtain convergence to a local minimizer even from remote starting points. This goal can be attained by selection $\mathbf{p}_k$ and α_k such that they satisfy the *Wolfe conditions*, which will be discussed in Section 3.5.

Conjugate-gradient (CG) methods are very useful for solving large problems and can be efficiently implemented on multiprocessor machines. The original CG method was proposed by Fletcher and Reeves [66], and was improved by Polak and Ribiere [161]. Powell [164] suggested a condition for restarting the CG algorithm, with the restart direction computed from a three-term recurrence iteration proposed by Beale [12]. The CG algorithms are attractive because they require only storage of a few vectors of memory and perform well for large systems. Shanno and Phua [188] proposed the Conjugate-Minimization (CONMIN) algorithm, which is a robust extension of the CG method requiring a few more vectors of storage. Navon and Legler [151] surveyed their properties and application to problems in meteorology. Davidon [47] introduced the quasi-Newton (Q-N) or variable metric methods, in which the initial approximation of the Hessian matrix is updated at every iteration by incorporating curvature information measured along each step. The most effective Q-N method to date appears to be the BFGS Q-N method, which was proposed in 1970 simultaneously by Broyden [20], Fletcher [63], Goldfarb [78], and Shanno et al. [187]. This method is fast, robust efficient, and provides a superlinear rate of convergence, but is not a good candidate for large-scale optimization since it requires storage of the approximation to the Hessian matrix, which is impractical for large-scale problems. Griewank and Toint [83] developed the partitioned quasi-Newton, which is designed for partially separable functions in order to exploit the structure of the cost functional in large-scale optimization; this method has been implemented by Conn et al. [34] in the LANCELOT package.

For large problems (e.g., as encountered in oceanography and meteorology, which involve $\approx 10^7$ degrees of freedom), it is not possible to store in memory information beyond that stemming from the first few iterations. Consequently, only the conjugate-gradient methods and the *limited-memory quasi-Newton* (LMQN) methods can be used in practice for large-scale problems.

Section 3.2 highlights the salient features of the CONMIN, E04DGF, L-BFGS, and BBVSCG methods, which all resemble the BFGS method but avoid storage of large matrices. The L-BFGS version of Liu and Nocedal [126] is implemented in the Harwell routine VA15 while a very similar version, M1QN3, was developed by Gilbert [74] at Institut National de Recherche en Informatique et en Automatique (INRIA) in the library Module Optimisation (MODULOPT). The studies of Gilbert and Lemaréchal [74] and of Liu and Nocedal [126] indicate that L-BFGS (LMQN) and its French equivalent M1QN3 are among the best LMQN methods available to date, and have been the main stay of 4–D VAR optimization in the last few years.

Section 3.3 presents *Truncated Newton* (T-N) methods, which require evaluations of functions and gradients only, and are therefore also suitable for large-scale minimization. In particular, this section will highlight the Truncated-Newton Package (TNPACK) algorithm of Schlick and Fogelson [179] and [180], as well as works using second-order adjoint Hessian vector products by Wang et al. ([212], [213], [214]) and by Le et al. [117]. *Hybrid methods*, combining the L-BFGS with the Truncated Newton method, have been proposed by Morales and Nocedal ([137], [138]), and tested by Daescu et al. [42] and Daley [46].

Section 3.4 highlights the use of information provided by the Hessian matrix for large-scale optimization. Section 3.5 discusses issues related to nonsmooth and nondifferentiable optimization, in view of the fact that precipitation and radiation parameterizations involve on-off processes. Section 3.6

addresses issues related to step-size searches, while Section 3.7 presents trust-region methods. Section 3.8 discusses scaling and preconditioning for linear and nonlinear problems. Nonlinear constrained optimization is discussed in Section 3.9. Since nonlinear optimization may involve cost functions characterized by presence of multiple minima. Section 3.10 briefly discusses two stochastic global minimization methods: simulated annealing and genetic algorithms.

3.2 Limited Memory Quasi-Newton(LMQN) Algorithms for Unconstrained Minimization

Limited-memory quasi-Newton (LMQN) algorithms combine the advantages of CG-algorithms (low storage requirements) with those of quasi-Newton (Q-N) methods (computational efficiency stemming from their superlinear convergence). The LMQN algorithms build several rank one or rank two matrix updates to the Hessian, thereby avoiding the need to store the approximate Hessian matrix, as required by full Q-N methods. Like the CG-methods, the LMQN methods require only a modest amount storage for generating the search directions. The basic structure of LMQN algorithms for minimizing a quadratic functional $J(\mathbf{x})$, $\mathbf{x} \in R^N$, comprises the following steps:

1. Choose an initial guess, $\mathbf{x}_0$, and a symmetric positive definite matrix, $\mathbf{H}_0$, as the initial approximation to the inverse Hessian matrix; the unit matrix may be chosen as $\mathbf{H}_0$.

2. Compute the gradient of $J(\mathbf{x})$ at $\mathbf{x}_0$:

$$\mathbf{g}_0 = \mathbf{g}(\mathbf{x}_0) = \nabla J(\mathbf{x}_0),\qquad(3.13)$$

   and set

$$\mathbf{d}_0 = -\mathbf{H}_0\mathbf{g}_0.\qquad(3.14)$$

3. For $k = 0, 1, \ldots, n+1$, set

$$\mathbf{x}_{k+1} = \mathbf{x}_k + \alpha_k\mathbf{d}_k,\qquad(3.15)$$

where α_k is the step-size obtained by a line-search ensuring sufficient descent.

4. Compute the updated gradient

$$\mathbf{g}_{k+1} = \nabla J(\mathbf{x}_{k+1}); \tag{3.16}$$

5. Check if a restart is needed (this step will be discussed in the sequel).

6. Generate a new search direction, $\mathbf{d}_{k+1}$, by setting

$$\mathbf{d}_{k+1} = -\mathbf{H}_{k+1}\mathbf{g}_{k+1}. \tag{3.17}$$

7. Check for convergence: if

$$\|\mathbf{g}_{k+1}\| \leq \varepsilon \ \max\{1, \|\mathbf{x}_{k+1}\|\}, \tag{3.18}$$

where ε is a small scalar (e.g., $\varepsilon = 10^{-5}$), stop; otherwise, go to step 3 and continue the computations by increasing the index k.

The LMNQ method with the BFGS update formula Liu and Nocedal [126], Nocedal [154] forms an approximate inverse Hessian from $\mathbf{H}_0$ and k pairs of vectors $(\mathbf{q}_i, \mathbf{p}_i)$, where $\mathbf{q}_i = \mathbf{g}_{i+1} - \mathbf{g}_i$ and $\mathbf{p}_i = \mathbf{x}_{i+1} - \mathbf{x}_i$ for $i \leq 0$. Since $\mathbf{H}_0$ is generally taken to be the unit matrix (or some other diagonal matrix), the pairs $(\mathbf{q}_i, \mathbf{p}_i)$ are stored (instead of storing $\mathbf{H}_k$) and the vector $\mathbf{H}_k\mathbf{g}_k$ is computed by an approximate algorithm. The LMQN methods CONMIN-BFGS, L-BFGS, E04DGF and BBVSCG, which will be presented in this section, all fall within this conceptual framework. They differ only in the selection of the vector pairs $(\mathbf{q}_i, \mathbf{p}_i)$, the choice of $\mathbf{H}_0$, the method for computing $\mathbf{H}_k\mathbf{g}_k$, the line-search implementation, and the handling of restarts. Details regarding the relative performance of CONMIN-BFGS, L-BFGS, E04DGF, and BBVSCG algorithms on two-dimensional 4–D VAR problems are provided in, e.g., refs. [227] and [234].

3.2.1 The CONMIN Algorithm

Shanno and Phua [188] proposed the CONMIN algorithm as a two-step LMQN-like CG method that incorporates Beale [12] restarts. Only seven vectors of storage are necessary. Step sizes are obtained by using Davidon's cubic interpolation method to satisfy the following Wolfe [217] conditions:

$$J\left(\mathbf{x}_k + \alpha_k \mathbf{d}_k\right) \leq J\left(\mathbf{x}_k\right) + \beta' \alpha_k \mathbf{g}_k^T \mathbf{d}_k, \tag{3.19}$$

$$\left| \frac{\nabla J(\mathbf{x}_k + \alpha_k \mathbf{d}_k)^T \mathbf{d}_k}{\mathbf{g}_k^T \mathbf{d}_k} \right| \leq \beta, \tag{3.20}$$

where $\beta' = 0.0001$ and $\beta = 0.9$. The restart criterion proposed by Powell [164] is used:

$$|\mathbf{g}_{k+1}^T \mathbf{g}_k| \geq 0.2 \|\mathbf{g}_{k+1}\|^2. \tag{3.21}$$

The new search direction $\mathbf{d}_{k+1}$, defined in Eq. (3.17), is obtained by setting

$$\mathbf{H}_{k+1} = \hat{\mathbf{H}}_k - \frac{\mathbf{p}_k \mathbf{q}_k^T \hat{\mathbf{H}}_k + \hat{\mathbf{H}}_k \mathbf{q}_k \mathbf{p}_k^T}{\mathbf{p}_k^T \mathbf{q}_k} + \left(1 + \frac{\mathbf{q}_k^T \hat{\mathbf{H}}_k \mathbf{q}_k}{\mathbf{p}_k^T \mathbf{q}_k}\right) \frac{\mathbf{p}_k \mathbf{p}_k^T}{\mathbf{p}_k^T \mathbf{q}_k}. \tag{3.22}$$

If a restart is satisfied, Eq. (3.17) is changed to

$$\mathbf{d}_{k+1} = -\hat{\mathbf{H}}_k \mathbf{g}_{k+1}, \tag{3.23}$$

where

$$\hat{\mathbf{H}}_k = \gamma_t \left(\mathbf{I} - \frac{\mathbf{p}_t \mathbf{q}_t^T + \mathbf{q}_t \mathbf{p}_t^T}{\mathbf{p}_t^T \mathbf{q}_t} + \frac{\mathbf{q}_t^T \mathbf{q}_t}{\mathbf{p}_t^T \mathbf{q}_t} \frac{\mathbf{p}_t \mathbf{p}_t^T}{\mathbf{p}_t^T \mathbf{q}_t} \right) + \frac{\mathbf{p}_t \mathbf{p}_t^T}{\mathbf{p}_t^T \mathbf{q}_t}. \tag{3.24}$$

In the expression above, the subscript t represents the last step of the previous cycle for which a line-search was performed. The parameter $\gamma_t = \mathbf{p}_t^T \mathbf{q}_t / \mathbf{q}_t^T \mathbf{q}_t$ is obtained by minimizing the condition number $\mathbf{H}_t^{-1} \mathbf{H}_{t+1}$ (see, e.g., ref. [188]).

As indicated by Eqs. (3.22) through (3.24), the current approximation of the Hessian matrix is constructed by using the pair of vectors $\mathbf{q}$ and $\mathbf{p}$. The advantage of the CONMIN algorithm is that it generates descent directions automatically without requiring exact line-searches as long as $(\mathbf{q}_k, \mathbf{p}_k)$ are positive at each iteration. This requirement can be ensured by satisfying the second Wolfe condition, cf. Eq. (3.20), in the line-search. However, CONMIN cannot take advantage of additional storage that might be available.

Shanno [185] and [186] showed that the CONMIN algorithm is globally convergent with inexact line-searches on strongly convex problems. Although it requires seven vectors of storage, the CONMIN algorithm turns out to be the most efficient amongst CG methods. Furthermore, Nocedal [154] indicates that increasing the storage available for CONMIN leads to fewer functions evaluations. CONMIN has provided the backbone algorithm for minimization applications in early works on adjoint parameter estimation, but has been gradually replaced by the L-BFGS algorithm (which will be discussed in a subsequent section). Typical uses of the CONMIN are provided in the works by Navon et al. [149] and [152] and Ramamurthy and Navon [167], as well as in the direct variational framework in oceanography by Legler and Navon [119] and Legler et al.[120].

3.2.2 The E04DGF Algorithm

The E04DGF routine of the NAG Software Library is a two-step LMQN method with preconditioning and restarts, as originally formulated in the nonlinear unconstrained minimization algorithm by Gill and Murray [75]. The amount of working storage required by this method is $12N$ real words of working space. The step-size is determined by considering a sequence of points $\{\alpha^j, j = 1, 2, \cdots, \}$, which tends in the limit to a local minimizer of the cost function $J(\mathbf{x})$ along a direction $\mathbf{d}_k$. This sequence can be computed by using a safeguarded polynomial interpolation algorithm. The initial step-length is chosen as suggested by Davidon:

$$\alpha^0 = \begin{cases} -2\left(J_k - J_{est}\right)/\mathbf{g}_k^T\mathbf{d}_k, & if -2\left(J_k - J_{est}\right)/\mathbf{g}_k^T\mathbf{d}_k \leq 1, \\ 1, & if -2\left(J_k - J_{est}\right)/\mathbf{g}_k^T\mathbf{d}_k > 1, \end{cases} \quad (3.25)$$

where J_{est} represents an estimate of the cost function at the solution point.

Consider now that t is the first index of the sequence α^j that satisfies the relation

$$\left|\nabla J\left(\mathbf{x}_k + \alpha^t\mathbf{d}_k\right)^T\mathbf{d}_k\right| \leq -\eta\mathbf{g}_k^T\mathbf{d}_k, \quad 0 \leq \eta \leq 1. \quad (3.26)$$

The E04DGF algorithm finds the smallest nonnegative integer r such that

$$J_k - J\left(\mathbf{x}_k + 2^{-r}\alpha^t\mathbf{d}_k\right) \geq -2^{-r}\alpha^t\mu\mathbf{g}_k^T\mathbf{d}_k, \quad 0 \leq \mu \leq \frac{1}{2}, \quad (3.27)$$

and subsequently sets $\alpha_k = s^{-r}\alpha^t$. A restart is required (see, e.g., ref. [164])

restart criteria

$$|\mathbf{g}_{k+1}^T \mathbf{g}_k| \geq 0.2\|\mathbf{g}_{k+1}\|^2, \ or$$
$$-1.2\,\|\mathbf{g}_{k+1}\|_2^2 \leq \mathbf{g}_{k+1}^T \mathbf{d}_{k+1} \leq -0.8\,\|\mathbf{g}_{k+1}\|_2^2. \tag{3.28}$$

is satisfied. The new search direction is generated by using Eq. (3.17), where $\mathbf{H}_{k+1}$ is calculated by the following two-step BFGS formula:

$$\mathbf{U}_2 = \mathbf{U}_1 - \frac{1}{\mathbf{q}_k^T \mathbf{p}_k}\left(\mathbf{U}_1\mathbf{q}_k\mathbf{p}_k^T + \mathbf{p}_k\mathbf{q}_k^T\mathbf{U}_1\right) + \frac{1}{\mathbf{q}_k^T \mathbf{p}_k}\left(1 + \frac{\mathbf{q}_k^T\mathbf{U}_1\mathbf{q}_k}{\mathbf{q}_k^T \mathbf{p}_k}\mathbf{p}_k\mathbf{p}_k^T\right), \tag{3.29}$$

$$\mathbf{H}_{k+1} = \mathbf{U}_2 - \frac{1}{\mathbf{q}_k^T \mathbf{p}_k}\left(\mathbf{U}_2\mathbf{q}_k\mathbf{p}_k^T + \mathbf{p}_k\mathbf{q}_k^T\mathbf{U}_2\right) + \frac{1}{\mathbf{q}_k^T \mathbf{p}_k}\left(1 + \frac{\mathbf{q}_k^T\mathbf{U}_2\mathbf{q}_k}{\mathbf{q}_k^T \mathbf{p}_k}\mathbf{p}_k\mathbf{p}_k^T\right). \tag{3.30}$$

If a restart is needed, the following self-scaling updating method due to Shanno [185] and [186] is used instead $\mathbf{U}_2$ in Eq. (3.29):

$$\hat{\mathbf{U}}_2 = \gamma\mathbf{U}_1 - \gamma\frac{1}{\mathbf{q}_t^T \mathbf{p}_t}\left(\mathbf{U}_1\mathbf{q}_t\mathbf{p}_t^T + \mathbf{p}_t\mathbf{q}_t^T\mathbf{U}_1\right) + \frac{1}{\mathbf{q}_t^T \mathbf{p}_t}\left(1 + \gamma\frac{\mathbf{q}_t^T\mathbf{U}_1\mathbf{q}_t}{\mathbf{q}_t^T \mathbf{p}_t}\mathbf{p}_t\mathbf{p}_t^T\right), \tag{3.31}$$

where $\gamma = \mathbf{q}_t^T\mathbf{p}_t/\mathbf{q}_t^T\mathbf{U}_1\mathbf{q}_t$, and $\mathbf{U}_1$ is a diagonal preconditioning matrix rather than the identity matrix.

3.2.3   The L-BFGS Quasi-Newton Algorithm

The LMQN algorithm L-BFGS by Liu and Nocedal [126] is implemented in the Harwell Software library and accommodates variable storage, which is very important in practice for the minimization of large-scale systems. Instead of a restart procedure, the L-BFGS algorithm generates updated matrices using information from the last m quasi-Newton (Q-N) iterations, where m is the number of Q-N updates supplied by the user (usually $3 \leq m \leq 7$). After $2Nm$ storage locations are exhausted, the Q-N matrix is updated by replacing the oldest by the newest information. This way, the Q-N approximation of the inverse Hessian matrix is continuously updated.

The matrix $\mathbf{H}_{k+1}$ in Eq. (3.30) is obtained by updating $\mathbf{H}_0$, $\hat{m} + 1$ times, where $\hat{m} = min\{k, m - 1\}$, by using the vector pairs $(\mathbf{q}_j, \mathbf{p}_j)_{j=k-\hat{m}}^k$, where $\mathbf{p}_k = \mathbf{x}_{k+1} - \mathbf{x}_k$, $\mathbf{q}_k = \mathbf{g}_{k+1} - \mathbf{g}_k$, such that

$$\mathbf{H}_{k+1} = \left(\mathbf{v}_k^T \cdots \mathbf{v}_{k-\hat{m}}^T\right) \mathbf{H}_0 \left(\mathbf{v}_{k-\hat{m}} \cdots \mathbf{v}_k\right)$$
$$+ \, \rho_{k-\hat{m}} \left(\mathbf{v}_k^T \cdots \mathbf{v}_{k-\hat{m}+1}^T\right) \mathbf{p}_{k-\hat{m}} \mathbf{p}_{k-\hat{m}}^T \left(\mathbf{v}_{k-\hat{m}+1} \cdots \mathbf{v}_k\right)$$
$$+ \, \rho_{k-\hat{m}+1} \left(\mathbf{v}_k^T \cdots \mathbf{v}_{k-\hat{m}+2}^T\right) \mathbf{p}_{k-\hat{m}+1} \mathbf{p}_{k-\hat{m}+1}^T \left(\mathbf{v}_{k-\hat{m}+2} \cdots \mathbf{v}_k\right)$$
$$\cdots$$
$$+ \, \rho_k \mathbf{p}_k \mathbf{p}_k^T, \tag{3.32}$$

where $\rho_k = 1/(\mathbf{q}_k^T \mathbf{p}_k)$, $\mathbf{v}_k = \mathbf{I} - \rho_k \mathbf{q}_k \mathbf{p}_k^T$, and $\mathbf{I}$ is the identity matrix.

Two options for the above procedure are implemented in the L-BFGS algorithm, as follows:

1. a more accurate line-search can be performed by using a small value for β (e.g., $\beta = 10^{-2}$ or $\beta = 10^{-3}$) in Eq. (3.20), which is advantageous when the evaluation of the functional and its gradient are inexpensive; and

2. a simple scaling can be used to reduce the number of iterations.

Within the algorithm, it is preferable to replace $\mathbf{H}_0$ in Eq. (3.14) by $\mathbf{H}_k^0$, so that $\mathbf{H}_0$ incorporates more up-to-date information, according to one of the following scaling options: (i) no scaling: $\mathbf{H}_k^0 = \mathbf{H}_0$; (ii) only initial scaling: $\mathbf{H}_k^0 = \gamma_0 \mathbf{H}_0$, $\gamma_0 = \mathbf{q}_0^T \mathbf{p}_0 / \|\mathbf{q}_0\|^2$; (iii) repeated scaling: $\mathbf{H}_k^0 = \gamma_k \mathbf{H}_0$, $\gamma_k = \mathbf{q}_k^T \mathbf{p}_k / \|\mathbf{q}_k\|^2$. Liu and Nocedal [126] have reported that the last option provides the most effective scaling.

3.2.4 The BBVSCG Algorithm

The BBVSCG algorithm implements the LMQN method of Buckley and Lenir [21], may be viewed as an extension of Shanno [188] CONMIN algorithm, since it accommodates extra storage. The BBVSCG algorithm begins by performing the BFGS Q-N update algorithm and retains the current BFGS approximation to the inverse Hessian matrix as a preconditioning matrix when all available storage is exhausted. The BBVSCG algorithm then continues by performing preconditioned memoryless Q-N steps, equivalent to the preconditioned CG method with exact line-searches. The memoryless Q-N steps are repeated until the criterion of Powell [164] indicates that a restart is required. At that time, all the BFGS corrections are discarded and a new approximation to the preconditioning matrix is initiated.

For the line-search, a step-size of $\alpha = 1$ is tried when $k \leq m$. A line-search using cubic interpolation is used only if the new point does not satisfy $\mathbf{p}_k^T \mathbf{q}_k > 0$. On the other hand, a step-size $\alpha_k = -\mathbf{g}_k^T \mathbf{d}_k / \mathbf{d}_k^T \mathbf{H}_k \mathbf{d}_k$ is chosen when $k > m$. At least one quadratic interpolation is performed before α_k is accepted. The search direction $\mathbf{d}_{k+1} = -\mathbf{H}_k \mathbf{g}_k$ is computed using a matrix $\mathbf{H}_k$ obtained as follows:

1. if $k = 1$, use a scaled Q-N BFGS formula

$$\mathbf{H}_1 = \Phi_0 - \frac{\Phi_0 \mathbf{q}_k \mathbf{p}_k^T + \mathbf{p}_k \mathbf{q}_k^T \Phi_0}{\mathbf{p}_k^T \mathbf{q}_k} + \left(1 + \frac{\mathbf{q}_k^T \Phi_0 \mathbf{q}_k}{\mathbf{p}_k^T \mathbf{q}_k}\right) \frac{\mathbf{p}_k \mathbf{p}_k^T}{\mathbf{p}_k^T \mathbf{q}_k}, \quad (3.33)$$

where $\Phi_0 = (\omega_0 / v_0) \mathbf{H}_0$, $\omega_0 = \mathbf{p}_0^T \mathbf{q}_0$ and $v_0 = \mathbf{q}_0^T \mathbf{H}_0 \mathbf{q}_0$;

2. if $1 < k \leq m$, use the Q-N BFGS formula

$$\mathbf{H}_k = \mathbf{H}_{k-1} - \frac{\mathbf{H}_{k-1} \mathbf{q}_k \mathbf{p}_k^T + \mathbf{p}_k \mathbf{q}_k^T \mathbf{H}_{k-1}}{\mathbf{p}_k^T \mathbf{q}_k} + \left(1 + \frac{\mathbf{q}_k^T \mathbf{H}_{k-1} \mathbf{q}_k}{\mathbf{p}_k^T \mathbf{q}_k}\right) \frac{\mathbf{p}_k \mathbf{p}_k^T}{\mathbf{p}_k^T \mathbf{q}_k};$$

$$(3.34)$$

3. if $k > m$, use the preconditioned memoryless Q-N formula

$$\mathbf{H}_k = \mathbf{H}_m - \frac{\mathbf{p}_k \mathbf{q}_k^T \mathbf{H}_m + \mathbf{H}_m \mathbf{q}_k \mathbf{p}_k^T}{\mathbf{p}_k \mathbf{q}_k^T} + \left(1 + \frac{\mathbf{q}_k^T \mathbf{H}_m \mathbf{q}_k}{\mathbf{p}_k \mathbf{q}_k^T}\right) \frac{\mathbf{p}_k \mathbf{p}_k^T}{\mathbf{p}_k \mathbf{q}_k^T},$$

$$(3.35)$$

where $\mathbf{H}_m$ is used as a preconditioning matrix. The matrix $\mathbf{H}_k$ need not be stored since only matrix-vector products $(\mathbf{H}_k \mathbf{v})$ are required, which are calculated from

$$\mathbf{H}_k \mathbf{v} = \mathbf{H}_q \mathbf{v} - \left[\frac{\mathbf{u}_k^T \mathbf{v}}{\omega_k} - \left(1 + \frac{v_k}{\omega_k}\right) \frac{\mathbf{p}_k^T \mathbf{v}}{\omega_k}\right] - \frac{\mathbf{p}_k^T \mathbf{v}}{\omega_k} \mathbf{u}_k, \quad (3.36)$$

where $v_k = \mathbf{q}_k^T \mathbf{H}_q \mathbf{q}_k$, $\omega_k = \mathbf{p}_k^T \mathbf{q}_k$, and $\mathbf{u}_k = \mathbf{H}_q \mathbf{q}_k$. The subscript q denotes either $k-1$ or m depending on whether $k \leq m$ or $k > m$. Recursive application of the above relation yields

$$\mathbf{H}_q\mathbf{v} = \mathbf{H}_0\mathbf{v} - \sum_{j=1}^{q}\left\{\left[\frac{\mathbf{u}_j^T\mathbf{v}}{\omega_j} - \left(1+\frac{v_j}{\omega_j}\right)\frac{\mathbf{p}_j^T\mathbf{v}}{\omega_j}\right] - \frac{\mathbf{p}_j^T\mathbf{v}}{\omega_j}\mathbf{u}_j\right\}. \qquad (3.37)$$

The total storage required for the matrices $\mathbf{H}_1, \ldots, \mathbf{H}_m$ amounts to $m(2N+2)$ locations. If $k > m$, a restart will take place if the Powell [164] criteria in Eqs. (3.21) and (3.18) are satisfied; in such as case, $\mathbf{H}_m$ is discarded, k is set to 1, and the algorithm is restarted from step 1.

Both the L-BFGS and Buckley and Lenir [21] methods allow the user to specify the number of Q-N updates m. When $m=1$, BBVSCG reduces to CONMIN, whereas when $m = \infty$ both L-BFGS and the Buckley-Lenir's method are identical to the Q-N BFGS method (implemented in the CONMIN-BFGS code).

Among the limited memory methods, L-BFGS has become the leading algorithm for large-scale applications, since it requires few vectors of memory, no matrix storage, and converges almost as rapidly as the Q-N BFGS algorithm but without requiring the storage of huge matrices. Currently, the L-BFGS algorithm is the widest used minimization algorithm at all operational centers where 4–D VAR is implemented for numerical weather prediction. On the other hand, the CONMIN algorithm is still widely used for minimizing 2-D problems.

3.3 Truncated-Newton (T-N) Methods

Just as LMQN methods attempt to combine the modest storage and computational requirements of CG COMIN methods with the convergence properties of the standard Q-N methods, *truncated-Newton* (T-N) methods attempt to retain the rapid (quadratic) convergence rate of classic Newton methods while making the storage and computational requirements feasible for large-scale applications (as shown, e.g., by Dembo and Steihaug [52]). Recall that Newton methods for minimizing a multivariate function $J(\mathbf{x}_k)$ are iterative techniques based on minimizing a local quadratic approximation to J at every step. The quadratic approximation of J at a point $\mathbf{x}_k$ along the direction of a vector $\mathbf{d}_k$ can be written as

$$J\left(\mathbf{x}_k + \mathbf{d}_k\right) \approx J\left(\mathbf{x}_k\right) + \mathbf{g}_k^T \mathbf{d}_k + \frac{1}{2}\mathbf{d}_k^T \mathbf{H}_k \mathbf{d}_k, \qquad (3.38)$$

where $\mathbf{g}_k$ and $\mathbf{H}_k$ denote the gradient and Hessian, respectively, of J at $\mathbf{x}_k$. Minimization of this quadratic approximation produces a linear system of equations (called the Newton equations) for the search vector $\mathbf{d}_k$:

$$\mathbf{H}_k \mathbf{d}_k = -\mathbf{g}_k. \qquad (3.39)$$

In the modified Newton framework, developed to guarantee global convergence (i.e., convergence to a local minimum from any starting point $\mathbf{x}_0$), a sequence of iterates is generated from the starting point $\mathbf{x}_0$ by using the relation $\mathbf{x}_{k+1} = \mathbf{x}_k + \alpha_k \mathbf{d}_k$. The vector $\mathbf{d}_k$ is a search vector that leads to a minimum, or an approximate minimum, of the right side of Eq. (3.38), while the scalar $\alpha_k > 0$ is computed to ensure sufficient decrease along $\mathbf{d}_k$. The vector $\mathbf{d}_k$ can be obtained as the solution (or approximate solution) of the system shown in Eq. (3.39) or, possibly, a modified version of it, where the Hessian $\mathbf{H}_k$ is replaced by some positive definite approximation, $\hat{\mathbf{H}}_k$. Note that $\mathbf{H}_k$ in Eq. (3.39) denotes the Hessian matrix rather than the inverse of the Hessian, as it customarily appears in the basic Q-N structure.

When an approximate solution is involved, the method is called a "truncated" Newton method, because the solution process of Eq. (3.39) is not carried to completion. In such a case, $\mathbf{d}_k$ is considered satisfactory when the residual vector $\mathbf{r}_k = \mathbf{H}_k \mathbf{d}_k + \mathbf{g}_k$ is sufficiently small. Truncation is justified by arguing that accurate search directions are not essential in regions far away from local minima. For such regions, any descent direction suffices, so the effort expanded in solving the system accurately is often unwarranted. However, as a solution of the optimization problem is being approached, the quadratic approximation in Eq. (3.38) is likely to become more accurate and a smaller residual is more important. Thus, the criterion for truncation should be chosen to enforce systematically a smaller residual as the minimization process advances. An effective strategy is to require that

$$\|\mathbf{r}_k\| \leq \eta_k \|\mathbf{g}_k\|, \quad where \ \eta_k = \min\left\{\frac{c}{k}, \|\mathbf{g}_k\|\right\}, \quad c \leq 1. \tag{3.40}$$

Dembo and Steihaug [52] have shown that the above strategy maintains quadratic convergence. Other truncation criteria have been discussed e.g., by Schlick and Fogelson [179] and by Nash [142] [143].

The quadratic subproblem of computing an approximate search direction at each step is accomplished by using an iterative scheme. This produces a nested iteration structure: an "outer" loop for updating $\mathbf{x}_k$, and an "inner" loop for computing $\mathbf{d}_k$. The linear CG method is attractive for the inner loop for large-scale problems because of its modest computational requirements and theoretical convergence in at most N iterations as shown by Golub and Van Loan [79]. However, since the original CG methods were developed for positive definite systems, they must be modified for minimizing Eq. (3.38), in which the Hessian may be indefinite. Typically, this situation is handled by terminating the inner loop (at iteration q) when a direction of negative curvature is detected (i.e., $\mathbf{d}_q^T \mathbf{H}_k \mathbf{d}_q < \xi$ where ξ denotes a small positive tolerance, e.g., 10^{-10}). Subsequently, a direction that is guaranteed to be a descent direction is chosen for the search direction, as proposed by Schlick and Fogelson in [179] and [180]. An alternative to the linear CG procedure for the inner loop can be constructed based on the Lanczos factorization, as proposed by Golub and Van Loan [79], which works for symmetric but not necessarily positive definite systems. It is important to note that different procedures for the inner-iteration loop can lead to very different overall minimization performances: different search directions may result, and the overall minimization performance is cumulative.

Note that the Newton equations in Eq. (3.39) constitute a linear system of the form $\mathbf{Ax} = \mathbf{b}$, where $\mathbf{A} = \nabla^2 J(\mathbf{x}_k)$ and $\mathbf{b} = -\nabla J(\mathbf{x}_k)$. To solve this system repeatedly, it is necessary to compute efficiently matrix-vector products of the type $\mathbf{Av} = \nabla^2 J(\mathbf{x}_k)\mathbf{v}$, for arbitrary vectors $\mathbf{v}$, while avoiding the need to store and/or compute the full Hessian matrix. One way of achieving this goal, even if just approximately, is to approximate the Hessian matrix-vector products by using values of the gradient in a first-order finite-difference scheme of the form

$$\nabla^2 J\left(\mathbf{x}_k\right) \mathbf{v} = \lim_{h \to 0} \frac{\nabla J\left(\mathbf{x}_k + h\mathbf{v}\right) - \nabla J\left(\mathbf{x}_k\right)}{h} \approx \frac{\nabla J\left(\mathbf{x}_k + h\mathbf{v}\right) - \nabla J\left(\mathbf{x}_k\right)}{h},$$
(3.41)

for some small values of h. The task of choosing an adequate h is difficult; Wang et al. [213] propose that h be taken as the square root of machine accuracy.

In-depth descriptions of the truncated-Newton (also referred to as the Hessian-free) method are provided in the works of Nash ([139], [140], [141], [142], [143]) and Nash and Sofer ([146], [147]), as well as Schlick and Fogelson in [179] and [180]. Nash and Nocedal [145] provide a comparison between the Limited Memory Quasi-Newton and Truncated-Newton methods. Zou et al. [230] present a comparison between LMQN and T-N methods applied to a 4–D VAR data assimilation meteorological problem. Work by Wang et al. ([212], [213]) showed the benefits and limitations of using the T-N methods with second-order adjoint Hessian vector information.

The T-N methods retain a quasi-quadratic convergence rate while needing only storage of vectors when using a finite difference approximation to the Hessian-vector products. The T-N algorithms achieve almost a quadratic rate of convergence when used in conjunction with second-order adjoint methods. The T-N methods offer competitive alternatives for two-dimensional (2-D) problems but are not yet competitive for 3-D operational problems, due to the high cost of the CG inner iterations, which require forward and backward adjoint iterations, thus offsetting the quadratic convergence rate advantage. For this reason, L-BFGS methods are preferred for minimization of large-scale systems.

3.4   Hessian Information in Optimization

Hessian information is essential for several fundamental aspects in both constrained and unconstrained minimization. All minimization methods start by assuming a quadratic model around the stationary point of the multivariate problem

$$\min_{\mathbf{x} \in R^n} F(\mathbf{x}). \tag{3.42}$$

The necessary condition for $\mathbf{x}^*$ to be a stationary point is

$$\nabla F(\mathbf{x}^*) = \mathbf{0}, \tag{3.43}$$

while the sufficient condition for the existence of the minimum of the multivariate unconstrained minimization problem is that the Hessian at $\mathbf{x}^*$ be positive definite.

3.4.1 Hessian's Spectrum: Convergence Rate in Unconstrained Minimization

The eigenvalues of the Hessian matrix determine the convergence rate for unconstrained minimization. This property can be demonstrated by considering $\mathbf{x}^*$ to denote a local minimizer of $F(\mathbf{x})$ satisfying the condition

$$F(\mathbf{x}^*) \leq F(\mathbf{x}), \tag{3.44}$$

for all $\mathbf{x}$ such that

$$|\mathbf{x} - \mathbf{x}^*| < \varepsilon, \tag{3.45}$$

where ε is typically a small positive number whose value may depend on the value of $\mathbf{x}^*$. If F is twice continuously differentiable, and $\mathbf{x}^*$ is an absolute minimum, then

$$\nabla F(\mathbf{x}^*) = \mathbf{0}, \tag{3.46}$$

and the Hessian $\mathbf{H}(\mathbf{x}^*)$ of F at $\mathbf{x}^*$ is positive definite, i.e.

$$\mathbf{p}^T \mathbf{H}(\mathbf{x}^*)\mathbf{p} > 0, \quad \forall \mathbf{p} \in R^n. \tag{3.47}$$

Expanding $F(\mathbf{x})$ in a Taylor series about $\mathbf{x}^*$ gives

$$F(\mathbf{x}) = F(\mathbf{x}^* + h\mathbf{p}) = F(\mathbf{x}^*) + \frac{1}{2}h^2\mathbf{p}^T\mathbf{H}(\mathbf{x}^*)\mathbf{p} + O(h^2), \tag{3.48}$$

where

$$\|\mathbf{p}\| = 1 \quad \text{and} \quad h = |\mathbf{x} - \mathbf{x}^*|. \tag{3.49}$$

For any acceptable solution, Eq. (3.48) indicates that

$$h^2 = |\mathbf{x} - \mathbf{x}^*|^2 \approx \frac{2\varepsilon}{\mathbf{p}^T \mathbf{H}(\mathbf{x}^*)\mathbf{p}}, \tag{3.50}$$

which implies that the Hessian substantially affects the size of $|\mathbf{x} - \mathbf{x}^*|$ and hence the rate of convergence of the unconstrained minimization problem. If $\mathbf{H}(\mathbf{x}^*)$ is ill conditioned, the error in $\mathbf{x}$ will vary with the direction of the perturbation $\mathbf{p}$.

If $\mathbf{p}$ is a linear combination of eigenvectors of $\mathbf{H}(\mathbf{x}^*)$ corresponding to the largest eigenvalues, the size of $|\mathbf{x} - \mathbf{x}^*|$ will be relatively small and consequently the convergence will be relatively fast. On the other hand, if $\mathbf{p}$ is a linear combination of eigenvectors of $\mathbf{H}(\mathbf{x}^*)$ corresponding to the smallest eigenvalues, the size of $|\mathbf{x} - \mathbf{x}^*|$ will be relatively large, and consequently the convergence will be slow.

The above considerations can be illustrated by considering the quadratic functional

$$f(\mathbf{x}) = \mathbf{c} + \mathbf{a}^T \mathbf{x} + \frac{1}{2}\mathbf{x}^T \mathbf{Q}\mathbf{x}, \tag{3.51}$$

where $\mathbf{Q}$ is a symmetric positive definite matrix. Luenberger [128] has shown that the rate of convergence of the method of steepest descent is governed by the relation

$$f(\mathbf{x}_{k+1}) - f(\mathbf{x}_k) \le \left(\frac{\sigma_1 - \sigma_n}{\sigma_1 + \sigma_n}\right)^2 [f(\mathbf{x}_k) - f(\mathbf{x}^*)], \tag{3.52}$$

where x^* is the exact solution of the quadratic function $f(\mathbf{x})$ where σ_1 and σ_n are the largest and the smallest eigenvalues of the matrix $\mathbf{Q}$, respectively. The larger the condition number of $\mathbf{Q}$, $\kappa(\mathbf{Q}) = \frac{\sigma_1}{\sigma_n}$, the smaller the ratio $\frac{\sigma_1 - \sigma_n}{\sigma_1 + \sigma_n}$, and consequently the slower the convergence rate of the method of steepest descent.

Alekseev and Navon have illustrated in [2] and [3] how to use information provided by the eigenvalues of the Hessian for the wavelet regularization of an ill-posed adjoint estimation of inflow parameters from down-flow data, in an inverse convection problem involving the two-dimensional parabolized Navier-Stokes equations. The wavelet method decomposed the problem into two subspaces, identifying both a well-posed and an ill-posed subspace. The scale of the ill-posed subspace was determined by finding the minimal eigenvalues of the Hessian of a cost functional measuring the discrepancy between model prediction and observed parameters. The control space was transformed into a wavelet space, and the Hessian of the cost functional was obtained from a discrete differentiation of the gradients obtained from the first-order adjoint model, and also by using the full second-order adjoint model. The minimum eigenvalues of the Hessian were obtained either by using the Rayleigh quotient or by employing a shifted iteration method, as shown in [231] and [232]. The numerical results thus obtained showed that if the Hessian minimal eigenvalue is greater or equal to the square of the data error dispersion, the problem can be considered to be well-posed. However, when the minimal Hessian eigenvalue is less than the square of the data error dispersion of the problem, the regularization fails.

3.4.2   Role of the Hessian in Constrained Minimization

The information provided by the Hessian also plays a very important role in constrained optimization, as can be illustrated by considering the minimization of an objective function F subject to linear equality constraints h_i. Such a problem can be stated as

$$\min_{\mathbf{x} \in R^n} F(\mathbf{x}), \qquad \text{subject to} \ \ \mathbf{Ax} = \mathbf{b}, \tag{3.53}$$

where F is assumed twice continuously differentiable, and $\mathbf{A}$ is a $m \times n$ matrix, $m \leq n$, with full row rank (i.e., the rows of $\mathbf{A}$ are independent). The feasible region consists of the set of points satisfying all constraints.

Any problem with linear constraints $\mathbf{Ax} = \mathbf{b}$ can be recast as an equivalent unconstrained problem, as follows: assume that a feasible point, $\bar{\mathbf{x}}$, exists, i.e., $\mathbf{A}\bar{\mathbf{x}} = \mathbf{b}$. Then any other feasible point can be expressed in the form $\mathbf{x} = \bar{\mathbf{x}} + \mathbf{p}$, where $\mathbf{p}$ is a feasible direction. Any feasible direction must lie in the null space of $\mathbf{A}$, i.e., the set of vectors $\mathbf{p}$ must satisfy $\mathbf{Ap} = \mathbf{0}$. In other words, the feasible region is given by $\{\mathbf{x} : \mathbf{x} = \bar{\mathbf{x}} + \mathbf{p}, \mathbf{p} \in N(\mathbf{A})\}$, where $N(\mathbf{A})$ denotes the null space of $\mathbf{A}$. Representing $N(\mathbf{A})$ by using the matrix $\mathbf{Z}$ of dimension $n \times r$ with $r \geq n - m$, the feasible region is defined by

$$\{\mathbf{x} : \mathbf{x} = \bar{\mathbf{x}} + \mathbf{Zv}, \text{ with } \ \mathbf{v} \in R^r\}, \quad \text{and} \quad \mathbf{AZ} = \mathbf{0}. \tag{3.54}$$

Consequently, the constrained minimization problem in $\mathbf{X}$ is equivalent to the unconstrained problem

$$\min_{\mathbf{v}\in R^r} \Phi\left(\mathbf{v}\right) = F\left(\mathbf{x} + \mathbf{Z}\mathbf{v}\right), \tag{3.55}$$

where $\mathbf{x}$ is a feasible point (see, e.g., refs. [76] and [148]). The function Φ is the restriction of F onto the feasible region, and is therefore called the *reduced function*. If $\mathbf{Z}$ is a basis matrix for the null space of $\mathbf{A}$, then Φ is a function of $n - m$ variables. The constrained problem has thus been transformed into an unconstrained problem with a reduced number of variables, involving the derivatives of the reduced function. If $\mathbf{x} = \bar{\mathbf{x}} + \mathbf{Z}\mathbf{v}$, then

$$\nabla\Phi\left(\mathbf{v}\right) = \mathbf{Z}^T\nabla F\left(\bar{\mathbf{x}} + \mathbf{Z}\mathbf{v}\right) = \mathbf{Z}^T\nabla F\left(\mathbf{x}\right), \tag{3.56}$$

and

$$\nabla^2\Phi\left(\mathbf{v}\right) = \mathbf{Z}^T\nabla^2 F\left(\bar{\mathbf{x}} + \mathbf{Z}\mathbf{v}\right)\mathbf{Z} = \mathbf{Z}^T\nabla^2 F\left(\mathbf{x}\right)\mathbf{Z}. \tag{3.57}$$

The vector $\nabla\Phi\left(\mathbf{v}\right) = \mathbf{Z}^T\nabla F\left(\mathbf{x}\right)$ is called the *reduced gradient* of F at $\mathbf{x}$, while the matrix $\nabla^2\Phi\left(\mathbf{v}\right) = \mathbf{Z}^T\nabla^2 F\left(\mathbf{x}\right)\mathbf{Z}$ is called the *reduced or projected Hessian* matrix. The reduced gradient and Hessian matrix are the gradient and Hessian, respectively, of the restriction of F onto the feasible region evaluated at $\mathbf{x}$. If $\mathbf{x}^*$ is a local solution of the constrained problem, then

$$\mathbf{x}^* = \bar{\mathbf{x}} + \mathbf{Z}\mathbf{v}^*, \qquad \text{for some } \mathbf{v}^*, \tag{3.58}$$

and therefore $\mathbf{v}^*$ is the local minimizer of Φ. It consequently follows that

$$\nabla\Phi\left(\mathbf{v}^*\right) = \mathbf{0}, \tag{3.59}$$

and $\nabla^2\Phi\left(\mathbf{v}^*\right)$ is positive semidefinite. Equivalently, if $\mathbf{x}^*$ is a local minimizer of F, and $\mathbf{Z}$ is the null-space matrix for $\mathbf{A}$, then the reduced Hessian matrix $\mathbf{Z}^T\nabla^2 F\left(\mathbf{x}^*\right)\mathbf{Z}$ is positive semidefinite and the reduced gradient vanishes at $\mathbf{x}^*$, i.e.,

$$\mathbf{Z}^T\nabla F\left(\mathbf{x}^*\right) = \mathbf{0}. \tag{3.60}$$

The second-order condition, which is used to distinguish local minimizers from other stationary points, is equivalent to the condition

$$\mathbf{v}^T \mathbf{Z}^T \nabla^2 F(\mathbf{x}^*) \mathbf{Z} \mathbf{v} \geq 0, \qquad \text{for all } \mathbf{v}. \tag{3.61}$$

Noting that $\mathbf{p} = \mathbf{Z}\mathbf{v}$ is a null space vector, the above inequality can also be written in the form

$$\mathbf{p}^T \nabla^2 F(\mathbf{x}^*) \mathbf{p} \geq 0, \qquad \text{for all } \mathbf{p} \in N(\mathbf{A}), \tag{3.62}$$

which indicates that the Hessian matrix at $\mathbf{x}^*$ must be positive semidefinite on the null space of $\mathbf{A}$.

3.5 Nondifferentiable Minimization: Bundle Methods

If the function F to be minimized is nonsmooth, then methods of nondifferentiable optimization are required. Since the gradient of a nonsmooth function F exists only almost anywhere, the customary gradient must be replaced by the *generalized gradient*, $\partial F(\mathbf{x})$, which is defined as the closure of the set that contains all convex linear combinations of subgradients (an element of the generalized gradient is called *subgradient*):

$$\partial F(\mathbf{x}) \equiv$$
$$\text{conv} \left\{ \begin{array}{l} g | \text{there exists a sequence } (\mathbf{x}_i)_{i \in \mathbf{N}} \text{ such that } \lim_{i \to \infty} \mathbf{x}_i = \mathbf{x}, \\ F \text{ differentiable at } \mathbf{x}_i, i \in \mathbf{N}, \text{ and } \lim_{i \to \infty} \nabla F(\mathbf{x}_i) = \mathbf{g}. \end{array} \right\} \tag{3.63}$$

where "*conv*" denotes *convex hull*. Nonsmooth optimization methods are based on the assumptions that: (i) the function F is locally Lipschitz continuous, and (ii) the function and its arbitrary subgradient can be evaluated at each point. Nonsmooth optimization methods can be divided into two main classes: *subgradient* methods and *bundle* methods.

The basic idea underlying subgradient methods is to replace the gradient $\nabla F(\mathbf{x}_k)$ for $\mathbf{d}_k$ by a normalized subgradient in the formula for generating search lines $\mathbf{d}_k$, by setting $\mathbf{d}_k = -\boldsymbol{\xi}_k / \|\boldsymbol{\xi}_k\|$, $\boldsymbol{\xi}_k \in \partial F(\mathbf{x}_k)$. This strategy of generating $\mathbf{d}_k$ does not ensure descent and hence minimizing line-searches becomes unrealistic. In addition, the standard stopping criterion can no longer be applied since an arbitrary subgradient contains no information on the optimality condition $0 \in \partial F(\mathbf{x})$. Due to these facts, it becomes necessary to choose *a priori* the step-sizes t_k in order to avoid line-searches and stopping criteria. Consequently, the successive iteration points are defined by the relation $\mathbf{x}_{k+1} = \mathbf{x}_k - t_k \boldsymbol{\xi}_k / \|\boldsymbol{\xi}_k\|$, for a suitably chosen $t_k > 0$. "Smoothing" the subgradient method would accelerate the rate of convergence. Currently, the most efficient subradient methods are based on generalized quasi-Newton methods, e.g., the ellipsoid and space dilation algorithms by Shor et al. [190] and the variable metric method by Uryasev [205].

The guiding principle underlying *bundle methods* is to gather the subgradient information from "previous iterations" into a *bundle* of subgradients. The pioneering bundle method, the ε-steepest descent method, was developed by Lemaréchal [121]. The main difficulty in Lemaréchal method is the *a priori* choice of an approximation tolerance which controls the radius of the ball in which the bundle model is considered to be a good approximation of the objective function.

A different approach was presented by Kiwiel [106], based on the idea of forming a convex piecewise linear approximation to the objective function by using the linearizations generated by subgradients. Kiwiel also presented two strategies to limit the number of stored subgradients: subgradient selection and aggregation. The main disadvantage of Kiwiel's method is its sensitivity to scaling of the objective function. In addition, the uncertain line-search may require, in general, many function evaluations compared to the number of iterations. Although they originated from different backgrounds, the methods of Lemaréchal [121] and Kiwiel [106] both generate a search direction by solving, at each direction, quadratic direction-finding problems. More recent methods aim at combining ideas of the bundle method with those of the trust region method, as proposed in the work by Schramm and Zowe [182], known as the bundle *trust region method*, and by Kiwiel [107], under the name of *proximal bundle method*.

The bundle methods have two characteristic features:

1. the gathering of subgradient information from past iterations into a bundle, and

2. the concept of a serious step and a null step in line-search.

These concepts can be illustrated by considering the search $\mathbf{y}_{k+1} = \mathbf{x}_k + t_k \mathbf{d}_k$ for some $t_k > 0$ and a subgradient $\boldsymbol{\xi}_{k+1} \in \partial F(\mathbf{y}_{k+1})$. Then, the following sequence is characteristic of bundle methods: (a) make a serious step $\mathbf{x}_{k+1} = \mathbf{y}_{k+1}$ if $F(\mathbf{y}_{k+1}) \leq F(\mathbf{x}_k) - \delta_k$ for a suitably chosen $\delta_k > 0$, and add $\boldsymbol{\xi}_{k+1}$ into the bundle; (b) otherwise, make a null step $\mathbf{x}_{k+1} = \mathbf{x}_k$, and add $\boldsymbol{\xi}_{k+1}$ into the bundle.

Several efficient globally convergent algorithms for nonconvex nonsmooth optimization have been developed based on versions of the bundle method, as exemplified in the works by Schramm and Zowe [182], Makela and Neittaanmäki [134], Bonnans et al. [17]. A hybrid method that combines the characteristics of the variable metric method and the bundle method has been proposed by Vlcek and Luksan [129] and [209]. This algorithm generates a sequence of basic points $(\mathbf{x}_k)_{k \in \mathbf{N}}$ and a sequence of trial points $(\mathbf{y}_k)_{k \in \mathbf{N}}$ satisfying the relations $\mathbf{x}_{k+1} = \mathbf{x}_k + t_L^k \mathbf{d}_k$, $\mathbf{y}_{k+1} = \mathbf{x}_k + t_R^k \mathbf{d}_k$, with starting values $\mathbf{y}_1 = \mathbf{x}_1$, where $t_R^k \in (0, t_{\max}]$ and $t_L^k \in (0, t_R^k]$ are appropriately chosen step-sizes, $\mathbf{d}_k = -\mathbf{H}_k \mathbf{g}_k$ is a direction vector, and $\mathbf{g}_k$ is an aggregate subgradient. The matrix $\mathbf{H}_k$ accumulates information about the previous subgradients and provides an approximation of the inverse Hessian matrix if the function F is smooth.

If the descent condition $F(\mathbf{y}_{k+1}) \leq F(\mathbf{x}_k) - c_L t_R^k \mathbf{w}_k$ is satisfied with a suitable step-size t_R^k, where $c_L \in (0, 0.5)$ is fixed and $-\mathbf{w}_k < 0$ represents the desirable amount of descent, then $\mathbf{x}_{k+1} = \mathbf{y}_{k+1}$ is taken as a descent step. Otherwise, a null step is taken, keeping the basic points unchanged but accumulating information about the minimized function.

The aggregate subgradient is constructed as follows: denote by m the lowest index j satisfying $\mathbf{x}_j = \mathbf{x}_k$ (index of the iteration after last descent step), and define $\tilde{\mathbf{g}}_{k+1} = \lambda_{k,1} \mathbf{g}_m + \lambda_{k,2} \mathbf{g}_{k+1} + \lambda_{k,3} \mathbf{g}_k$ as a convex combination of the known basic subgradient $\mathbf{g}_m \in \partial f(\mathbf{x}_k)$, the trial subgradient $\mathbf{g}_{k+1} \in \partial f(\mathbf{y}_{k+1})$, and the current aggregate subgradient $\tilde{\mathbf{g}}_k$. The multipliers λ_k can be determined by minimizing a simple quadratic function that depends on these three subgradients and two subgradient locality measures; this approach replaces the solution of a rather complicated quadratic programming problem, which appears in the standard bundle method Lemaréchal [122].

The matrices $\mathbf{H}_k$ are generated using either a symmetric quasi-Newton rank-one update after the null steps (to preserve the property of being bounded and other characteristics required for the global convergence), or using the standard BFGS update after the descent steps. Fletcher [64] presents an in-depth discussion regarding both types of updates.

Although the additional computational cost for building the subgradient is several times larger than that for L-BFGS, bundle nonsmooth optimization methods may work advantageously for problems with discontinuities, where L-BFGS methods usually fail. Bundle nonsmooth optimization methods have not been tested yet on operational 4–D VAR systems, but investigations as described by Karmitsa et al. ([102], [103]) are needed to assess the applicability of such methods to realistic large-scale problems.

3.6   Step-Size Search

The unconstrained minimization of a smooth function $F(\mathbf{x})$, $\mathbf{x} \in \mathbb{R}^n$, when its gradient $\mathbf{g} \equiv \nabla F^T$ is available, typically involves iterations of the form $\mathbf{x}_{k+1} = \mathbf{x}_k + \alpha_k \mathbf{d}_k$, where $\mathbf{d}_k$ denotes a search direction and α_k denotes the step-length obtained using a one-dimensional search. In CG methods, for instance, the search direction is $\mathbf{d}_k = -\mathbf{g}_k + \beta_k \mathbf{d}_{k-1}$, where $\mathbf{g} \equiv \nabla F_k^T$ and where the scalar β_k is chosen so that the CG method reduces to the linear CG for a quadratic $F(\mathbf{x})$ for which the line-search is exact. On the other hand, for the (full) Newton-method, the Q-N method, and the steepest descent methods, the search direction is defined as $\mathbf{d}_k = -\mathbf{B}_k^{-1}\mathbf{g}_k$, where $\mathbf{B}_k$ is a symmetric non-singular matrix. Specifically, the choices are $\mathbf{B}_k = \nabla^2 F(\mathbf{x})$ for the Newton method, and $\mathbf{B}_k = \mathbf{I}$ for the steepest descent methods, respectively. For the Q-N method, $\mathbf{B}_k$ depends on $\mathbf{x}_k$, $\mathbf{x}_{k-1}$, and $\mathbf{B}_{k-1}$. In all these methods, $\mathbf{d}_k$ must be a descent direction, i.e., it must satisfy the inequality $\mathbf{d}_k^T\mathbf{g}_k < 0$. For Newton-type methods, $\mathbf{d}_k$ is assured to be a descent direction by ensuring that $\mathbf{B}_k$ is positive definite. For CG methods, however, it is necessary to choose an adequate line-search, which depends on resolving two questions:

1. How good is the search direction?

2. What is the best choice for the length of the step along the search direction?

One indicator of the quality of a search direction is the *angle* between the steepest descent direction $-\mathbf{g}_k$ and the search direction, defined as

$$cos\theta_k = - \left(\mathbf{g}_k^T \mathbf{d}_k\right) / \|\mathbf{g}_k\| \, \|\mathbf{d}_k\| \, . \tag{3.64}$$

The *step-size* length α_k must satisfy the inequality $F\left(\mathbf{x}_k + \alpha_k\mathbf{d}_k\right) < F\left(\mathbf{x}_k\right)$, and is determined by minimizing the function $\phi_k\left(\alpha_k\right) = F\left(\mathbf{x}_k + \alpha_k\mathbf{d}_k\right)$ for a positive α. The step-size α_k is optimal when the quantity $\mathbf{x}_{k+1} = \mathbf{x}_k + \alpha_k\mathbf{d}_k$ satisfies the inequality

$$F\left(\mathbf{x}_{k+1}\right) \leq F\left(\mathbf{x}_k + \bar{\alpha}_k\mathbf{d}_k\right), \tag{3.65}$$

where $\bar{\alpha}_k$ is the smallest positive stationary point of the function ϕ_k. By definition, a step-size α_k is called "exact" if it is a stationary point of ϕ_k. Finding such an "exact" step-size is itself a nonlinear minimization problem whose solution is approximate except when F is quadratic. For this reason, an alternative strategy for accepting a positive step-length α_k has evolved based on solving two inequalities which are jointly known as the *Wolfe conditions* [217]. The *first Wolfe* condition or inequality is designed to ensure that the function is reduced sufficiently, i.e.,

$$F\left(\mathbf{x}_k + \alpha_k\mathbf{d}_k\right) < F\left(\mathbf{x}_k\right) + \sigma_1\alpha_k\mathbf{g}_k^T\mathbf{d}_k, \quad \sigma_1 = 10^{-4}. \tag{3.66}$$

Typically, the above inequality is always satisfied for some small positive α_k. The *second Wolfe* condition is designed to prevent the step-lengths from becoming too small, by means of the inequality:

$$\mathbf{g}_k^T\left(\mathbf{x}_k + \alpha_k\mathbf{d}_k\right)\mathbf{d}_k \geq \sigma_2\mathbf{g}_k^T\mathbf{d}_k, \quad 0 < \sigma_1 < \sigma_2 \simeq 0.9 < 1. \tag{3.67}$$

Wolfe has shown in [217] and [218] that if $\mathbf{d}_k$ is a descent direction and if $F\left(\mathbf{x}\right)$ is continuously differentiable and bounded along the set of directions $\{\mathbf{x}_k + \alpha_k\mathbf{d}_k | \alpha_k > 0\}$, then there always exist step-lengths satisfying the two

Wolfe inequality conditions. Other line-search strategies (e.g., backtracking) successively decrease the step-length, starting from an initial guess, until a sufficient reduction in $F(\mathbf{x})$ is obtained.

3.7   Trust Region Methods

The line-search methods reviewed in the foregoing sections used the quadratic model $F(\mathbf{x}_c + \mathbf{d}) = F(\mathbf{x}_c) + \mathbf{g}^T(\mathbf{x}_c)\mathbf{d} + (\mathbf{d}^T\mathbf{H}_c\mathbf{d})/2$ to find a search direction $\mathbf{d}_c = -\mathbf{H}_c^{-1}\mathbf{g}_c$ and subsequently choose a step-length. The procedures for choosing a step-length did not make further use of the Hessian approximation $\mathbf{H}_c$. In addition to line-search methods, *trust region methods* also seek global convergence while retaining fast local convergence of optimization algorithms. The trust region methods follow a reverse sequence of operations, by first choosing a trial step-length Δ_c, and subsequently using a quadratic model to select the best step-length by solving

$$\min_{\mathbf{s} \in R^n} F(\mathbf{x}_c + \mathbf{s}) = F(\mathbf{x}_c) + \mathbf{g}^T(\mathbf{x}_c)\mathbf{s} + (\mathbf{s}^T\mathbf{H}_c\mathbf{s})/2, \text{ subject to } \|\mathbf{s}_c\| \leq \Delta_c.$$
$$(3.68)$$

The trial step Δ_c is called a *trust radius*, indicating the extent to which the quadratic model could be trusted. The trust radius Δ_c is related to the length of the successful step at the previous iteration and may be adjusted as the current iteration proceeds. The works of Gay [72] and Sorensen [191] are the early references on trust region methods, although Celis et al. [30] later coined the name for these methods. The book by Conn et al. [35] provides a comprehensive presentation of all aspects of trust region methods.

3.8   Scaling and Preconditioning

3.8.1   Preconditioning for Linear Problems

The goal of preconditioning is to improve the performance of conjugate-gradient type minimization methods, by reducing the number of iterations required to achieve a prescribed accuracy. The convergence properties of the algorithm should be considered when selecting a preconditioner. This selection process can be illustrated by considering the linear conjugate-gradient (CG) method, which converges theoretically in a finite number of iterations, equal to the number of distinct eigenvalues of the Hessian matrix of the cost function (a cluster of eigenvalues is viewed practically as a single eigenvalue). Prior to attaining the exact solution, this algorithm displays a linear rate of convergence proportional to

$$\frac{[cond\ (\mathbf{H})]^{\frac{1}{2}} - 1}{[cond\ (\mathbf{H})]^{\frac{1}{2}} + 1}, \tag{3.69}$$

where $cond\ (\mathbf{H})$ denotes the condition number of the Hessian matrix $\mathbf{H}$. The preconditioning transformation aims at clustering (i.e., reducing the spread) and reducing the number of distinct eigenvalues of $\mathbf{H}$, and/or reducing the condition number of $\mathbf{H}$. Specifically, if the conjugate-gradient method is employed to solve the linear system

$$\mathbf{G}\mathbf{x} = \mathbf{c}, \tag{3.70}$$

where $\mathbf{G}$ is symmetric positive definite matrix, the number of iterations required is equal to the number of distinct eigenvalues of $\mathbf{G}$. The rate of convergence could be improved by replacing the original system with an equivalent system having a matrix with many unit eigenvalues. This can be achieved by introducing a positive definite and symmetric matrix $\mathbf{W}$ to recast Eq. (3.70) into the equivalent form

$$\mathbf{W}^{-1/2}\mathbf{G}\mathbf{W}^{-1/2}\mathbf{y} = -\mathbf{W}^{-1/2}\mathbf{c}, \tag{3.71}$$

which is solved for $\mathbf{x} = \mathbf{W}^{-1/2}\mathbf{y}$. The matrix $\mathbf{R} = \mathbf{W}^{-1/2}\mathbf{G}\mathbf{W}^{-1/2}$ has the same eigenvalues as $\mathbf{W}^{-1}\mathbf{G}$, since $\mathbf{W}^{-1/2}\mathbf{R}\mathbf{W}^{-1/2} = \mathbf{W}^{-1}\mathbf{G}$. The matrix $\mathbf{W}$ should be chosen so that the matrix $\mathbf{W}^{-1}\mathbf{G}$ has as many eigenvalues close to unity as possible, or, equivalently, so that the condition number *cond* $\left(\mathbf{W}^{-1}\mathbf{G}\right)$ of $\mathbf{W}^{-1}\mathbf{G}$ is as small as possible. A suitable matrix $\mathbf{W}$ can be found by performing r (where $r << n$) steps of a LMQN method. If exact line-searches are made, the LMQN matrix $\mathbf{M}$ satisfies the Q-N condition $\mathbf{s}_j = \mathbf{M}\mathbf{y}_j$ for the space r pairs of vectors $\{\mathbf{s}_j, \mathbf{y}_j\}$. If F is a quadratic function, the relation $\mathbf{G}\mathbf{s}_j = \mathbf{y}_j$ holds, which implies that $\mathbf{s}_j = \mathbf{M}\mathbf{G}\mathbf{s}_j$, which in turn implies that the matrix $\mathbf{M}\mathbf{G}$ has r unit eigenvalues with eigenvectors $\{\mathbf{s}_j\}$. Therefore, the matrix $\mathbf{M}$ can be used as $\mathbf{W}^{-1}$.

3.8.2 Preconditioning for Nonlinear Problems

For nonlinear problems, the preconditioning matrix $\mathbf{W}_k$ will vary from iteration to iteration. In many situations, a limited-memory approximate Hessian matrix $\mathbf{B}_k$ can play the role of $\mathbf{W}_k^{-1}$. A relatively simple and successful method is to use diagonal preconditioning based upon a diagonal scaling. Such methods assume that the direction of search is obtained from $\mathbf{B}_k\mathbf{d}_k = -\mathbf{g}_k$, while the off-diagonal elements of the approximate Hessian matrix $\mathbf{B}_k$ are unknown. The LMQN methods can be modified to accept a diagonal preconditioning matrix rather than the unit matrix. Nash [144] describes an effective automatic preconditioner for the inner conjugate-gradient iterations within the truncated-Newton method. Additional ways to construct preconditioners include: diagonal approximations from Q-N update formulas, sparse approximations from incomplete Cholesky, and polynomial approximations.

Scaling can also substantially improve the performance of minimization algorithms. An effective automatic scaling could also improve the condition number of the Hessian matrix for well-scaled problems, thus facilitating their solution. On the other hand, badly scaled nonlinear problems can become extremely difficult to solve, as exemplified by Navon and De Villiers [150], and Courtier and Talagrand [37]. In meteorological problems, for example, the variables in the control vector have widely different magnitudes, spanning a range of eight orders of magnitude. Scaling by variable transformations converts the variables from units that reflect the physical nature of the problem to units that display desirable properties for improving the efficiency of the minimization algorithms. The general form of a scaling procedure is

$$\mathbf{x}^s = \mathbf{S}^{-1}\mathbf{x}, \quad \mathbf{g}^s = \mathbf{Sg}, \quad \mathbf{H}^s = \mathbf{SHS}, \tag{3.72}$$

where $\mathbf{S}$ is a diagonal scaling matrix, $\mathbf{x}$ and $\mathbf{g}$ denote the state variable and its gradient, respectively, while $\mathbf{H}$ denotes the Hessian matrix. There is no general rule for choosing scaling factors. Furthermore, scaling is problem dependent, since convergence tolerances and other criteria are necessarily based upon an implicit definition of "small" and "large," so that variables with widely varying orders of magnitude may cause difficulties, as discussed by Gill et al. [76]. Scaling the variables so they become of similar magnitudes and of order unity improves the computational performance. A simple strategy is to use values that are typically for the various fields (e.g., a scaling factor of 10^{-5} can be used for the vorticity field).

3.9 Nonlinearly Constrained Minimization

The inequality-constrained problem typically assumes the form

$$\min_{\mathbf{x} \in R^n} F(\mathbf{x}), \quad subject\ to\ C_i(\mathbf{x}) \geq 0, \quad i = 1, \ldots, m. \tag{3.73}$$

In the derivations to follow in this section, the gradient of the constraint function $C_i(\mathbf{x})$ will be denoted by $\mathbf{a}_i(\mathbf{x})$, and the Hessian of $C_i(\mathbf{x})$ will be denoted by $\mathbf{G}_i(\mathbf{x})$. Furthermore, the matrix $\mathbf{A}(\mathbf{x})$ will denote a matrix whose i^{th}-row is the vector $\mathbf{a}_i^T(\mathbf{x})$.

3.9.1 Penalty and Barrier Function Methods

The *penalty method*, attributed to Courant and Hilbert [36], replaces a constrained optimization problem by a series of unconstrained problems whose solutions should converge to the solution of the original constrained problem. The unconstrained problems minimize an objective function $P(\mathbf{x}, \rho)$, which is constructed by adding to the original objective function $F(\mathbf{x})$ a term that comprises a *penalty parameter multiplying a measure of the violation of the constraints*, i.e.,

$$P\left(\mathbf{x}, \rho\right) = F\left(\mathbf{x}\right) + \frac{\rho}{2}\mathbf{C}^{T}\left(\mathbf{x}\right)\mathbf{C}\left(\mathbf{x}\right), \quad \rho = penalty\ parameter, \qquad (3.74)$$

where

$$\lim_{\rho \to \infty} \mathbf{x}^{*}(\rho) = \mathbf{x}^{*}. \qquad (3.75)$$

The measure of violation is nonzero when the constraints are not satisfied, and is zero in the region where the constraints are satisfied. This procedure makes it possible to construct a function whose constrained minimum is either $\mathbf{x}^{*}$ itself or is related to $\mathbf{x}^{*}$ in a known way. The original problem can thus be solved by formulating a sequence of unconstrained subproblems.

Barrier methods are an alternative class of algorithms for constrained optimization. These methods also use a penalty-like term added to the objective function, but the iterates within the barrier methods are forced by the barrier to remain interior to and away from the boundary of the feasible solution domain. The functional $B\left(\mathbf{x}, \gamma\right)$ to be minimized in barrier methods has the form

$$B\left(\mathbf{x}, \gamma\right) = F\left(\mathbf{x}\right) - \gamma \sum_{i=1}^{m} \ln C_{i}\left(\mathbf{x}\right), \quad \gamma = barrier\ parameter, \qquad (3.76)$$

where

$$\lim_{\gamma \to \infty} \mathbf{x}^{*}(\rho) = \mathbf{x}^{*}. \qquad (3.77)$$

3.9.2   Augmented Lagrangian Methods

Augmented Lagrangian methods transform the constrained minimization problem defined in Eq. (3.73), into the unconstrained minimization of the functional $L\left(\mathbf{x}, \lambda, \rho\right)$ defined as

$$L\left(\mathbf{x}, \lambda, \rho\right) = F\left(\mathbf{x}\right) - \lambda^{T}\mathbf{C}\left(\mathbf{x}\right) + \frac{\rho}{2}\mathbf{C}^{T}\left(\mathbf{x}\right)\mathbf{C}\left(\mathbf{x}\right). \qquad (3.78)$$

The critical point $\mathbf{x}^*$ denotes the solution of $\nabla L(\mathbf{x}^*, \lambda, \rho) = \mathbf{0}$. The Hessian of $L(\mathbf{x}, \lambda, \rho)$ is

$$\nabla^2 L(\mathbf{x}, \lambda, \rho) = \nabla^2 F(\mathbf{x}) - \sum_{i=1}^{m} [\lambda_i - C_i(\mathbf{x})] \nabla^2 C_i(\mathbf{x}) + \rho \mathbf{A}^T(\mathbf{x}) \mathbf{A}(\mathbf{x}). \quad (3.79)$$

The Hessian of the penalty term evaluated at $\mathbf{x}^*$ is $\rho \mathbf{A}^T(\mathbf{x}^*) \mathbf{A}(\mathbf{x}^*)$, a semi-definite matrix with strictly positive eigenvalues corresponding to eigenvectors in the range of $\mathbf{A}^T(\mathbf{x}^*)$. Augmented Lagrangian methods were developed independently by Powell [163] and Hestenes [87], and are sometimes called *multiplier methods*, or *Powell-Rockafellar penalty function methods*. The effect of the penalty term in L is to increase the (possibly negative) eigenvalues of $\nabla^2 L(\mathbf{x}, \lambda, \rho)$ corresponding to eigenvectors in the range of $\mathbf{A}(\mathbf{x}^*)$, but to leave the other eigenvalues unchanged. This property makes it possible to show that, under mild conditions, there exists a finite penalty parameter $\bar{\rho}$, such that the minimum point x^* is an unconstrained minimizer of $L(\mathbf{x}, \lambda, \rho)$ for all $\rho > \bar{\rho}$. Navon and De Villiers [150] discuss the use of augmented Lagrangian methods for variational data assimilation.

3.9.3 Sequential Quadratic Programming (SQP) Methods

The SQP method can be used within either a line-search or a trust region framework, and is very efficient for solving both small and large problems. For a nonlinear equality-constrained problem, SQP-methods use a quadratic model of the Lagrangian function, of the form

$$L(\mathbf{x}, \lambda) = F(\mathbf{x}) - \lambda^T \mathbf{C}(\mathbf{x}). \quad (3.80)$$

The SQP method solves a sequence of sub-problems designed to minimize a quadratic model of the objective functional $L(\mathbf{x}, \lambda)$ subject to linearization of the constraints. Thus, the SQP method comprises an algorithm of the form $\mathbf{x}_{k+1} = \mathbf{x}_k + \alpha_k \mathbf{p}_k$, where $\mathbf{p}_k$ denotes a search direction and α_k denotes a non-negative step-length. The optimal point $\mathbf{x}^*$ should be feasible, i.e., $\mathbf{C}(\mathbf{x}^*) = \mathbf{0}$. If the minimization problem is unconstrained, the SQP method reduces to Newton's method for finding the point where the gradient of the objective vanishes. If the minimization problem comprises solely equality constraints, the SQP method is equivalent to applying Newton's method to the first-order optimality conditions, for the minimization problem.

Expanding $\mathbf{C}(\mathbf{x})$ in Taylor series about $\mathbf{x}_k$ along a general vector $\mathbf{p}$ yields

$$\mathbf{C}(\mathbf{x}_k + \mathbf{p}) = \mathbf{C}_k + \mathbf{A}_k \mathbf{p} + O\left(\|\mathbf{p}\|^2\right), \tag{3.81}$$

where $\mathbf{C}_k \equiv \mathbf{C}(\mathbf{x}_k)$ and $\mathbf{A}_k \equiv \mathbf{A}(\mathbf{x}_k)$. The desired search direction $\mathbf{p}_k$ will be a step towards a zero of a local linear approximation to $\mathbf{C}(\mathbf{x})$ if $\mathbf{C}_k + \mathbf{A}_k \mathbf{p}_k = \mathbf{0}$; this relation defines a set of linear equality constraints to be satisfied by $\mathbf{p}_k$. Hence, $\mathbf{p}_k$ can be determined by solving the following equality-constrained quadratic program:

$$\min_{\mathbf{p} \in R^n} \mathbf{g}_k^T \mathbf{p} + \frac{1}{2} \mathbf{p}^T \mathbf{B}_k \mathbf{p}, \quad \text{subject to} \quad \mathbf{A}_k \mathbf{p} = -\mathbf{C}_k. \tag{3.82}$$

The step-length α_k for a nonlinearly constrained problem is chosen to yield "sufficient decrease," in the sense of Ortega and Rheinboldt [156], in a *merit function* measuring progress toward the solution of the *nonlinear equality constrained* problem. Typical merit functions are the L_1-penalty function

$$M_1(\mathbf{x}, \rho) \equiv F(\mathbf{x}) + \rho \|C(\mathbf{x})\|_1, \tag{3.83}$$

and the augmented Lagrangian functional

$$M(\mathbf{x}, \lambda, \rho) = F(\mathbf{x}) - \lambda^T C(\mathbf{x}) + \frac{\rho}{2} \mathbf{C}^T(\mathbf{x}) \mathbf{C}(\mathbf{x}). \tag{3.84}$$

The principle underlying SQP can be illustrated by considering the equality-constrained problem

$$\min F(\mathbf{x}), \quad \text{subject to } \mathbf{C}(\mathbf{x}) = \mathbf{0}, \tag{3.85}$$

where $F : \mathbb{R}^n \to \mathbb{R}$ and $\mathbf{C} : \mathbb{R}^n \to \mathbb{R}^m$ are smooth functions of $\mathbf{x}$. The Lagrangian for the problem defined in Eq. (3.85) is defined as

$$L(\mathbf{x}, \lambda) \equiv F(\mathbf{x}) - \lambda^T \mathbf{C}(\mathbf{x}). \tag{3.86}$$

The essential idea underlying SQP is to model the problem in Eq. (3.86) at the current iterate $(\mathbf{x}_k, \lambda_k)$ by a quadratic programming subproblem, and to use the minimizer of this subproblem to define a new iterate $\mathbf{x}_{k+1}$. The challenge is to design the quadratic subproblem so that it yields a good step for the underlying constrained optimization problem while ensuring that the overall SQP algorithm has good convergence properties and good practical performance. Let us denote the Hessian of the Lagrangian by $\mathbf{W}(\mathbf{x}, \lambda) = \nabla^2_{\mathbf{xx}} L(\mathbf{x}, \lambda)$. Let us also denote the Jacobian matrix of the constraints C_i, $i = 1, \ldots, m$, by $\mathbf{A}^T(\mathbf{x}) = [\nabla C_1(\mathbf{x}), \nabla C_2(\mathbf{x}), \ldots, \nabla C_m(\mathbf{x})]$. Thus, we solve the quadratic subproblem

$$\min_{p} \left(\frac{1}{2} \mathbf{p}^T \mathbf{W}_k \mathbf{p} + \nabla F_k^T \mathbf{p} \right), \quad \text{subject to } \mathbf{A}_k \mathbf{p} + \mathbf{C}_k = \mathbf{0}, \qquad (3.87)$$

at iteration $(\mathbf{x}_k, \lambda_k)$, where $\mathbf{A}_k$ and $\mathbf{W}_k$ denote approximations for $\mathbf{A}$ and $\mathbf{W}$, respectively. If the Jacobian $\mathbf{A}_k$ of the constraints has full row rank, and the matrix $\mathbf{W}_k$ satisfies the inequality $\mathbf{d}^T \mathbf{W}_k \mathbf{d} > 0$ on the tangent space of constraints (i.e., for all $\mathbf{d} \neq \mathbf{0}$ such that $\mathbf{A}_k \mathbf{d} = \mathbf{0}$), then the problem in Eq. (3.87) has a unique solution $(\mathbf{p}_k, \lambda_{k+1})$ which satisfies the system of equations

$$\begin{bmatrix} \mathbf{W}_k & -\mathbf{A}_k^T \\ \mathbf{A}_k & \mathbf{0} \end{bmatrix} \begin{bmatrix} \mathbf{p}_k \\ \lambda_{k+1} \end{bmatrix} = \begin{bmatrix} -\nabla F_k \\ -\mathbf{C}_k \end{bmatrix}. \qquad (3.88)$$

To be practical, a SQP method must be able to converge on nonconvex problems, starting from remote points. When $\mathbf{W}_k$ is positive definite on the tangent space of constraints, the quadratic subproblem in Eq. (3.87) can be solved without any additional requirements. When $\mathbf{W}_k$ does not have this property, line-search methods either replace $\mathbf{W}_k$ by a positive definite approximation $\mathbf{B}_k$ or modify $\mathbf{W}_k$ directly during the process of matrix factorization. Trust region methods provide yet another approach by adding a constraint to the subproblem, limiting the step to a region where the model in Eq. (3.87) is considered to be reliable. Difficulties may arise, however, because the inclusion of the trust region may cause the subproblem to become infeasible, thereby forcing a relaxation of the constraints at some iterations, which complicates the algorithm and increases its computational cost. Due to these trade-offs, neither one of the two SQP approaches (line-search or trust region) can be regarded as clearly superior to the other. There is no unique recommendation for optimally choosing the matrix $\mathbf{W}_k$ in the approximate quadratic model. As discussed by Nocedal and Wright [155], choices of $\mathbf{W}_k$ which have performed well on some problems have performed poorly or even failed on other problems.

The most popular choices for computing $\mathbf{W}_k$ (or suitable replacements thereof) are as follows:

1. Maintain a quasi-Newton approximation $\mathbf{B}_k$ to the full Hessian $\nabla^2_{\mathbf{xx}} L(\mathbf{x}, \lambda)$ of the Lagrangian $L(\mathbf{x}, \lambda)$ using a BFGS update. The update for $\mathbf{B}_k$ uses vectors $\mathbf{s}_k$ and $\mathbf{y}_k$ defined via the relations

$$\mathbf{s}_k = \mathbf{x}_{k+1} - \mathbf{x}_k \mathbf{y}_k = \nabla_{\mathbf{x}} L(\mathbf{x}_{k+1}, \lambda_{k+1}) - \nabla_{\mathbf{x}} L(\mathbf{x}_k, \lambda_{k+1}). \quad (3.89)$$

   The next approximation, $\mathbf{B}_{k+1}$, is subsequently computed using the BFGS formula. These iterations will converge robustly and rapidly if $\nabla^2_{\mathbf{xx}} L(\mathbf{x}, \lambda)$ is positive definite. However, if $\nabla^2_{\mathbf{xx}} L(\mathbf{x}, \lambda)$ contains negative eigenvalues, the BFGS algorithm of approximating it with a positive matrix may be ineffective.

2. A more effective modification is the damped BFGS updating; this approach ensures that the update, $\mathbf{B}_{k+1}$, is always well defined by modifying the definition of $\mathbf{y}_k$. Defining $\mathbf{s}_k$ and $\mathbf{y}_k$ as in Eq. (3.89), the matrix $\mathbf{B}_k$ is updated by using the formula

$$\mathbf{B}_{k+1} = \mathbf{B}_k - \frac{\mathbf{B}_k \mathbf{s}_k (\mathbf{B}_k \mathbf{s}_k)^T}{\mathbf{s}_k^T \mathbf{B}_k \mathbf{s}_k} + \frac{\mathbf{r}_k \mathbf{r}_k^T}{\mathbf{s}_k^T \mathbf{r}_k}, \quad (3.90)$$

   where the vector $\mathbf{r}_k$ is defined as

$$\mathbf{r}_k \equiv \theta_k \mathbf{y}_k + (1 - \theta_k) \mathbf{B}_k \mathbf{s}_k, \quad (3.91)$$

   with the scalar θ_k is defined as

$$\theta_k = \begin{cases} 1; & \text{if } \mathbf{s}_k^T \mathbf{y}_k \geq 0.2 \mathbf{s}_k^T \mathbf{B}_k \mathbf{s}_k; \\ (0.8 \mathbf{s}_k^T \mathbf{B}_k \mathbf{s}_k) / (\mathbf{s}_k^T \mathbf{B}_k \mathbf{s}_k - \mathbf{s}_k^T \mathbf{y}_k); & \text{if } \mathbf{s}_k^T \mathbf{y}_k < 0.2 \mathbf{s}_k^T \mathbf{B}_k \mathbf{s}_k; \end{cases}$$
$$(3.92)$$

   Although Eq. (3.90) guarantees that $\mathbf{B}_{k+1}$ is positive definite, the foregoing algorithm nevertheless fails to address directly the difficulties caused by a nonpositive definite Lagrangian Hessian.

3. A third approach modifies the Lagrangian Hessian directly by adding terms to the Lagrangian function to ensure positive definiteness. The modified Lagrangian reads

$$L_{modif}(\mathbf{x}, \lambda; \mu) = F(\mathbf{x}) - \lambda^T \mathbf{C}(\mathbf{x}) + \frac{1}{2\mu} ||\mathbf{C}(\mathbf{x})||^2, \qquad (3.93)$$

for some positive scalar $0 < \mu < \mu^*$, where μ^* is chosen such that the Hessian of the modified Lagrangian is positive definite. The matrix $\mathbf{W}_k$ can be chosen to be $\nabla^2_{\mathbf{xx}} L_{modif}$ or a quasi-Newton approximation $\mathbf{B}_k$ to this matrix. The main difficulty here stems from the choice of μ^*, which depends on quantities that are not routinely known (e.g., bounds on the second derivatives of the problem's functions).

To ensure that the SQP method converges from remote starting points, a *merit function* Φ is employed to control the size of the steps (in line-search methods) and to determine whether a step is acceptable or whether the trust region radius needs to be modified (in trust region methods). The merit function plays the role of the objective function in unconstrained optimization, since each step must reduce it. The most widely employed is the L_1-merit function proposed by Fletcher et al. [65], which is defined as

$$\Phi_1(\mathbf{x}; \mu) \equiv F(\mathbf{x}) - \lambda^T C(\mathbf{x}) + \frac{1}{2\mu} \sum C_i^2(\mathbf{x}). \qquad (3.94)$$

Reduced-Hessian quasi-Newton methods are designed for solving problems in which second derivatives are difficult to compute, and for which the number $(n - m)$ of degrees of freedom is small. This approach is employed if only the reduced Hessian of the Lagrangian $\mathbf{Z}_k^T \mathbf{W}_k \mathbf{Z}_k$ is to be approximated, where $\mathbf{Z}_k$ is a matrix which spans the range of $\mathbf{A}_k$. The update will be an $(n - m) \times (n - m)$ matrix, $\mathbf{M}_k$, of the reduced-Hessian approximation. Since $(n - m)$ is small, $\mathbf{M}_k$ will be of high quality and the line-search computation will be inexpensive. In addition, the reduced Hessian is much more likely to be positive definite, even when the current iterate is still some distance away from the solution, so that the safeguarding mechanism in the quasi-Newton update will be required less often than in the line-search implementation.

The *trust region method* is implemented by using the modified model

$$\min_{\mathbf{p}} \left(\frac{1}{2}\mathbf{p}^T\mathbf{W}_k\mathbf{p} + \nabla F_k^T\mathbf{p} \right), \text{ subject to } \mathbf{A}_k\mathbf{p} + \mathbf{C}_k = \mathbf{0}, \text{ with } ||\mathbf{p}|| \leq \Delta_k.$$

(3.95)

The trust region radius Δ_k is updated based on the agreement between the predicted and actual reductions in the merit function. If the agreement is good, the trust region radius is unaltered or increased; if the agreement is poor, the radius is decreased. In principle, it is possible to simply increase Δ_k until the set of steps $\mathbf{p}$ satisfying the linear constraints in Eq. (3.95) intersect the trust region; however, this approach is not likely to resolve the conflict between the linear and the trust-region constraints. A more appropriate approach is to improve the feasibility of these constraints at each step and to satisfy them exactly only in the limit.

In an operational 4–D VAR setting, the simplest methods for solving non-linearly constrained minimization problems are the barrier and penalty methods, followed by the augmented Lagrangian methods, as shown, for example, in the works by Zou et al. ([230], [231], [232]) and Zhu and Navon [225]. The SQP method is currently the most advanced, robust, and reliable method for constrained optimization but its implementation is more demanding. Dennis et al. [53] implemented a version of SQP coupled with trust region methods and interior-point techniques in the package **TRICE** (Trust-Region Interior-Point algorithms for optimal control and engineering design problems). The book by Conn et al. [35] provides further details on advanced SQP methods, which are most often used in optimal control problems. Only recently have Fisher et al. [62] attempted to implement SQP in a 4–D VAR setting.

3.10   Global Optimization

The aim of global optimization is to determine all of the critical points of a function, F, particularly if several local optima exist where the corresponding function values differ substantially from one another. Global optimization methods can be classified into two major categories, namely deterministic and stochastic methods. Deterministic methods attempt to compute all of the critical points with probability one (i.e., with absolute success). The global optimization and sensitivity analysis algorithm conceived by Cacuci [24] and Cacuci et al. [29] typifies this aim. On the other hand, stochastic methods sacrifice the possibility of an absolute guarantee of success, attempting to minimize F in a random sample of points from a set $S \in R^n$, which is assumed to be convex, compact, and to contain the global minimum as an interior point. Five different underlying philosophies for global optimization can be identified from the literature Rinnooy, Kan and Timmer [169], as follows:

1. *Partition and search*: The set S is partitioned into successively smaller subregions among which global minimum is sought.

2. *Approximation and search*: F is replaced by an increasingly better approximation which is computationally easier to work with.

3. *Global decrease*: Search for a permanent improvement in F values, culminating in arrival at the global minimum.

4. *Improvement of local minima*: Exploit the availability of an efficient local search routine to generate a sequence of local minima in decreasing order, with the global minima being the last one encountered in this sequence.

5. *Enumeration of local minima*: Enumerate all of the local minima (or at least a subset thereof) as a way to solve the global minimization problem.

This section will present two stochastic global minimization algorithms, namely the *simulated annealing* and the *genetic* algorithms, which have recently been implemented in variational data assimilation in geophysical sciences applications. It is important and to note, however, that the convergence rate of these global minimization methods does not outperform the convergence rate of LMQN methods for large-scale operational 4–D VAR models.

3.10.1   Simulated Annealing

The *simulated annealing* (SA) algorithm exploits the analogy between the search for a global minimum and the annealing process (i.e., the way in which a metal cools and freezes) into a minimum energy crystalline structure. The SA algorithm uses the Monte Carlo algorithm originally proposed by Metropolis et al. [135] to find the equilibrium configuration of a thermodynamic system (i.e., of a collection of atoms at a given temperature). Pincus [160] first noted the connection between the Metropolis-algorithm and mathematical minimization, while Kirkpatrick et al. [172] proposed it as an alternative for the optimization of combinatorial and other minimization problems. Kirkpatrick et al. [105], proposed the SA algorithm as an adaptation of the Metropolis-Hastings algorithm, which is a Monte Carlo method for generating sample states of the annealing of a thermodynamic system.

The major advantage of the SA algorithm (over other methods) is its ability to avoid becoming trapped at local minima. The SA algorithm employs a random search that accepts not only changes that decrease objective function F but also some changes that increase it. The latter are accepted with a probability $p = \exp\left(-\delta F/T\right)$, where δF denotes the increase in F, and T is a control parameter, which is known (by analogy to the original application) as the system's "temperature," irrespective of the objective function involved. The implementation of the SA comprises the following basic structure and elements:

1. representation of possible solutions;

2. a generator of random changes in solutions;

3. means of evaluating the problem functions; and

4. an annealing schedule, comprising an initial temperature and rules for lowering it as the search progresses.

3.10.1.1   Annealing Schedule

The annealing schedule determines the degree of uphill movement permitted during the search using the relation $p = \exp\left(-\delta F/T\right)$, which is therefore critical to the algorithm's performance. The annealing schedule should commence at an initial temperature that is sufficiently high to "melt" the system completely; the temperature should subsequently be reduced towards the system's "freezing point" as the search progresses. However, as Bounds [18] observed, "choosing an annealing schedule for practical purposes is still something of a black art."

The standard implementation of the SA algorithm involves the generation of homogeneous Markov chains of finite length, at decreasing temperatures. Such an implementation requires the specification of the initial temperature T_0, the final temperature (or stopping criterion) T_f, the lengths of the respective Markov chains, and a rule for decrementing the temperature.

3.10.1.2   Choice of Initial and Final Temperatures

The value of the initial T_0 depends on the scaling of F and is therefore problem-specific. It can be estimated by first conducting an initial search in which all increases are accepted, calculating the average objective increase observed, δF^+, and subsequently computing $T_0 = (\delta F^+)/\ln(\chi_0)$ by using a value $\chi_0 \simeq 0.8$, as suggested by Kirkpatrick [104]. In some simple implementations of SA algorithms, the final temperature is determined by specifying a limit on the number of temperature values to be used, or the total number of solutions to be generated. Alternatively, the search can be halted when it ceases to make progress. Lack of progress can be defined in several ways; a working definition combines two criteria, namely:

1. "no improvement" (i.e., no new best solution) being found in an entire Markov chain at a certain temperature, and

2. the acceptance ratio falling below a given (small) value χ_f.

3.10.1.3   Computational Considerations

Since the procedures controlling the generation and acceptance of new solutions are simple, the computational cost of implementing the SA algorithm is usually dominated by the costs of evaluating the problem's functions. It is therefore essential that such evaluations be performed as efficiently as possible. In general, efforts to improve performance (reduce Central Processing Unit (CPU) time) should be aimed at exploiting the vectorization or parallelization capabilities of computational platforms in order to accelerate the function evaluations. Due to its recursive structure, SA is an intrinsically sequential algorithm. Aarts and Korst [1] have reviewed parallel designs that use of the idea of *multiple trial parallelism*, in which several different trial solutions are simultaneously generated, evaluated, and tested on individual processors. Whenever one of these processors accepts its solution, it becomes the new solution from which other solutions are generated. Although such a strategy yields N times as many solutions being investigated per unit time (where N is the number of processors), it has been found that the total (elapsed) time required for convergence is not proportionally reduced. This is due to the fact that the instantaneous concurrent efficiency varies as the search progresses. (In this context, the instantaneous concurrent efficiency can be defined as $\eta \equiv \delta t_S / N \delta t_P$, where δt is the time taken for a new solution to be accepted by the serial or parallel algorithm). Initially, the vast majority of the solutions generated are accepted, so that $(N-1)$ processors are redundant and $\eta \simeq 1/N$. As the annealing temperature is reduced and the solution acceptance probability falls, η increases, approaching 100% as T nears zero. However, the overall incentive for parallelizing the optimization scheme is not great, especially since, in many instances, the function evaluation procedure can be multitasked with greater ease and effect.

A significant component of an SA code is the *random number generator*, which is used both for generating random changes in the control variables and for the temperature-dependent increase acceptance test. Random number generators are often provided within standard function libraries or as machine specific functions. The random number generator must have good spectral properties, particularly when analyzing large-scale problems requiring thousands of iterations.

3.10.2 Genetic Algorithms

Although Kirkpatrick et al. [105] has described SA as "an example of an evolutionary process modeled accurately by purely stochastic means," this description is more literally applicable to another class of optimization routines, which are known collectively as *genetic algorithms* (GAs). The GAs attempt to simulate the phenomenon of natural evolution first observed by Darwin [45] and elaborated by Dawkins [48]. In natural evolution, each species searches for beneficial adaptations in an ever-changing environment. As species evolve, new attributes are encoded in the chromosomes of individual members. This information changes by random mutation, but the actual driving force behind evolutionary development is the combination and exchange of chromosomal material during breeding. Although sporadic attempts at incorporating these principles in optimization routines have been made since the early 1960s, as reviewed in Chapter 4 of Goldberg [77], GAs were first established on a sound theoretical basis by Holland [89]. The two key axioms underlying this work are: (i) complicated nonbiological structures could be described by simple bit strings; and (ii) these structures could be improved by the application of simple transformations to these strings. GAs differ from traditional optimization algorithms in four important respects:

1. GAs work using an encoding of the control variables, rather than the variables themselves;

2. GAs search from one population of solutions to another, rather than from individual to individual;

3. GAs use only objective function information, not derivatives;

4. GAs use probabilistic, rather than deterministic, transition rules.

GAs share the last two attributes with SA and, not surprisingly, have found applications in many of the same areas as SA. A more significant difference between GAs and SA is the replacement of the usual operation of generating a new solution by three separate activities in GAs, namely: population selection, recombination, and mutation.

3.10.2.1 Solution Representation

In order to solve a problem using a GA, candidate solutions must be encoded in a suitable form. In the traditional GA, solutions are represented by binary bit strings (chromosomes). The different stages of the genetic algorithm are as follows:

1. generate initial population;

2. assess initial population;

3. select population;

4. recombine new population;

5. mutate new population;

6. assess new population;

7. terminate search?

8. stop? (no/yes)

While integer and decision variables can readily be encoded in this form, the representation of continuous control variables is not simple. In general, the only option is to approximate the continuous control variables, rescaled if necessary, by equivalent integer variables. Consequently, the accuracy with which an optimum solution can be resolved depends on the encoded bit length of these integers, leading to an inevitable compromise between precision and execution time. For combinatorial optimization problem, problem-specific solution encodings, such as ordered lists, are necessary. For example, a solution to the "*Traveling Salesman Problem*" can be represented by a string that lists the cities in the order in which they are to be visited. Problem-specific operators are also required to manipulate correctly such strings.

3.10.2.2   Population Selection

The initial population for a GA search is usually selected randomly, although there may be occasions when heuristic selection is appropriate, as noted by Grefenstette [82]. Within the GA, population selection is based on the principle of "survival of the fittest." If the objective function F is to be maximized, the standard procedure is to define the probability $P_{Si} = F_i/F_{\Sigma}$ that a particular solution i survives, where F_i is the fitness (objective value) of solution i, and $F_{\Sigma} = \sum_{i=1}^{N} F_i$ is the total fitness of the population of size N. On the other hand, if F is to be minimized, then a particular solution i is defined to survive with a probability $P_{Si} = 1 - F/F_{\Sigma}$. The new population is subsequently selected by simulating the spinning of a suitably weighted roulette wheel N times. This scheme can be used only if F is always positive. The range and scaling of F are also important. Early in a search, it is possible for a few superindividuals (solutions with fitness values significantly better than the average) to dominate the selection process. Various procedures have been suggested to overcome this potential drawback; the simplest procedure is   linear scaling, whereby F is rescaled through an equation of the form $\tilde{F} = af + b$. The coefficients a and b are chosen for each generation so that:

1. the average values of F and $\tilde{F}$ are equal, and

2. the maximum value of $\tilde{F}$ is a specified multiple of (usually twice) the average. Caution should be exercised when using linear scaling because, for low performance solutions, there exists a risk of introducing negative values for $\tilde{f}$.

Baker and Daley [7] suggested that P_{Si} should be taken as a (linear) function of the solution's rank within the population. For example, a survival probability of $2/N$ could be allocated to the best solution. In such a case, the survival probability for the worst solution is consequently constrained to be zero by the normalization condition that the survival probabilities sum to unity. This ranking procedure has been found to overcome the difficulties of over or under selection, without displaying any obvious drawbacks.

Roulette wheel selection suffers from the disadvantage of being a high-variance process with the result that there are often large differences between the actual and expected numbers of copies made; there is no guarantee that the best solution will be copied. On the other hand, De Jong [49] tested an elitist scheme, which gave just such a guarantee by enlarging the population to include a copy of the best solution if it had not been retained. He found that on problems with just one maximum or minimum the algorithm performance was much improved, but on multimodal problems it was degraded.

Numerous schemes, which introduce various levels of determinism into the selection process, have been investigated. Overall, it seems that a procedure entitled *stochastic remainder selection without replacement* offers the best performance. Within this procedure, the expected number of copies of each solution is calculated as $E_i = NP_{Si}$. Each solution is then copied I_i times, where I_i is the integer part of E_i. The fractional remainder, $R_i = E_i - I_i$, is treated as the probability of further duplication. For example, a solution for which $E_i = 1.8$ would certainly be copied once and would be copied again with probability 0.8. Each solution is successively subjected to an appropriately weighted simulated coin toss until the new population is complete.

3.10.2.3   Advanced GA Operators

The simple operators and representations described in the foregoing form the backbone of all GAs but, because natural genetics is actually a considerably more complex phenomenon than portrayed so far, it is possible to conceive of several alternative representations and operators could bring particular advantages for specific GA applications. As discussed in detail in the book by Goldberg [77], it is possible to introduce more sophisticated concepts, including:

1. *Diploidy* and *dominance*, whereby solutions are represented by (several) pairs of chromosomes. The decoding of these chromosomes then depends on whether individual bits are dominant or recessive. Such a representation allows alternative solutions to be held in abeyance, which may prove particularly useful for optimization problems where the solution space is time dependent.

2. *Niche and speciation* in multimodal problems, whereby diversity to breed different species by exploiting different niches in the environment, are deliberately maintained by elaborating the selection and recombination rules described above, in order to locate several of the local optima.

3. *Intelligent control* over the selection of mating partners, such as the "inbreeding with intermittent crossbreeding" procedure described by Hollstein [91]. In this procedure, similar individuals are mated with each other as long as the "family" fitness continues to improve. When this improvement ceases, new genetic material is added by crossbreeding with other families.

3.10.2.4   Population Assessment

Like the SA algorithms, GAs do not use derivative information; they only need to be supplied with a fitness value for each member of each population. Thus, the evaluation of the problem proceeds essentially as a "black box" operation, as far as the GAs are concerned. Just as in the case of SA algorithms, the overall computational efficiency of GAs depends strongly on the efficient performance of problem-function evaluations.

The procedures for solving constrained minimization problems with GAs are similar to those used for SA algorithms. As long as there are no equality constraints and the feasible space is not disjoint, infeasible solutions can simply be "rejected." In a GA, this means ensuring that rejected particular solutions are not selected as parents in the next generation. Such a rejection procedure can be implemented, for example, by allocating a zero survival probability to the unwanted solutions. If these conditions on the constraints are not met, then a penalty function method can be used. A suitable form for a GA is a functional of the form $F_A(\mathbf{x}) = F(\mathbf{x}) + M^k \mathbf{w}^T \mathbf{c}_V(\mathbf{x})$, where $\mathbf{w}$ is a vector of nonnegative weighting coefficients, the vector $\mathbf{c}_V$ quantifies the magnitudes of any constraint violations, M denotes the number of the current generation, and k is a suitable exponent. The dependence of the penalty term on the generation number M biases the search increasingly more towards the feasible space as the search progresses.

3.10.2.5 Control Parameters

The efficiency of a GA is highly dependent on the values of the algorithm's control parameters. Assuming that the selection procedure is predetermined, the control parameters available for adjustment are the population size N, the crossover probability P_C, and the mutation probability P_M. De Jong [49] analyzed the performance of GAs on several test problems, which included examples with difficult characteristics such as discontinuities, high dimensionality, noise, multimodality, and recommended the settings $(N, P_C, P_M) = (50, 0.60, 0.001)$ for satisfactory performance over a wide range of problems. Grefenstette [82] optimized these parameters concluding that the settings $(N, P_C, P_M) = (30, 0.95, 0.010)$ gave the best performance when the average fitness of each generation was used as the indicator, while $(N, P_C, P_M) = (80, 0.45, 0.010)$ yielded the best performance when the fitness of the best individual member in each generation (which is the more usual performance measure for optimization routines) was monitored. In general, the population size should not be smaller than 25 or 30; for problems of high dimensionality, larger populations (of the order of hundreds) are appropriate.

3.10.2.6 GA Computational Considerations

A significant component of a GA code is the random number generator, which is essential for the processes of selection, crossover, and mutation. As with the SA algorithms, the procedures controlling the generation of new solutions are relatively simple, so that the computational cost of implementing a GA is dominated by the cost of evaluating the problem's functions. It is therefore important that these evaluations should be performed efficiently. However, unlike the SA algorithm, which is intrinsically a sequential algorithm, GAs are particularly well suited for implementation on parallel computers. Evaluation of the objective function and constraints can be done simultaneously for the entire population; the production of the new population by mutation and crossover can also be parallelized. On highly parallel machines, therefore, a GA can be expected to run nearly N times faster than on nonparallel machines, where N is the population size. If it is possible to parallelize the evaluation of individual problem functions effectively, experimentation will be needed to determine the level at which multitasking should be performed. This level will depend on the number of processors available, the intended population size, and the potential speed-ups available. If the number of processors exceeds the population size, as could be the case on a highly parallel machine, multilevel parallelization may be possible.

3.10.2.7 GA Operators in Detail

Reproduction is the first operator applied to a population. The ith string in the population is selected for mating with a probability proportional to f_i. The cumulative probability for all strings must be one. Hence, the probability of selecting the ith string is $F_i \left(\sum_{j=1}^{N} F_j \right)^{-1}$, where N is the population size. A binary tournament selection is used in which two strings are selected at random for the tournament and the better of the two is retained by choosing the string with the smaller objective function value.

The crossover operation is performed in order to search the parameter space. In this operation, two strings are randomly selected from the mating pool and some portions of the strings are exchanged between the strings. In a two-point crossover operator, two random sites are chosen and the contents bracketed by these sites are exchanged between two parents.

3.10.2.8 Extensions of GA Methods to Constrained Optimization

Many large-scale search and optimization problems involve inequality and/or equality constraints. Solving constrained optimization problems using GAs usually employs penalty function methods, due to their simplicity and ease of implementation. However, since the penalty function methods are generic and applicable to any type of constraint (linear or nonlinear), their performance is not always satisfactory. The most difficult aspect of the penalty function method is to find appropriate penalty parameters to guide the search towards the constrained optimum. Deb [50] exploited the GA's ability to make pairwise comparisons in the tournament selection operator to devise a penalty function method that does not require any penalty parameter. Comparisons among feasible and infeasible solutions provide a search direction towards the feasible region. When sufficient feasible solutions are found, a "niching" method with a controlled mutation operator is used to maintain diversity among feasible solutions. This allows a real parameter in the GA's crossover operator to find continuously better feasible solutions, leading the search to the vicinity of the true optimum solution.

4

—————

Basic Principles of 4–D VAR

—————

CONTENTS

When first introduced, data assimilation methods were referred to as "objective analyses" (see, e.g., ref. [40] and [11]), to contrast them to "subjective analyses," in which numerical weather predictions (NWP) forecasts were adjusted "by hand" by meteorologists, using their professional expertise. Subsequently, methods called "nudging" were introduced based on the simple idea of Newtonian relaxation. In nudging, the right side of the model's dynamical equations is augmented with a term which is proportional to the difference between the calculated meteorological variable and the observation value. This term keeps the calculated state vector closer to the observations. Nudging can be interpreted as a simplified Kalman-Bucy filter with the gain matrix being prescribed rather than obtained from covariances. The nudging method is used in simple operational global-scale and meso-scale models for assimilating small-scale observations when lacking statistical data. The recent advances in nudging methods are briefly presented in Section 4.1.

Section 4.2 briefly mentions the "optimal interpolation" (OI) method, "three-dimensional variational data assimilation" (3–D VAR), and the physical space statistical analysis (PSAS) methods. These methods were introduced independently, but were shown to be formally equivalent; in particular, PSAS is a dual formulation of 3–D VAR.

Data assimilation requires the explicit specification of the error statistics for model forecast and the current observations, which are the primary quantities needed for producing an analysis. A correct specification of observation and background error covariances are essential for ensuring the quality of the analysis, because these covariances determine to what extent background fields will be corrected to match the observations. The essential parameters are the variances, but the correlations are also very important because they specify the manner in which the observed information will be smoothed in the model space if the resolution of the model does not match the density of the observations. Section 4.3 briefly outlines the prevailing operational practices employed for the practical estimation of observation error covariance matrices and background error covariance matrices.

The goal of the four dimensional data assimilation (4–D VAR) formalism is to find the solution of a numerical forecast or numerical weather prediction (NWP) model that best fits sequences of observational fields distributed in space over a finite time interval. Section 4.4 discusses the basic framework of (4–D VAR) methods utilizing optimal control theory (variational approach). The advance brought by the variational approaches is that the meteorological fields satisfy the dynamical equations of the forecast model while simultaneously minimizing a cost functional, which measures the differences between the computed and the observed fields, by solving a constrained minimization problem. The 4–D VAR formalism is first presented without taking the modeling errors into account; subsequently, the functional to be minimized is extended to include model errors. This section concludes with a discussion of the consistent optimality and transferable optimality properties of the 4–D VAR procedure.

Section 4.5 presents results of numerical experiments with unconstrained minimization methods for 4–D VAR using the shallow water equations, which are widely used in meteorology and oceanography for testing new algorithms since they contain most of the physical degrees of freedom (including gravity waves) present in the more sophisticated operational models. The numerical experiments were performed with four limited-memory quasi-Newton (LMQN) methods (CONMIN-CG, E04DGF, L-BFGS, and BBVSCG) and two truncated Newton (T-N) methods. The CONMIN-CG and BBVSCG algorithms failed after the first iteration, even when both gradient scaling and nondimensional scaling were applied. The L-BFGS algorithm was successful only with gradient scaling. On the other hand, the E04DGF algorithm worked only with the nondimensional shallow water equations model. This indicates that using additional scaling is essential for the success of LMQN minimization algorithms when applied to large-scale minimization problems. On the other hand, T-N methods appear to perform best for large-scale minimization problems, especially in conjunction with a suitable preconditioner. The importance of preconditioning increases with increasing dimensionality of the minimization problem under consideration. Furthermore, for the Shallow-Water-Equations (SWE) numerical experiments, the T-N methods required far fewer iterations and function calls than the LMQN methods.

In the so-called strong constraint variational data assimilation (VDA) (or classical VDA), it is assumed that the forecast model perfectly represents the evolution of the actual atmosphere. The best fit model trajectory is obtained by adjusting only the initial conditions via the minimization of a cost functional that is subject to the model equations as strong constraints. However, numerical weather prediction (NWP) models are imperfect since subgrid processes are not included. Furthermore, numerical discretizations produce additional dissipative and dispersion errors. Modeling errors also arise from the incomplete mathematical modeling of the boundary conditions and forcing terms, and from the simplified representation of physical processes and their interactions in the atmosphere. Usually, all of these modeling imperfections are collectively called *model error* (ME). Model error is formally introduced as a correction to the time derivatives of model variables. Section 4.6 highlights the treatment of model errors (ME) in (VDA). Taking into account numerical errors explicitly as additional terms in the cost functional to be minimized at doubles (and can even triple) the size of the system to be optimized by comparison to minimizing the cost functional when the model errors are neglected.

This chapter discusses only aspects of 4–D VAR that were researched by the authors and will refer the reader to recent literature concerning aspects that were previously addressed in books such as refs. [43], [101], and [123].

_____

4.1   Nudging Methods (Newtonian Relaxation)

Nudging (also called *Newtonian relaxation*) methods are characterized by the addition of empirical forcing terms in a pre-forecast period to drive the model variables towards the observations, which are introduced into the data assimilation system at each time step of model integration during the assimilation time period (see, e.g., refs. [4], [73], [110], [166], [231], [232]). The concepts underlying nudging can be illustrated by considering a spatially discretized time-evolution model of the form

$$\frac{\partial \mathbf{X}}{\partial t} = \mathbf{F}(\mathbf{X}), \quad \mathbf{X}(0) = \mathbf{V}, \tag{4.1}$$

where $\mathbf{X}$ represents the discretized state variables of the model, t denotes time, and $\mathbf{V}$ represents the initial conditions for the model. Denoting a set of experimental observations by $\mathbf{X}^o(t)$, the nudging data assimilation (NDA) strives to achieve a compromise between the model and the observations by considering the state of the atmosphere to be defined by

$$\frac{\partial \mathbf{X}}{\partial t} = \mathbf{F}(\mathbf{X}) + \mathbf{G}(\mathbf{X}^o - \mathbf{X}); \quad \mathbf{X}(0) = \mathbf{V}, \tag{4.2}$$

where $\mathbf{G}$ is a diagonal matrix. For a given matrix $\mathbf{G}$, Eq. (4.2) has a unique solution $\mathbf{X}(\mathbf{V}, \mathbf{G})$.

The main difficulty underlying NDA resides in the estimation of the nudging coefficient matrix $\mathbf{G}$. If $\mathbf{G}$ is too large, the diffusion term will completely dominate the time tendency and will have an effect similar to replacing the model data by the observations at each time step. If a particular observation has a large error that prevents attaining a dynamic balance, an exact fit to the observation is not required since it may lead to a false amplification of observational errors. On the other hand, if $\mathbf{G}$ is too small, the observation will have little effect on the solution. In general, $\mathbf{G}$ decreases with: (i) increasing observation error, (ii) increasing horizontal and vertical distance separation, and (iii) increasing time separation. In the experiment reported by Anthes [4], a nudging coefficient of 10^{-3} was used for all the fields for a hurricane model and was applied on all the domain of integration. In the experiment of Krishnamurti et al. [110], the relaxation coefficients for the estimated NDA experiment were kept invariant in space and time and their values were simply determined by numerical experience. The implicit dynamic constraints of the model then spread the updated information to the other variables (temperature and moisture) resulting eventually in a set of balanced conditions at the end of the nudging period. In the work of Zou et al. ([231], [232]), adjoint functions were used to obtain optimal nudging coefficients by minimizing the difference between the model solution and the observations.

A more sophisticated nudging method is the *backward nudging* method, in which the state equations of the model are solved backwards in time, starting from the observation of the state of the system at the final instant. A nudging term, with the opposite sign compared to the standard nudging algorithm, is added to the state equations. Thus, the system to be solved has the form

$$\frac{\partial \tilde{\mathbf{X}}}{\partial t} = \mathbf{F}\left(\tilde{\mathbf{X}}\right) - \mathbf{G}'\left(\mathbf{X}^o - \tilde{\mathbf{X}}\right), \quad T > t > 0, \quad \tilde{\mathbf{X}}\left(T\right) = \mathbf{V}_T, \qquad (4.3)$$

where $\mathbf{V}_T$ denotes a "final condition" at the final-time T. Nudging is applied to the above backward model with the opposite sign of the feedback term, in order to have a well-posed problem. The state thus obtained at $t = T$ is considered to represent the initial state of the real pyhsical system.

The *back and forth nudging* (BFN) method was introduced by Auroux et al. [5], and requires solving first the forward nudging equation and then the direct system, backwards in time, with a feedback term whose sign is opposite to the one introduced in the forward equation. The "initial" condition for the backward resolution is the final state obtained by the standard nudging method. The resolution of this backward equation yields an estimate of the initial state of the system. These forward and backward computations, with the added feedback terms, are repeated until attaining convergence. The forms of the BFN equations are:

$$\frac{\partial \mathbf{X}_k}{\partial t} = \mathbf{F}(\mathbf{X}_k) + \mathbf{G}(\mathbf{X}^o - \mathbf{X}_k), \quad \mathbf{X}_k(0) = \tilde{\mathbf{X}}_{k-1}(0), \qquad (4.4)$$

$$\frac{\partial \tilde{\mathbf{X}}_k}{\partial t} = \mathbf{F}\left(\tilde{\mathbf{X}}_k\right) - \mathbf{G}'\left(\mathbf{X}^o - \tilde{\mathbf{X}}_k\right), \quad \tilde{\mathbf{X}}_k(T) = \mathbf{X}_k(T), \qquad (4.5)$$

with $\tilde{\mathbf{X}}_0(0) = \mathbf{V}$. This implies that $\mathbf{X}_1(0) = \mathbf{V}$, and the resolution of the direct model gives $\mathbf{X}_1(T)$ and hence, $\tilde{\mathbf{X}}_1(T)$. A resolution of the backward model provides $\tilde{\mathbf{X}}_1(0) = \mathbf{X}_2(0)$, which becomes the new initial condition of the system, and so on. Auroux et al. [5] proved that the BFN algorithm convergences for a linear model, provided that the feedback term is sufficiently large. The BFN method can be implemented with relative ease, and may provide a useful initial state under certain circumstances.

4.2   Optimal Interpolation, Three-Dimensional Variational, and Physical Space Statistical Analysis Methods

Historically, these methods were developed independently, but Lorenc [127] has shown that they are formally equivalent. Since these methods have been described in detail in the books by Kalnay [101] and Lewis et al. [123], only the equivalence relations among them will be highlighted succinctly in this section. The introduction of variational calculus to meteorology is attributed to Sasaki [177]. In subsequent works, Sasaki ([173], [174], [175], [176], [177]) included dynamic model laws, proposing three basic types of variational formalisms:

1. a "time-wise localized" formalism;

2. a formalism with strong constraints;

3. a formalism with weak constraints.

   In the first two formalisms, Sasaki introduced a cost functional which was to be minimized subject to constraints representing the exact satisfaction of selected prognostic equations. This time-independent variational framework approach is now generically referred to as three-dimensional variational data assimilation (3–D VAR). Specifically, the 3–D VAR framework involves the minimization of a "cost function" proportional to the square of the distance between analysis and both background and observations, of the form

$$J\left(\mathbf{x}\right) = \frac{1}{2}\left[\mathbf{y}^{o} - \mathbf{H}\left(\mathbf{x}\right)\right]^{T}\mathbf{R}^{-1}\left[\mathbf{y}^{o} - \mathbf{H}\left(\mathbf{x}\right)\right] + \left(\mathbf{x} - \mathbf{x}^{b}\right)^{T}\mathbf{B}^{-1}\left(\mathbf{x} - \mathbf{x}^{b}\right), \quad (4.6)$$

where $\mathbf{x}$ denotes the field of model variables, $\mathbf{B} \equiv E\left\{\varepsilon_{b}\varepsilon_{b}{}^{T}\right\}$ denotes the background error covariance, $\mathbf{R} \equiv E\left\{\varepsilon_{o}\varepsilon_{o}{}^{T}\right\}$ denotes the observation error covariance, $\mathbf{H}$ denotes an interpolation (or observation) operator, $\mathbf{x}^{b}$ is an estimate of the background state, $\mathbf{y}^{o}$ denotes the vector of observations, and $\left[\mathbf{y}^{o} - \mathbf{H}\left(\mathbf{x}^{b}\right)\right]$ are the observational increments satisfying the relation

$$\mathbf{x}^{a} = \mathbf{x}^{b} + \mathbf{W}\left[\mathbf{y}^{o} - \mathbf{H}\left(\mathbf{x}^{b}\right)\right], \quad (4.7)$$

with $\mathbf{W}$ representing a weight matrix based on statistical error covariances of forecast and observations. The observations and background errors are assumed to be uncorrelated, i.e., $E\left\{\varepsilon_{0}\varepsilon_{b}{}^{T}\right\} = 0$. The background term contains *a priori* information, which is required because the number of degrees of freedom of the model (in 3–D and time) exceeds by far the number of observations taken within the same time window. In numerical weather prediction (NWP), the previous model forecast provides information for filling data voids, as well as the *a priori* information for the current forecast. The background covariance matrix $\mathbf{B}$ is too large to be stored, so various proposals have been put forward to approximate it (e.g., to construct it using univariate variables that model vertical and horizontal covariances separately); its construction and properties will be discussed in the next section.

Statistical interpolation can be traced back to Kolmogorov et al. [108] and Weiner [216]. The terminology of "optimal interpolation" (OI) appears to stem from Weiner [216]; these early works are reviewed in the book by Yaglom [222] on stochastic processes. Krige [109] used statistical interpolation in the mining industry. Eliassen [58] introduced statistical interpolation in atmospheric sciences, while Gandin [70] used it in meteorology under the name of "objective analysis," to distinguish it from "subjective analysis," in which data are manipulated according to the opinions of experts. Gandin's algorithm is actually a reduced version of the Kalman filtering (KF) algorithm, *in which the covariance matrices are not calculated from the dynamical equations but are predetermined.*

Da Silva et al. [41] proposed the *physical space statistical analysis system* (PSAS), in which the minimization of a cost functional equivalent to Eq. (4.6) is performed in the space of observations ("physical space") rather than in the model space, as in OI and 3–D VAR procedures, which is more efficient when the space of observations is smaller than the number of degrees of freedom in the model. The formal equivalences among the OI, 3–D VAR, and PSAS procedures can be summarized as follows:

1. The optimal least-squares estimator, also called *Best Linear Unbiased Estimator* (BLUE), is defined by the following interpolation equations:

$$\mathbf{x}^a = \mathbf{x}^b + \mathbf{K}\left[\mathbf{y}^o - \mathbf{H}\left(\mathbf{x}^b\right)\right], \qquad (4.8)$$

$$\mathbf{K} = \mathbf{BH}^T\left(\mathbf{HBH}^T + \mathbf{R}\right)^{-1}, \qquad (4.9)$$

   where the linear operator $\mathbf{K}$ is called the *gain*, or *weight matrix*, of the BLUE analysis.

2. For any gain $\mathbf{K}$, the *analysis error covariance matrix*, $\mathbf{A} \equiv E\left\{\varepsilon_a\varepsilon_a{}^T\right\}$, is given by

$$\mathbf{A} = (\mathbf{I} - \mathbf{KH})\,\mathbf{B}(\mathbf{I} - \mathbf{KH})^T + \mathbf{KRK}^T. \qquad (4.10)$$

   If $\mathbf{K}$ is the optimal least-squares gain, the above expression becomes

$$\mathbf{A} = (\mathbf{I} - \mathbf{KH})\,\mathbf{B}. \qquad (4.11)$$

3. Equivalently, the BLUE results can be obtained as the solution of the *variational optimization* of the cost functional $J(\mathbf{x})$ defined in Eq. (4.6).

The equivalence between propositions (1) and (3) stems from the requirement that the gradient of $J(\mathbf{x})$ vanishes at the optimum $\mathbf{x}^a$, which gives

$$\nabla J(\mathbf{x}^a) = 2\mathbf{B}^{-1}\left(\mathbf{x}^a - \mathbf{x}^b\right) - 2\mathbf{H}^{\mathbf{T}}\mathbf{R}^{-1}\left[\mathbf{y} - \mathbf{H}\left(\mathbf{x}^a\right)\right],$$

$$0 = \mathbf{B}^{-1}\left(\mathbf{x}^a - \mathbf{x}^b\right) - \mathbf{H}^{\mathbf{T}}\mathbf{R}^{-1}\left[\mathbf{y} - \mathbf{H}\left(\mathbf{x}^b\right)\right] - \mathbf{H}^{\mathbf{T}}\mathbf{R}^{-1}\mathbf{H}\left(\mathbf{x}^a - \mathbf{x}^b\right),$$

$$\left(\mathbf{x}^a - \mathbf{x}^b\right) = \left(\mathbf{B}^{-1} + \mathbf{H}^{\mathbf{T}}R^{-1}\mathbf{H}\right)^{-1}\mathbf{H}^{\mathbf{T}}\mathbf{R}^{-1}\left[\mathbf{y} - \mathbf{H}\left(\mathbf{x}^b\right)\right].$$

$$(4.12)$$

The equivalence among Eqs. (4.12), (4.8), and (4.9) follows from the sequence of equalities

$$\mathbf{H}^{\mathbf{T}}\mathbf{R}^{-1}\left(\mathbf{HBH}^{\mathbf{T}} + \mathbf{R}\right) = \left(\mathbf{B}^{-1} + \mathbf{H}^{\mathbf{T}}\mathbf{R}^{-1}\mathbf{H}\right)\mathbf{BH}^{\mathbf{T}} = \mathbf{H}^{\mathbf{T}} + \mathbf{H}^{\mathbf{T}}\mathbf{R}^{-1}\mathbf{HBH}^{\mathbf{T}},$$

which imply that $\left(\mathbf{B}^{-1} + \mathbf{H}^{\mathbf{T}}\mathbf{R}^{-1}\mathbf{H}\right)^{-1}\mathbf{H}^{\mathbf{T}}\mathbf{R}^{-1} = \mathbf{B}\mathbf{H}^{T}\left(\mathbf{H}\mathbf{B}\mathbf{H}^{T} + \mathbf{R}\right)^{-1}$.

The expressions for the analysis error covariance matrix $\mathbf{A}$ shown in Eq. (4.10) can be obtained by recalling the definition of the various errors involved, namely: $\varepsilon_b = \mathbf{x}^b - \mathbf{x}^t$, $\varepsilon_a = \mathbf{x}^a - \mathbf{x}^t$, $\varepsilon_o = \mathbf{y} - \mathbf{H}(\mathbf{x}^t)$, and $\varepsilon_a - \varepsilon_b = \mathbf{K}(\varepsilon_o - \mathbf{H}\varepsilon_b)$, which imply that $\varepsilon_a = (\mathbf{I} - \mathbf{KH})\varepsilon_b + \mathbf{K}\varepsilon_o$. The expression of $\mathbf{A}$ shown in Eq. (4.10) can be obtained by developing the expression of $\varepsilon_a\varepsilon_a^{T}$, taking its expectation, and using the linearity of the expectation operator (recall that the errors are considered to be uncorrelated, so their cross-correlations are zero). The simpler form of analysis error covariance $\mathbf{A}$ shown in Eq. (4.11) is derived by substituting Eq. (4.9) for the optimal gain $\mathbf{K}$, and simplifying the resulting expression. The expression for the gain $\mathbf{K}$ shown in Eq. (4.9) can be proven by minimizing the trace of the analysis error covariance matrix $\mathbf{A}$ given by Eq. (4.10), while noting that $\mathbf{B}^{T} = \mathbf{B}$ and $\mathbf{R}^{\mathbf{T}} = \mathbf{R}$, as follows:

$$\mathrm{Tr}(\mathbf{A}) = \mathrm{Tr}(\mathbf{B}) + \mathrm{Tr}\left(\mathbf{KHBH}^{\mathbf{T}}\mathbf{K}^{\mathbf{T}}\right) - 2\mathrm{Tr}\left(\mathbf{BH}^{\mathbf{T}}\mathbf{K}^{\mathbf{T}}\right) + \mathrm{Tr}\left(\mathbf{KRK}^{\mathbf{T}}\right),$$

The above expression is a continuous and differentiable scalar function of the coefficients of $\mathbf{K}$. Therefore, the derivative d_K of $\mathrm{Tr}(\mathbf{A})$ can be written as the first-order terms in $\mathbf{K}$ of the difference $\mathrm{Tr}(\mathbf{A})(\mathbf{K} + \mathbf{L}) - \mathrm{Tr}(\mathbf{A})(\mathbf{K})$, where $\mathbf{L}$ is an arbitrary test matrix. It follows that

$$d_K[\mathrm{Tr}(\mathbf{A})]\mathbf{L} = 2\mathrm{Tr}\ (\mathbf{KHBH}^\mathrm{T}\mathbf{L}^\mathrm{T}) - 2\mathrm{Tr}\ (\mathbf{BH}^\mathrm{T}\mathbf{L}^\mathrm{T}) + 2\mathrm{Tr}\ (\mathbf{KRL}^\mathrm{T})$$

$$= 2\mathrm{Tr}\ (\mathbf{KHBH}^\mathrm{T}\mathbf{L}^\mathrm{T} - \mathbf{BH}^\mathrm{T}\mathbf{L}^\mathrm{T} + \mathbf{KRL}^\mathrm{T})$$

$$= 2\mathrm{Tr}\ \left\{[\mathbf{K}\ (\mathbf{HBH}^\mathrm{T} + \mathbf{R}) - \mathbf{BH}^\mathrm{T}\ ]\mathbf{L}^\mathrm{T}\right\}$$

The last line above indicates that the derivative d_K vanishes for any choice of $\mathbf{L}$ if and only if $(\mathbf{HBH}^\mathrm{T} + \mathbf{R})\,\mathbf{K}^\mathrm{T} - \mathbf{HB} = 0$, which is equivalent to $\mathbf{K} = \mathbf{BH}^\mathrm{T}(\mathbf{HBH}^\mathrm{T} + \mathbf{R})^{-1}$.

Note that that the variational methods lead to results that are closest in a "root mean square" sense to the true state. Furthermore, if the background and observation error probability distribution functions are Gaussian, then $\mathbf{x}^a$ is also the maximum likelihood estimation of $\mathbf{x}^t$ (true state). The minimization of the cost functional J in 3–D VAR (and PSAS) can be performed using global unconstrained minimization algorithms, which indicates that the approximations made in OI are not necessary. Relative advantages and disadvantages of 3–D VAR, PSAS, and OI are discussed in Kalnay [101], and Lewis et al. [123].

4.3   Estimation of Error Covariance Matrices

Data assimilation requires the explicit specification of the error statistics for model forecast and the current observations, which are the primary quantities needed for producing an analysis. A correct specification of observation and background error covariances are essential for ensuring the quality of the analysis, because these covariances determine to what extent background fields will be corrected to match the observations. The essential parameters are the variances, but the correlations are also very important because they specify the manner in which the observed information will be smoothed in the model space if the resolution of the model does not match the density of the observations.

In practice, most models of *observation error covariances* $\mathbf{R} \equiv E\left\{\boldsymbol{\varepsilon}_o \boldsymbol{\varepsilon}_o{}^T\right\}$ are almost, if not entirely, diagonal. The observation error correlations are minimized in practice by using bias corrections, avoiding unnecessary observation preprocessing, thinning dense data, and improving the design of the model and observation operators.

Background error covariances $\mathbf{B} \equiv E\left\{\varepsilon_b \varepsilon_b{}^T\right\}$ also play an essential role for:

1. Information spreading: in data-sparse areas, the shape of the analysis increment is completely determined by the covariance structures. Hence, the correlations in $\mathbf{B}$ will induce a spatial spreading of information from the observation points to a finite domain surrounding it.

2. Information smoothing: in data-dense areas, in the presence of discrete observations, the amount of smoothing of the observed information is governed by the correlations in $\mathbf{B}$. Smoothing of the increments is important to ensure that the analysis contains scales that are statistically compatible with the smoothness properties of the physical fields. For instance, when analyzing stratospheric or anti-cyclonic air masses, it is desirable to smooth the increments substantially in the horizontal directions in order to average and spread efficiently the measurements. On the contrary, when performing low level analysis in frontal, coastal or mountainous areas, it is desirable to limit the extent of the increments in order to avoid producing an analysis which may be too smooth. Such properties need to be reflected in the specification of background error correlations.

3. Information balancing: often, there are more degrees of freedom in a model than in reality. For instance, the large-scale atmosphere is usually hydrostatic and it is almost geostrophic, thus reflecting *balance* properties. Such balance properties introduce constraints on the analysis, which could be enforced through a posteriori normal-mode initialization (which eliminates short gravity waves) or through statistical properties linking various model variables. In other words, the existence of a balance in reality and also in the model state implies the existence of a (linearized) version of the corresponding balance in the background error covariances.

This is important when using observed information: observing one model variable yields information about all variables that are balanced with it, e.g., a low-level wind observation allows corrections in the surface pressure field by assuming a certain amount of geostrophy. When combined with spatial smoothing of increments, correlations impact considerably the quality of the analysis. For example, a temperature observation at one point can be smoothed to produce a correction to the geopotential height surrounding the respective point.

The choice of the covariance matrix $\mathbf{B}$ greatly influences the data assimilation process, since its properties determine the components of the solution that correspond to areas with few or no observations. By altering the weight of the background term relative to the observation terms, it is possible to adjust the degree to which the analysis will hew to the observations. The relative amplitude of the increments in terms of the various model fields will depend directly on the specified amount of correlation as well as on the assumed error variance in the concerned parameters. It is therefore essential to represent realistically the background error covariance matrix $\mathbf{B}$, which must reflect (in the analysis) the following features:

1. correlations must be smooth in physical space, on sensible scales;

2. correlations should tend to zero for very large separations if it is believed that observations should only have only a local effect on the increments;

3. correlations should not exhibit physically unreasonable variations according to direction or location;

4. correlations should not lead to unreasonable effective background error variances for any observed parameter which is used in the subsequent model forecast; and

5. the most fundamental balance properties (e.g., geostrophy) must be reasonably well enforced.

The most popular and universally adopted method for estimating the background error covariance matrix for 3–D VAR was introduced by Parrish and

Derber [157], and is now known as the "NCEP (National Center for Environmental Prediction) method." This method does not use measurements for constructing the covariance matrix; instead, it uses differences between forecasts of different time lengths of the form

$$\mathbf{B} \approx \alpha E \left\{ [\mathbf{X}_f(48h) - \mathbf{X}_f(24h)] [\mathbf{X}_f(48h) - \mathbf{X}_f(24h)]^T \right\}. \qquad (4.13)$$

The above expression represents a multivariate global forecast difference covariance. If the time interval is longer than the forecast used to generate background fields, the covariances of the forecast difference will be broader than those of the background error. Differences between forecasts of different lengths are produced by data assimilation in the period between the starting times of the two forecasts. In poorly-observed regions, the available observations may be too few to effect a change in the initial condition of the subsequent forecast, yielding two forecasts with very similar characteristics. In such data-sparse regions, Eq. (4.13) is likely to underestimate the variance of background error. Furthermore, since the length of the forecasts (typically between 12 and 48 hours) is significantly longer than the forecast used to generate the background fields, covariances of forecast differences are likely to be broader in the horizontal and the vertical directions than those of background error.

The European Centre for Medium-Range Weather Forecasts (ECMWF) uses the so-called *ensemble of analyses* to estimate the background errors. In this method (see, e.g., Fisher [59]), the ensemble is generated by perturbing all inputs to the analysis system, yielding a perturbed analysis, and a forecast is run from this perturbed analysis. By running the analysis forecast twice for the same period and perturbing both runs using statistically independent perturbations, the difference between these pairs of background fields acquire statistical characteristics of differences between background error fields. A typical analysis system for numerical weather prediction (NWP) has a state vector comprising some 10^7 components. Consequently, the background error covariance matrix would contain roughly 10^{14} elements, which cannot be stored in the memory of current computers. Moreover, specifying the elements of such a matrix would require at least 10^7 background or forecast differences, far more than available. To reduce the storage problem to manageable proportions, the background covariance matrix is constructed from a set of very sparse matrices, in the form

$$\mathbf{B} = \mathbf{L}^T \mathbf{\Sigma}^T \mathbf{C} \mathbf{\Sigma} \mathbf{L}, \tag{4.14}$$

where:

1. $\mathbf{L}$ denotes a balance operator which accounts for correlations between the mass field (e.g., temperature and surface pressure) and the wind field; $\mathbf{L}$ is sparse.

2. $\mathbf{\Sigma}$ denotes a matrix that is implemented as the inverse spectral transform of the spectral coefficients of the model state vector, followed by a multiplication at each gridpoint by the standard deviation of background error, followed by a spectral transform; these operations are also sparse. The advantage of this spectral approach is that it reduces the horizontal correlation matrix to a diagonal matrix. The disadvantage is that, by assuming the correlations to be equivalent to a convolution, the resulting correlations are homogeneous and isotropic. Nevertheless, the spectral method remains attractive due to its efficiency, the ease of computing coefficients from forecast or background differences, and the absence of polar discontinuities. It is also possible to relax the restrictions of homogeneity and isotropy while retaining the advantages of the spectral approach.

3. $\mathbf{C}$ is a block diagonal matrix, with diagonal blocks of the form $h_n \mathbf{V}_n$, where $\mathbf{V}_n$ represents the vertical correlation for a particular variable and wave number. By specifying different vertical correlation matrices for different wave numbers, it is possible to produce vertical correlations that depend on horizontal scales, so that features with a large horizontal scale have deeper vertical correlations than features with small horizontal scales. This nonseparability of vertical and horizontal correlations is necessary for specifying simultaneously and correctly mass and wind correlations.

A square-root factorization of the background error covariance of the form shown in Eq. (4.15) is based on formulations provided by Bouttier and Courtier

[19], and Weaver and Courtier [215]; this procedure avoids the need for computing explicitly the inverse of $\mathbf{B}$, via the transformation

$$J_b = \frac{1}{2}\delta_X^T \mathbf{B}^{-1}\delta_X = \frac{1}{2}\delta_X^T \left(\mathbf{B}^{\frac{1}{2}}\mathbf{B}^{\frac{T}{2}}\right)^{-1}\delta_X = \frac{1}{2}\mathbf{V}^T\mathbf{V}, \qquad (4.15)$$

where $\delta_X \equiv \mathbf{X}(t_0) - \mathbf{X}_b$, and $\mathbf{V} \equiv \mathbf{B}^{-\frac{1}{2}}\delta_X$; these relations implies that $\delta_X = \mathbf{B}^{\frac{1}{2}}\mathbf{V}$.

4.4   Framework of Time-Dependent ("Four-Dimensional") Variational Data Assimilation (4–D VAR)

The pioneering work of Cacuci ([22], [23]) laid the rigorous mathematical foundation for the application of adjoint-operator based methods to analyze nonlinear systems. An application of this work was to introduce the use of adjoint operators and functions to nonlinear and discontinuous problems in the atmospheric sciences by Cacuci and workers ([25], [26], [27]), which opened the way for constructing the modern techniques nowadays known as *four-dimensional variational data assimilation* (4–D VAR). The technical report by Le Dimet [116] and the work of Thompson [195] were used by Derber [54] in the first application of the adjoint method to data assimilation in atmospheric sciences. Penenko and Obraztsov [158] used adjoint operators to perform simple data assimilation experiments on a *linear model* (see Talagrand and Courtier [193]), while Derber [54] used adjoint functions in his Ph.D. thesis to adjust analysis to a multi-level quasi-geostrophic model. Hoffman [88] also contributed to the development of the 4–D VAR concepts, even though he used a simplified primitive equation model and he perturbed successively all the components of the initial state in order to estimate the gradient. Talagrand and Courtier [193] presented a more in-depth general exposition of the theory of adjoint equations in the framework of variational assimilation and applied it to the inviscid vorticity equation and to the Haurwitz wave (see also [37]). Related work by Navon and De Villiers [150] on augmented Lagrangian methods presented procedures for enforcing conservation of integral invariants, as mentioned in the early work of Le Dimet and Talagrand [118]. A detailed account of state of theory in data assimilation for the period 1982 by 1986 was presented Lorenc [127]. A 3–D VAR methodology was placed in operational use at the National Center for Environment Prediction (NCEP), replacing OI, in 1991 by Derber et al. [56]. Since then, most of meteorological centers in the world have implemented 3–D VAR methodologies, while intensively developing 4–D VAR in conjunction with operational models. Thus, 4–D VAR was implemented for operational use at the ECMWF in 1997. Several countries (France, Japan, and Canada) have since also implemented 4–D VAR methodologies using numerical weather prediction (NWP) models comprising sophisticated physical processes (see, e.g., references [133], [165], [228], [229], [235]).

The goal of the 4–D VAR formalism is to find the solution of a numerical forecast or numerical weather prediction (NWP) model that best fits sequences of observational fields distributed in space over a finite time interval. Formally, the forecast model can be written as an evolution equation of the form

$$\mathbf{I}_0 \frac{d\mathbf{X}}{dt} + \mathbf{A}\left(\mathbf{X}\right) = \mathbf{0}, \qquad (4.16)$$

where $\mathbf{I}_0$ denotes the identity operator for a dynamical model or the null operator for a steady state model, $\mathbf{A}$ is a linear or nonlinear operator acting on the unknown vector of state variables $\mathbf{X}\left(\mathbf{U}\right)$, and $\mathbf{U}$ denotes a vector of "control variables," which may consist of initial conditions, boundary conditions and/or model parameters. The control $\mathbf{U}$ should belong to a class of admissible controls, $\mathbf{U}_{ad}$. In discretized form, the NWP model defined by Eq. (4.16) can be represented as

$$\mathbf{X}_{i+1} = \mathbf{M}_{i+1,i}\left(\mathbf{X}_i\right), \qquad (4.17)$$

where the operator $\mathbf{M}_{i+1,i}$ represents the evolution of the (nonlinear) NWP model from time t_i to time t_{i+1}. In 4–D VAR methods, the typical cost functional J to be minimized measures the distance between the model's trajectory and observations during a finite time interval (called the "time window"), and includes the discrepancies between the model's initial state $\mathbf{X}_0$ and the background at the initial time t_0. The mathematical representation of such a typical cost functional is:

$$J\left(\mathbf{X}_0\right) = \frac{1}{2}\left(\mathbf{X}_0 - \mathbf{X}_b\right)^T \mathbf{B}^{-1}\left(\mathbf{X}_0 - \mathbf{X}_b\right)$$

$$+ \frac{1}{2}\sum_{i=0}^{N}\left[\mathbf{H}_i\left(\mathbf{X}_i\right) - \mathbf{y}_i\right]^T \mathbf{R_i}^{-1}\left[\mathbf{H}_i\left(\mathbf{X}_i\right) - \mathbf{y}_i\right], \qquad (4.18)$$

where:

$\mathbf{X}_0$ represents the NWP model state at time t_0;

$\mathbf{X}_b$ represents the background state at time t_0 (typically a $6h$ forecast from a previous analysis);

$\mathbf{B}$ represents the background error covariance matrix;

$\mathbf{y}_i$ represents the observation vector at time t_i;

$\mathbf{H}_i$ represents the observation operator;

$\mathbf{X}_i = \mathbf{M}_{i,0}\left(\mathbf{X}_0\right)$ represents the model state at time t_i;

$\mathbf{R_i}$ represents the observation error covariance matrix at time t_i.

The minimization of the above cost functional requires the computation of its gradient, which can be performed in three equivalents ways, namely within: (a) a Lagrangian framework; (b) an adjoint-operator framework; and (c) a synthesis of optimality conditions within the framework of optimal control theory.

The conceptual steps underlying 4–D VAR methods can be illustrated by considering a *linear* discrete stochastic dynamical system described by the stochastic vector difference equation,

$$\mathbf{x}_{k+1} = \mathbf{L}\left(k+1, k\right)\mathbf{x}_k + \mathbf{\Gamma}\left(k\right)\mathbf{w}_{k+1}, \quad k = 0, 1, \cdots, \qquad (4.19)$$

where the n-dimensional vector $\mathbf{x}_k$ denotes the state of the system at t_k, $\mathbf{L}\left(k+1, k\right)$ is a linear operator which advances the system from t_k to t_{k+1}, $\mathbf{\Gamma}\left(k\right)$ is a $n \times r$-dimensional matrix, while the r-dimensional vector $\mathbf{w}_k \sim N\left(0, \mathbf{Q}_k\right)$, $k = 1, 2, \cdots$, denotes a white-noise Gaussian sequence, with $N\left(0, \mathbf{Q}_k\right)$ denoting the Gaussian (normal) distribution with zero mean and covariance (of $\mathbf{w}_k$) $\mathbf{Q}_k$. The distribution of the initial condition $\mathbf{x}_0$ is assumed to be a known Gaussian of the form

$$\mathbf{x}_0 \sim N\left(\mathbf{x}_0^b, \mathbf{B}_0\right), \tag{4.20}$$

where $\mathbf{x}_0$ is independent of the sequence $\{\mathbf{w}_k\}$. In view of the above conditions, the random sequence generated by Eq. (4.19) is a Markov Chain (see, e.g., Jazwinski [100]), since $\mathbf{x}_{k+1}$ is completely determined by the previous state ($\mathbf{x}_k$ and $\mathbf{w}_k$), independently of the sequence $\{\mathbf{x}_{k-1}, \cdots, \mathbf{x}_0\}$. The noisy discrete observations (measurements) are denoted by the m-dimensional vector $\mathbf{y}_k$, and are related to the state vector $\mathbf{x}_k$ through the linear stochastic equation

$$\mathbf{y}_k = \mathbf{h}_k \mathbf{x}_k + \mathbf{v}_k, \quad k = 1, \dots; \tag{4.21}$$

where $\mathbf{h}_k$ is a $m \times n$ matrix and $\mathbf{v}_k \sim N\left(0, \mathbf{R}_k\right)$ is an m-dimensional vector denoting a white-noise Gaussian with covariance matrix $\mathbf{R}_k$. The sequences $\{\mathbf{w}_k\}$ and $\{\mathbf{v}_k\}$ are assumed to be independent of each other, and $\{\mathbf{w}_k\}$ is assumed to be independent of $\mathbf{x}_0$. Under these assumptions, the joint sequence $\{\mathbf{x}_k, \mathbf{y}_k\}$ is also a Markov process.

The goal of data assimilation is to obtain an optimal estimate $\hat{\mathbf{x}}_k$ of $\mathbf{x}_k$, given a sequence of observations $\mathbf{Y}_l = \{\mathbf{y}_1, \dots, \mathbf{y}_l\}$. Typically, the optimal estimate $\hat{\mathbf{x}}_k$ is chosen to minimize the error $\delta \mathbf{x}_k \equiv \mathbf{x}_k^t - \hat{\mathbf{x}}_k$ of estimation of the true state $\mathbf{x}_k^t$. Depending on the specific values of the indices k and l, the data assimilation problem is called a *smoothing* problem when $k < l$, or a *filtering* problem if $k = l$, or a *prediction* problem when $k > l$.

Estimation theory (see, e.g., ref. [33]) shows that the "minimum mean square error" (or the "minimum variance") estimate is the mean conditioned by observations ("conditional mean"). It is noteworthy that this result is general, independent of the nature of the conditional probability density $p(\mathbf{x}_k|\mathbf{Y})_l$. Thus, the conditional mean is a reasonable optimal estimate. The alternative "optimal estimate" is provided by the maximum likelihood (Bayesian) estimate obtained by maximizing the conditional probability density function $p(\mathbf{x}_k|\mathbf{Y})_l$. When $p(\mathbf{x}_k|\mathbf{Y})_l$ is a Gaussian, the maximum likelihood (Bayesian) estimate is identical to the minimum variance estimate. In the following sections, the terminology associated with the maximum likelihood (Bayesian) estimate will be used to analyze both the variational and sequential data assimilation methods.

4.4.1 Perfect Model

The standard 4–D VAR procedure (see, e.g., ref. [127]) minimizes a cost (objective) functional that measures the weighted sum of squares of distances to the background state $\mathbf{x}^b$ and to the observations $\mathbf{y}$ distributed over a time interval $[t_0, t_l]$, of the form

$$J^p = \frac{1}{2}\left(\mathbf{x}_0 - \mathbf{x}_0^b\right)^T \mathbf{B}_0^{-1}\left(\mathbf{x}_0 - \mathbf{x}_0^b\right) + \frac{1}{2}\sum_{k=1}^{l}\left(\mathbf{h}_k\mathbf{x}_k - \mathbf{y}_k\right)^T \mathbf{R}_k^{-1}\left(\mathbf{h}_k\mathbf{x}_k - \mathbf{y}_k\right).$$

$$(4.22)$$

In the above functional, $\mathbf{B}_0$ denotes the analysis error covariance matrix at time t_0. The observations are considered to start from time t_1, which is convenient for describing "cycling" 4–D VAR procedures. In the *standard* 4–D VAR, the minimization of the above cost functional is carried out with respect to the initial state $\mathbf{x}_0$. In such a case, the relation

$$\mathbf{L}\left(k, i\right) = \mathbf{L}\left(k, k-1\right)\cdots\mathbf{L}\left(i+1, i\right), \quad \text{for} \quad i \leq k, \qquad (4.23)$$

can be used to rewrite the cost function in Eq. (4.22) in order to show specifically the dependence on the initial state $\mathbf{x}_0$, in the form

$$J^p\left(\mathbf{x}_0\right) = \frac{1}{2}\left(\mathbf{x}_0 - \mathbf{x}_0^b\right)^T \mathbf{B}_0^{-1}\left(\mathbf{x}_0 - \mathbf{x}_0^b\right)$$

$$+ \frac{1}{2}\sum_{k=1}^{l}\left[\mathbf{h}_k\mathbf{L}\left(k, 0\right)\mathbf{x}_0 - \mathbf{y}_k\right]^T \mathbf{R}_k^{-1}\left[\mathbf{h}_k\mathbf{L}\left(k, 0\right)\mathbf{x}_0 - \mathbf{y}_k\right]. \qquad (4.24)$$

The gradient of $J^p\left(\mathbf{x}_0\right)$ is obtained by differentiating the above expression with respect to $\mathbf{x}_0$, which yields

$$\nabla_{\mathbf{x}_0} J^p\left(\mathbf{x}_0\right) = \mathbf{B}_0^{-1}\left(\mathbf{x}_0 - \mathbf{x}_0^b\right) + \sum_{k=1}^{l}\mathbf{L}^T\left(k, 0\right)\mathbf{h}_k^T\mathbf{R}_k^{-1}\left[\mathbf{h}_k\mathbf{L}\left(k, 0\right)\mathbf{x}_0 - \mathbf{y}_k\right].$$

$$(4.25)$$

The optimal solution $\hat{\mathbf{x}}_0$ is obtained by imposing the (necessary) condition

$\nabla_{\mathbf{x}_0} J^p(\mathbf{x}_0) = 0$, which gives

$$\hat{\mathbf{x}}_0 = \mathbf{x}_0^b - \mathbf{H}_{0,l}^{-1} \nabla_{\mathbf{x}_0} J\left(\mathbf{x}_0^b\right), \tag{4.26}$$

where

$$\mathbf{H}_{0,l} = \mathbf{B}_0^{-1} + \sum_{k=1}^{l} \mathbf{L}^T\left(k, 0\right) \mathbf{h}_k^T \mathbf{R}_k^{-1} \mathbf{h}_k \mathbf{L}\left(k, 0\right), \tag{4.27}$$

is the Hessian semipositive symmetric matrix (of second-order derivatives) of the cost function $J^p\left(\mathbf{x}_0\right)$ at $\mathbf{x}_0$, while the quantity $\nabla_{\mathbf{x}_0} J\left(\mathbf{x}_0^b\right)$, defined as

$$\nabla_{\mathbf{x}_0} J^p\left(\mathbf{x}_0^b\right) \equiv \sum_{k=1}^{l} \mathbf{L}^T\left(k, 0\right) \mathbf{h}_k^T \mathbf{R}_k^{-1}\left(\mathbf{y}_k^b - \mathbf{y}_k\right), \tag{4.28}$$

is the gradient of the cost function $J^p\left(\mathbf{x}_0\right)$ at $\mathbf{x}_0^b$. In the above expression, the quantity $\mathbf{y}_k^b$ is defined as

$$\mathbf{y}_k^b \equiv \mathbf{h}_k \mathbf{x}_k^b = \mathbf{h}_k \mathbf{L}(k, 0) \mathbf{x}_0^b. \tag{4.29}$$

The analysis error covariance matrix for the optimal estimate is

$$\mathbf{P}_0^a \equiv \left\langle \left(\hat{\mathbf{x}}_0 - \mathbf{x}_0^t\right) \left(\hat{\mathbf{x}}_0 - \mathbf{x}_0^t\right)^T \right\rangle = \mathbf{H}_{0,l}^{-1}, \tag{4.30}$$

where $\mathbf{x}_0^t$ denotes the true value of $\mathbf{x}_0$. The above result shows that the analysis error covariance matrix is the inverse $\mathbf{H}_{0,l}^{-1}$ of the Hessian matrix, which is of significant practical importance. The equivalent expressions, in terms of the Kalman gain, derived for the optimal estimate $\hat{\mathbf{x}}_0$ by several authors (Lorenc [127], Thépaut and Courtier [194], and Cohn [33]), can be obtained by re-writing $\mathbf{H}_{0,l}^{-1}$ using the Sherman-Morrison-Woodbury formula.

To show the optimality of each state on the trajectory determined by $\hat{\mathbf{x}}_0$ over the entire time window, it is necessary to prove that $\mathbf{L}\left(k, 0\right) \hat{\mathbf{x}}_0$ minimizes $J^p\left(\mathbf{x}_0\right)$. This can be demonstrated by first showing that the sequence $\hat{\mathbf{x}}_k = \mathbf{L}\left(k, 0\right) \hat{\mathbf{x}}_0$, $k = 1, \ldots, l$, minimizes the following cost function:

$$J^p\left(\mathbf{x}_k\right) = \frac{1}{2}\left(\mathbf{x}_k - \mathbf{x}_k^b\right)^T \mathbf{L}^{-T}\left(k,0\right) \mathbf{B}_0^{-1} \mathbf{L}^{-1}\left(k,0\right)\left(\mathbf{x}_k - \mathbf{x}_k^b\right)$$

$$+ \frac{1}{2}\sum_{i=1}^{k}\left[\mathbf{h}_i \mathbf{L}^{-1}\left(k,i\right)\mathbf{x}_k - \mathbf{y}_i\right]^T \mathbf{R}_i^{-1}\left[\mathbf{h}_i \mathbf{L}^{-1}\left(k,i\right)\mathbf{x}_k - \mathbf{y}_i\right]$$

$$+ \frac{1}{2}\sum_{i=k+1}^{l}\left[\mathbf{h}_i \mathbf{L}^{-1}\left(k,i\right)\mathbf{x}_k - \mathbf{y}_i\right]^T \mathbf{R}_i^{-1}\left[\mathbf{h}_i \mathbf{L}^{-1}\left(k,i\right)\mathbf{x}_k - \mathbf{y}_i\right], \quad (4.31)$$

In Eq. (4.31), the quantity $\mathbf{L}^{-T}$ denotes $\left(\mathbf{L}^T\right)^{-1}$. The above cost functional involves the inverse of $\mathbf{L}\left(k,0\right)$, and some (e.g., geophysical) models may not necessarily be invertible. However, all (i.e., both the invertible and noninvertible) cases can be treated by using in Eq. (4.31) the singular value decomposition (SVD)

$$\mathbf{L}\left(k,0\right) = \mathbf{U}\left(k,0\right)\mathbf{\Sigma}\left(k,0\right)\mathbf{V}^T\left(k,0\right), \quad (4.32)$$

where the diagonal matrix $\mathbf{\Sigma}\left(k,0\right) \equiv \mathrm{diag}\left(\sigma_1,\cdots,\sigma_i,\cdots,0,\cdots,0\right)$ contains the singular eigenvalues of $\mathbf{L}\left(k,0\right)$, while $\mathbf{U}\left(k,0\right)$ and $\mathbf{V}\left(k,0\right)$ are orthonormal matrices containing the left and right singular eigenvectors, respectively. Consequently, the generalized (or Moore-Penrose) inverse, $\tilde{\mathbf{L}}^{-1}\left(k,0\right)$, defined as

$$\tilde{\mathbf{L}}^{-1}\left(k,0\right) = \mathbf{V}\left(k,0\right)\tilde{\mathbf{\Sigma}}^{-1}\left(k,0\right)\mathbf{U}^T\left(k,0\right), \quad (4.33)$$

with

$$\tilde{\mathbf{\Sigma}}^{-1}\left(k,0\right) = \mathrm{diag}\left(\sigma_1^{-1},\cdots,\sigma_i^{-1},\cdots,0,\cdots,0\right), \quad (4.34)$$

can be used in Eq. (4.31) to rewrite that cost function in the form

$$J^p\left(\mathbf{x}_k\right) = \frac{1}{2}\left(\mathbf{x}_k - \mathbf{x}_k^b\right)^T \tilde{\mathbf{L}}^{-T}\left(k,0\right)\mathbf{B}_0^{-1}\tilde{\mathbf{L}}^{-1}\left(k,0\right)\left(\mathbf{x}_k - \mathbf{x}_k^b\right)$$

$$+ \frac{1}{2}\sum_{i=1}^{k}\left[\mathbf{h}_i\tilde{\mathbf{L}}^{-1}\left(k,i\right)\mathbf{x}_k - \mathbf{y}_i\right]^T \mathbf{R}_i^{-1}\left[\mathbf{h}_i\tilde{\mathbf{L}}^{-1}\left(k,i\right)\mathbf{x}_k - \mathbf{y}_i\right]$$

$$+ \frac{1}{2}\sum_{i=k+1}^{l}\left[\mathbf{h}_i\mathbf{L}\left(i,k\right)\mathbf{x}_k - \mathbf{y}_i\right]^T \mathbf{R}_i^{-1}\left[\mathbf{h}_i\mathbf{L}\left(i,k\right)\mathbf{x}_k - \mathbf{y}_i\right]. \quad (4.35)$$

Using the relations $\mathbf{x}_0 = \tilde{\mathbf{L}}^{-T}\left(k,0\right)\mathbf{x}_k$ and $\tilde{\mathbf{L}}^{-1}\left(k,0\right)\mathbf{L}\left(k,0\right) = \mathbf{I}$ in Eq.

(4.35) shows that $J^p\left(\mathbf{x}_k\right)$ is actually the same as $J^p\left(\mathbf{x}_0\right)$. The gradient of $J^p\left(\mathbf{x}_k\right)$ with respect to $\mathbf{x}_k$ is obtained using the chain rule:

$$\nabla_{\mathbf{x}_k} J^p\left(\mathbf{x}_k\right) = \tilde{\mathbf{L}}^{-1}\left(k,0\right)\nabla_{\mathbf{x}_0} J^p\left(\mathbf{x}_0\right). \tag{4.36}$$

Using the optimality condition $\nabla_{\mathbf{x}_k} J^p\left(\mathbf{x}_k\right) = 0$ together with Eq. (4.25) in Eq. (4.36) yields, after some algebraic manipulations, the following expression for the optimal estimate $\hat{\mathbf{x}}_k$ (of $\mathbf{x}_k$):

$$\hat{\mathbf{x}}_k = \mathbf{x}_k^b - \mathbf{H}_{k,l}^{-1}\nabla_{\mathbf{x}_k} J\left(\mathbf{x}_k^b\right), \tag{4.37}$$

where

$$\mathbf{H}_{k,l} = \tilde{\mathbf{L}}^{-T}\left(k,0\right)\mathbf{B}_0^{-1}\tilde{\mathbf{L}}^{-1}\left(k,0\right) + \sum_{i=1}^{k}\tilde{\mathbf{L}}^{-T}\left(k,i\right)\mathbf{h}_i^T\mathbf{R}_i^{-1}\mathbf{h}_i\tilde{\mathbf{L}}^{-1}\left(k,i\right)$$

$$+ \sum_{i=k+1}^{l}\mathbf{L}^T\left(i,k\right)\mathbf{h}_i^T\mathbf{R}_i^{-1}\mathbf{h}_i\mathbf{L}\left(i,k\right) \tag{4.38}$$

is actually the Hessian matrix of the cost function $J^p\left(\mathbf{x}_k\right)$ at $\mathbf{x}_k$, and where

$$\nabla_{\mathbf{x}_k} J^p\left(\mathbf{x}_k^b\right) = \tilde{\mathbf{L}}^{-1}\left(k,0\right)\nabla_{\mathbf{x}_0} J^p\left(\mathbf{x}_0^b\right) \tag{4.39}$$

is the gradient of $J^p\left(\mathbf{x}_k\right)$ at $\mathbf{x}_k^b$. Denoting the true value of $\mathbf{x}_k$ by $\mathbf{x}_k^t$, we define the analysis error covariance matrix $\mathbf{P}_k^a$ for the optimal estimate $\mathbf{x}_k$ is defined analogously to the initial analysis covariance matrix $\mathbf{P}_0^a$, cf. Eq. (4.30), namely

$$\mathbf{P}_k^a \equiv \left\langle\left(\hat{\mathbf{x}}_k - \mathbf{x}_k^t\right)\left(\hat{\mathbf{x}}_k - \mathbf{x}_k^t\right)^T\right\rangle = \mathbf{H}_{k,l}^{-1}. \tag{4.40}$$

Applying the matrix $\mathbf{L}\left(k,0\right)$ to Eqs. (4.37) through (4.40) yields the following counterpart expressions to the optimal solution $\mathbf{x}_0$:

$$\hat{\mathbf{x}}_k = \mathbf{L}\left(k,0\right)\hat{\mathbf{x}}_0, \tag{4.41}$$

$$\mathbf{L}^T\left(k,0\right)\mathbf{H}_{k,l}\mathbf{L}\left(k,0\right) = \mathbf{H}_{0,l}, \tag{4.42}$$

$$\mathbf{L}^T\left(k,0\right)\nabla_{\mathbf{x}_k} J^p\left(\mathbf{x}_k^b\right) = \nabla_{\mathbf{x}_0} J^p\left(\mathbf{x}_0^b\right), \tag{4.43}$$

$$\mathbf{P}_k^a = \mathbf{L}\left(k,0\right)\mathbf{P}_0^a\mathbf{L}^T\left(k,0\right), \tag{4.44}$$

which show that the optimal solution associated with the cost function $J^p(\mathbf{x}_k)$ is on the trajectory initiated with $\hat{\mathbf{x}}_0$, while the associated analysis error covariance is propagated by the model. This is true regardless of whether the model is invertible or not.

4.4.2 Model with Errors

When model errors that are uncorrelated in time are included in the analysis, the cost function J^p defined in (4.22) is augmented by the term $\frac{1}{2}\sum_{k=1}^{l}\mathbf{w}_k^T\mathbf{Q}_k^{-1}\mathbf{w}_k$. Therefore, the cost functional to be optimized in this case becomes

$$J^e \equiv J^p + \frac{1}{2}\sum_{k=1}^{l}\mathbf{w}_k^T\mathbf{Q}_k^{-1}\mathbf{w}_k$$

$$= \frac{1}{2}\left(\mathbf{x}_0 - \mathbf{x}_0^b\right)^T\mathbf{B}_0^{-1}\left(\mathbf{x}_0 - \mathbf{x}_0^b\right) + \frac{1}{2}\sum_{k=1}^{l}\left(\mathbf{h}_k\mathbf{x}_k - \mathbf{y}_k\right)^T\mathbf{R}_k^{-1}\left(\mathbf{h}_k\mathbf{x}_k - \mathbf{y}_k\right)$$

$$+ \frac{1}{2}\sum_{k=1}^{l}\mathbf{w}_k^T\mathbf{Q}_k^{-1}\mathbf{w}_k. \tag{4.45}$$

Several methods can be used for minimizing the above cost functional, as follows:

1. Performing a constrained minimization J^e with respect to the set $\{\mathbf{x}_0, \cdots, \mathbf{x}_l, \mathbf{w}_1, \cdots, \mathbf{w}_l\}$, subject to the constraints expressed in Eq. (4.19). However, the optimization is computationally prohibitive for this formulation in meteorological data assimilation, although it has been applied by using so-called "representers reduction approximation" in oceanic data assimilation (see, e.g., Bennett [14]). This formulation is usually employed for theoretical analysis, and it has been exclusively used to prove the equivalence between Kalman smoothers and 4–D VAR.

2. Performing the unconstrained minimization of the following cost

functional:

$$J_1^e \equiv \frac{1}{2} (\mathbf{x}_0 - \mathbf{x}_0^b)^T \mathbf{B}_0^{-1} (\mathbf{x}_0 - \mathbf{x}_0^b)$$

$$= \frac{1}{2} \sum_{k=1}^{l} (\mathbf{h}_k \mathbf{x}_k - \mathbf{y}_k)^T \mathbf{R}_k^{-1} (\mathbf{h}_k \mathbf{x}_k - \mathbf{y}_k)$$

$$+ \frac{1}{2} \sum_{k=1}^{l} [\mathbf{x}_k - \mathbf{L}(k, k-1) \mathbf{x}_{k-1}]^T$$

$$\mathbf{\Gamma}^T (k-1) \mathbf{Q}_k^{-1} \mathbf{\Gamma} (k-1) [\mathbf{x}_k - \mathbf{L}(k, k-1) \mathbf{x}_{k-1}] , \qquad (4.46)$$

with respect to $\{\mathbf{x}_0, \cdots, \mathbf{x}_l\}$. In this case, the 4–D VAR procedure can be considered as a method for searching an optimal trajectory segment, rather than searching for the optimal initial condition.

3. Performing the unconstrained minimization of the following cost functional:

$$J_2^e \equiv \frac{1}{2} (\mathbf{x}_0 - \mathbf{x}_0^b)^T \mathbf{B}_0^{-1} (\mathbf{x}_0 - \mathbf{x}_0^b)$$

$$+ \frac{1}{2} \sum_{k=1}^{l} \left\{ \mathbf{h}_k \left[\mathbf{L}(k, 0) \mathbf{x}_0 + \sum_{j=1}^{k} \mathbf{L}(k, j) \mathbf{\Gamma}(j-1) \mathbf{w}_j \right] - \mathbf{y}_k \right\}^T$$

$$\times \mathbf{R}_k^{-1} \left\{ \mathbf{h}_k \left[\mathbf{L}(k, 0) \mathbf{x}_0 + \sum_{j=1}^{k} \mathbf{L}(k, j) \mathbf{\Gamma}(j-1) \mathbf{w}_j \right] - \mathbf{y}_k \right\}$$

$$+ \frac{1}{2} \sum_{k=1}^{l} \mathbf{w}_k^T \mathbf{Q}_k^{-1} \mathbf{w}_k. \qquad (4.47)$$

with respect to $\{\mathbf{x}_0, \mathbf{w}_1 \cdots, \mathbf{w}_l\}$. In the derivation of J_2^e, the time-integration solution of the difference equation (4.19), was used in the form

$$\mathbf{x}_k = \mathbf{L}(k, 0) \mathbf{x}_0 + \sum_{j=1}^{k} \mathbf{L}(k, j) \mathbf{\Gamma}(j-1) \mathbf{w}_j. \qquad (4.48)$$

Since the vectors $\mathbf{w}_k$ are r-dimensional, and since r may be smaller in practice than the dimension n of $\mathbf{x}_k$, the functional J_2^e is especially useful for approximately representing model errors with a small r. Such an approximation has been implemented by Zupanski [235]. On the other hand, in the older variational continuous assimilation technique due to Derber [55], the noise $\mathbf{w}_k$ is considered to be time independent (i.e., k is fixed).

The derivations in this section will use the functional J_2^e, so the results are computational and algorithm-oriented. Since the set $\{\mathbf{x}_0, \mathbf{w}_1 \cdots, \mathbf{w}_l\}$ is treated jointly, it is convenient to introduce the augmented $(n + lr)$-dimensional vector $\mathbf{Z}_0^T = (\mathbf{x}_0^T, \mathbf{w}_1^T, \cdots, \mathbf{w}_l^T)$ to rewrite J_2^e in the form

$$J_2^e = \frac{1}{2}(\mathbf{Z}_0 - \mathbf{Z}_0^b)^T \mathbf{B}_{Z_0}^{-1}(\mathbf{Z}_0 - \mathbf{Z}_0^b) + \frac{1}{2}\sum_{k=1}^{l}(\mathbf{h}_k\mathbf{C}_k\mathbf{Z}_0 - \mathbf{y}_k)^T \mathbf{R}_k^{-1}(\mathbf{h}_k\mathbf{C}_k\mathbf{Z}_0 - \mathbf{y}_k),$$

(4.49)

where $(\mathbf{Z}_0^b)^T = \left[(\mathbf{x}_0^b)^T, \mathbf{0}, \cdots, \mathbf{0}\right]$, $\mathbf{B}_{Z_0}$ is a $(1 + l) \times (1 + l)$ block diagonal matrix of total dimension $[(n + rl) \times (n + rl)]$ with diagonal blocks $\mathbf{B}_0, \mathbf{Q}_1, \cdots, \mathbf{Q}_l$, while

$$\mathbf{C}_k \equiv [\mathbf{L}(k, 0), \mathbf{L}(k, 1)\mathbf{\Gamma}(0), \cdots, \mathbf{L}(k, k)\mathbf{\Gamma}(k - 1), \mathbf{0}, \cdots, \mathbf{0}].$$ (4.50)

are $n \times (n + rl)$-dimensional matrices. When the model is perfect, J_2^e reduces, as it should, to J^p. Following the same conceptual procedure as for J^p in the previous section leads to the optimal solution

$$\hat{\mathbf{Z}}_0 = \mathbf{Z}_0^b + \mathbf{H}_{Z_0}^{-1}\nabla_{Z_0}J_2^e(\mathbf{Z}_0^b),$$ (4.51)

where the Hessian matrix $\mathbf{H}_{Z_0}$ is given by

$$\mathbf{H}_{Z_0} = \mathbf{B}_{Z_0}^{-1} + \sum_{k=1}^{l}\mathbf{C}_k^T\mathbf{h}_k^T\mathbf{R}_k^{-1}\mathbf{h}_k\mathbf{C}_k,$$ (4.52)

while the gradient of J_2^e at $\mathbf{Z}_0^b$ has the expression

$$\nabla_{Z_0}J_2^e(\mathbf{Z}_0^b) = \sum_{k=1}^{l}\mathbf{C}_k^T\mathbf{h}_k^T\mathbf{R}_k^{-1}(\mathbf{h}_k\mathbf{C}_k\mathbf{Z}_0^b - \mathbf{y}_k)$$

$$= \sum_{k=1}^{l}\mathbf{C}_k^T\mathbf{h}_k^T\mathbf{R}_k^{-1}\left[\mathbf{h}_k\mathbf{L}(k, 0)\mathbf{x}_0^b - \mathbf{y}_k\right].$$ (4.53)

Furthermore, the error covariance of $\hat{\mathbf{Z}}_0$ can be shown to be given by the inverse of the Hessian matrix $\mathbf{H}_{Z_0}$, i.e.,

$$\mathbf{P}^a_{Z_0} = \mathbf{H}^{-1}_{Z_0}. \tag{4.54}$$

Analogous to the previous section, where the model was considered to be "perfect" (i.e., free of modeling errors), every model state on the trajectory determined by the optimal estimate $\hat{\mathbf{Z}}_0 = (\hat{\mathbf{x}}_0^T, \hat{\mathbf{w}}_1^T, \ldots, \hat{\mathbf{w}}_l^T)$ minimizes the functional J_2^e in the presence of white-noise-type modeling errors. The proof is analogous to that for a perfect model, which led to Eqs. (4.41) through (4.44), by showing that $\hat{\mathbf{Z}}_k = \left[\mathbf{L}\,(k,0)\,\hat{\mathbf{x}}_0^T, \hat{\mathbf{w}}_1^T, \ldots, \hat{\mathbf{w}}_l^T \right]$ also minimizes J_2^e. This can be done by:

1. Employing the quantity $\mathbf{Z}_k = \mathbf{A}\,(k,0)\,\mathbf{Z}_0$, where $\mathbf{A}\,(k,0)$ is a $(n+rl)\times(n+rl)$ matrix having its first n rows occupied by $\mathbf{C}\,(k,0)$, and having all other rows with diagonal entries set to 1 and non-diagonal entries set to 0;

2. noting from Eq. (4.48) that the model state on this trajectory is

$$\hat{\mathbf{x}}_k = \mathbf{L}\,(k,0)\,\hat{\mathbf{x}}_0 + \sum_{j=1}^{k} \mathbf{L}\,(k,j)\,\mathbf{\Gamma}\,(j-1)\,\hat{\mathbf{w}}_j = \mathbf{C}_k \hat{\mathbf{Z}}_0. \tag{4.55}$$

Also analogously to the case of a "perfect" model, one can obtain the following results

$$\mathbf{A}^T\,(k,0)\,\mathbf{H}_{Z_k}\,\mathbf{A}\,(k,0) = \mathbf{H}_{Z_0}, \tag{4.56}$$

$$\mathbf{P}^a_{Z_k} = \mathbf{A}\,(k,0)\,\mathbf{P}^a_{Z_0}\,\mathbf{A}^T\,(k,0). \tag{4.57}$$

The 4–D VAR procedure performed on models where errors are present also retains an "additive" property similar to that for perfect models. This property can be proven by considering two cost functions, $J_{2,1}^e\,(\mathbf{Z}_0)$ and $J_{2,2}^e\,(\mathbf{Z}_0)$, defined as follows:

$$J_{2,1}^e\,(\mathbf{Z}_0) = \frac{1}{2}(\mathbf{x}_0 - \mathbf{x}_0^b)^T \mathbf{B}_0^{-1}(\mathbf{x}_0 - \mathbf{x}_0^b) + \frac{1}{2}\sum_{i=1}^{l_1}\left(\mathbf{h}_{k_1(i)}\mathbf{x}_{k_1(i)} - \mathbf{y}_{k_1(i)}\right)^T$$

$$\mathbf{R}^{-1}_{k_1(i)}\left(\mathbf{h}_{k_1(i)}\mathbf{x}_{k_1(i)} - \mathbf{y}_{k_1(i)}\right) + \sum_{k=1}^{l} \mathbf{w}_k^T \mathbf{Q}_k^{-1} \mathbf{w}_k, \tag{4.58}$$

and

$$J_{2,2}^e (\mathbf{Z}_0) = \frac{1}{2}\left(\mathbf{Z} - \hat{\mathbf{Z}}_0\right)^T \hat{\mathbf{B}}_Z^{-1} \left(\mathbf{Z} - \hat{\mathbf{Z}}_0\right) + \frac{1}{2}\sum_{i=1}^{l_2} \left(\mathbf{h}_{k_2(i)}\mathbf{x}_{k_2(i)} - \mathbf{y}_{k_2(i)}\right)^T$$

$$\mathbf{R}_{k_2(i)}^{-1} \left(\mathbf{h}_{k_2(i)}\mathbf{x}_{k_2(i)} - \mathbf{y}_{k_2(i)}\right). \tag{4.59}$$

Consider that $\hat{\mathbf{Z}}_0$ is the minimization solution of $J_{2,1}^e (\mathbf{Z}_0)$ and $\hat{\mathbf{B}}_Z^{-1}$ is taken as $\mathbf{H}_{Z_0}$. Expanding $J_{2,1}^e (\mathbf{Z}_0)$ in a Taylor series around $\hat{\mathbf{Z}}_0$ yields

$$J_{2,1}^e (\mathbf{Z}_0) = J_{2,1}^e \left(\hat{\mathbf{Z}}_0\right) + \frac{1}{2}\left(\mathbf{Z} - \hat{\mathbf{Z}}_0\right)^T \hat{\mathbf{B}}_Z^{-1} \left(\mathbf{Z} - \hat{\mathbf{Z}}_0\right). \tag{4.60}$$

Combining Eqs. (4.59) and (4.60) gives

$$J_2^e (\mathbf{Z}_0) = J_{2,2}^e (\mathbf{Z}_0) + J_{2,1}^e (\mathbf{Z}_0), \tag{4.61}$$

indicating that the minimization solution of $J_{2,2}^e (\mathbf{Z})$ is identical to that of $J_2^e (\mathbf{Z})$, where all observations are processed simultaneously.

The foregoing derivations assumed that the model errors can be represented by uncorrelated Gaussian white noise. Correlated model errors were investigated by Zupanski [235].

4.4.3 Optimality Properties of 4–D VAR

a) **Perfect model**

For a perfect model, the joint conditional probability density is

$$p(\mathbf{x}_0, \mathbf{x}_1, \cdots, \mathbf{x}_l | \mathbf{y}_1, \cdots, \mathbf{y}_l) = C\, exp\left\{-\frac{1}{2}(\mathbf{x}_0 - \mathbf{x}_0^b)^T \mathbf{B}_0^{-1} (\mathbf{x}_0 - \mathbf{x}_0^b)\right.$$

$$\left. -\frac{1}{2}\sum_{k=1}^{n} (\mathbf{y}_k - \mathbf{h}_k\mathbf{x}_k)^T \mathbf{R}_k^{-1} (\mathbf{y}_k - \mathbf{h}_k\mathbf{x}_k)]\right\}, \tag{4.62}$$

where C is a normalization constant. Maximizing $p(\mathbf{x}_0, \mathbf{x}_1, \cdots, \mathbf{x}_l | \mathbf{y}_1, \cdots, \mathbf{y}_l)$ with respect to $\{\mathbf{x}_0, \cdots, \mathbf{x}_l\}$ is thus equivalent to minimizing the cost functional J^p defined in Eq. (4.22). Hence, the minimum of J^p is also the *joint maximum likelihood* (Bayesian) estimate (or the most probable estimate). In particular, the optimal solution with respect to $\mathbf{x}_k$ $(0 \leq k \leq l)$ is the solution that maximizes the marginal density $p(\mathbf{x}_k|\mathbf{y}_1, \cdots, \mathbf{y}_l)$, rather than the joint density.

Consider first the marginal density $p(\mathbf{x}_0|\mathbf{y}_1, \cdots, \mathbf{y}_l)$ for the initial state $\mathbf{x}_0$ (i.e., $k = 0$). According to Bayes' theorem, this distribution is given by

$$p(\mathbf{x}_0|\mathbf{y}_1, \cdots, \mathbf{y}_l) = \frac{p(\mathbf{y}_1, \cdots, \mathbf{y}_l|\mathbf{x}_0)\, p(\mathbf{x}_0)}{p(\mathbf{y}_1, \cdots, \mathbf{y}_l)}. \tag{4.63}$$

With $\mathbf{x}_0$ given, the relation $\mathbf{x}_k = \mathbf{L}(k, 0)\mathbf{x}_0$ shows that the set $\{\mathbf{x}_1 \cdots, \mathbf{x}_l\}$ contains independent quantities. Furthermore, using Eq. (4.21) yields

$$\mathbf{y}_i - \mathbf{h}_i\mathbf{L}(i, 0)\mathbf{x}_0 = \mathbf{v}_i \quad i = 1, \ldots, l, \tag{4.64}$$

which indicates that $\{\mathbf{y}_1, \cdots, \mathbf{y}_l\}$ are also independent variates. It therefore follows that

$$p(\mathbf{y}_1, \ldots, \mathbf{y}_l|\mathbf{x}_0) = C'p[\mathbf{y}_1|\mathbf{L}(1, 0)\mathbf{x}_0] \ldots p[\mathbf{y}_l|\mathbf{L}(l, 0)\mathbf{x}_0]$$
$$= p_{v_1}[\mathbf{y}_1 - \mathbf{h}_1\mathbf{L}(1, 0)\mathbf{x}_0] \ldots p_{v_l}[\mathbf{y}_l - \mathbf{h}_l\mathbf{L}(l, 0)\mathbf{x}_0], \tag{4.65}$$

where C' is a normalization constant. Since the variates $\mathbf{x}_0$ and $\mathbf{v}_i$ are assumed to follow Gaussian distributions, Eq. (4.63) becomes

$$p(\mathbf{x}_0|\mathbf{y}_1, \cdots, \mathbf{y}_l) = C'' exp \left\{ -\frac{1}{2}\left(\mathbf{x}_0 - \mathbf{x}_0^b\right)^T \mathbf{B}_0^{-1}\left(\mathbf{x}_0 - \mathbf{x}_0^b\right) \right.$$
$$\left. -\frac{1}{2}\sum_{i=1}^{l}[\mathbf{h}_i\mathbf{L}(i, 0)\mathbf{x}_0 - \mathbf{y}_i]^T \mathbf{R}_i^{-1}[\mathbf{h}_i\mathbf{L}(i, 0)\mathbf{x}_0 - \mathbf{y}_i] \right\}, \tag{4.66}$$

where C'' is another normalization constant which depends only on C' and $p(\mathbf{y}_1, \cdots, \mathbf{y}_l)$. Thus, maximizing $p(\mathbf{x}_0|\mathbf{y}_1, \cdots, \mathbf{y}_l)$ turns out to be equivalent to minimizing the cost functional $J^p(\mathbf{x}_0)$ defined in Eq. (4.24). Due to the consistency property of 4–D VAR, maximizing $p(\mathbf{x}_0|\mathbf{y}_1, \cdots, \mathbf{y}_l)$ is equivalent to maximizing $p(\mathbf{x}_0, \mathbf{x}_1, \cdots, \mathbf{x}_l|\mathbf{y}_1, \cdots, \mathbf{y}_l)$.

Since $\mathbf{x}_k = \mathbf{L}(k, 0)\mathbf{x}_0$, it follows (see, e.g., ref. [100]) that

$$\|\mathbf{L}(k, 0)\|\, p(\mathbf{x}_0|\mathbf{y}_1, \cdots, \mathbf{y}_l) = p(\mathbf{x}_k|\mathbf{y}_1, \cdots, \mathbf{y}_l), \tag{4.67}$$

where $\|\mathbf{L}\,(k,0)\|$ is the absolute value of the determinant of $\mathbf{L}\,(k,0)$. Thus, maximizing $p\,(\mathbf{x}_k|\mathbf{y}_1,\cdots,\mathbf{y}_l)$ is equivalent to maximizing $p\,(\mathbf{x}_0|\mathbf{y}_1,\cdots,\mathbf{y}_l)$, which implies that the 4–D VAR solution is optimal not only with respect to the model trajectory comprising the set $\{\mathbf{x}_0,\cdots,\mathbf{x}_l\}$, but also with respect to $\mathbf{x}_k$ at a single observation time.

b) Model with errors

In the presence of errors, the maximum likelihood estimate is obtained by maximizing the distribution $p\,(\mathbf{x}_0,\mathbf{w}_1,\cdots,\mathbf{w}_l|\mathbf{y}_1,\cdots,\mathbf{y}_l)$ with respect to $\{\mathbf{x}_0,\mathbf{w}_1,\cdots,\mathbf{w}_l\}$. The solution thus obtained can be shown (following the derivations in [100], p. 153) to be the same as the 4–D VAR solution minimizing the cost functional J_2^e defined in Eq. (4.47). According to Bayes' theorem, the density $p\,(\mathbf{x}_0,\mathbf{w}_1,\cdots,\mathbf{w}_l|\mathbf{y}_1,\cdots,\mathbf{y}_l)$ can be written in the form

$$p(\mathbf{x}_0,\mathbf{w}_1,\cdots,\mathbf{w}_l|\mathbf{y}_1,\cdots,\mathbf{y}_l) =$$
$$\frac{p\,(\mathbf{y}_1,\cdots,\mathbf{y}_l|\mathbf{x}_0,\mathbf{w}_1,\cdots,\mathbf{w}_l)\,p\,(\mathbf{x}_0|\mathbf{w}_1,\cdots,\mathbf{w}_l)\,p\,(\mathbf{w}_1,\cdots,\mathbf{w}_l)}{p\,(\mathbf{y}_1,\cdots,\mathbf{y}_l)}. \tag{4.68}$$

When the set $\{\mathbf{x}_0,\mathbf{w}_1,\cdots,\mathbf{w}_l\}$ is given, $\mathbf{x}_k$ is determined from Eq. (4.48). With $\mathbf{x}_k$ and the white-noise $\mathbf{v}_k$ known, Eq. (4.21) indicates that the variates $\mathbf{y}_k$ are independent Gaussian distributions. It therefore follows that

$$p\,(\mathbf{y}_1,\cdots,\mathbf{y}_l|\mathbf{x}_0,\mathbf{w}_1,\cdots,\mathbf{w}_l) = p\,(\mathbf{y}_1|\mathbf{x}_0,\mathbf{w}_1,\cdots,\mathbf{w}_l)\cdots p\,(\mathbf{y}_l|\mathbf{x}_0,\mathbf{w}_1,\cdots,\mathbf{w}_l)$$
$$= p\,(\mathbf{y}_1|\mathbf{x}_1)\cdots p\,(\mathbf{y}_l|\mathbf{x}_l)$$
$$= p_{v_1}\left\{\mathbf{y}_1 - \mathbf{h}_1\left[\mathbf{L}\,(k,0)\,x_0 + \sum_{j=1}^{1}\mathbf{L}\,(1,j)\,\mathbf{\Gamma}\,(j-1)\,\mathbf{w}_j\right]\right\}$$
$$\cdots p_{v_l}\left\{\mathbf{y}_l - \mathbf{h}_l\left[\mathbf{L}\,(k,0)\,x_0 + \sum_{j=1}^{1}\mathbf{L}\,(l,j)\,\mathbf{\Gamma}\,(j-1)\,\mathbf{w}_j\right]\right\}. \tag{4.69}$$

Since $\mathbf{x}_0$ is independent of the variates $\{\mathbf{w}_1,\cdots,\mathbf{w}_l\}$, it follows that

$$p\,(\mathbf{x}_0|\mathbf{w}_1,\cdots,\mathbf{w}_l) = p_{x_0}\,(\mathbf{x}_0), \tag{4.70}$$

and since the variates $\{\mathbf{w}_1, \cdots, \mathbf{w}_l\}$ are also independent, it follows that

$$p(\mathbf{w}_1, \cdots, \mathbf{w}_l) = p_{w_1}(\mathbf{w}_1) \cdots p_{w_l}(\mathbf{w}_l). \qquad (4.71)$$

Substituting Eqs. (4.69) through (4.71) into (4.68) yields

$$p(\mathbf{x}_0, \mathbf{w}_1, \cdots, \mathbf{w}_l | \mathbf{y}_1, \cdots, \mathbf{y}_l) = C' exp\left(-J_2^e\right), \qquad (4.72)$$

where J_2^e is defined in Eq. (4.47), and the quantity C' only depends on the variates $\{\mathbf{y}_1, \cdots, \mathbf{y}_l\}$ but is independent of the variates $\{\mathbf{x}_0, \mathbf{w}_1, \cdots, \mathbf{w}_l\}$. Thus, maximizing $p(\mathbf{x}_0, \mathbf{w}_1, \cdots, \mathbf{w}_l | \mathbf{y}_1, \cdots, \mathbf{y}_l)$ is equivalent to minimizing the cost functional J_2^e as defined in Eq. (4.47). Furthermore, Eq. (4.72) also indicates that the distribution $p(\mathbf{x}_0, \mathbf{w}_1, \cdots, \mathbf{w}_l | \mathbf{y}_l)$ is Gaussian, and hence the distributions $p(\mathbf{x}_0 | \mathbf{y}_l)$, $p(\mathbf{w}_1 | \mathbf{y}_l), \ldots, p(\mathbf{w}_l | \mathbf{y}_l)$ are also Gaussian. Therefore, the minimum solution of J_2^e, namely $\hat{\mathbf{Z}}_0 = \left(\hat{\mathbf{x}}_0^T, \hat{\mathbf{w}}_1^T, \ldots, \hat{\mathbf{w}}_l^T\right)$, also provides the maximum likelihood estimates, namely the solution that simultaneously maximizes $p(\mathbf{x}_0 | \mathbf{y}_l)$, $p(\mathbf{w}_1 | \mathbf{y}_l), \ldots, p(\mathbf{w}_l | \mathbf{y}_l)$, respectively.

On the other hand, the optimal estimation with respect to the set of variates $\{\mathbf{x}_k, \mathbf{w}_1, \cdots, \mathbf{w}_l\}$, $0 \leq k \leq l$, is given by the estimate which maximizes the conditional density $p(\mathbf{x}_k, \mathbf{w}_1, \cdots, \mathbf{w}_l | \mathbf{y}_1, \cdots, \mathbf{y}_l)$. Recalling the definition $\mathbf{Z}_0^T = \left(\mathbf{x}_0^T, \mathbf{w}_1^T, \cdots, \mathbf{w}_l^T\right)$, and recalling that $\mathbf{Z}_k = \mathbf{A}(k, 0)\mathbf{Z}_0$, it follows that the vector $\mathbf{Z}_k^T = \left(\mathbf{x}_k, \mathbf{w}_1, \cdots, \mathbf{w}_l\right)^T$ satisfies the transformation

$$\mathbf{Z}_k^T = \mathbf{A}(k, 0)\mathbf{Z}_0^T. \qquad (4.73)$$

By applying the Theorem 2.7 in [100] to the above relation gives

$$\|\mathbf{A}\| \, p(\mathbf{x}_k, \mathbf{w}_1, \cdots, \mathbf{w}_l | \mathbf{y}_1, \cdots, \mathbf{y}_l) = p(\mathbf{x}_0, \mathbf{w}_1, \cdots, \mathbf{w}_l | \mathbf{y}_1, \cdots, \mathbf{y}_l), \qquad (4.74)$$

where $\|\mathbf{A}\|$ denotes the absolute value of the determinant of $\mathbf{A}(k, 0)$. Hence, maximizing $p(\mathbf{x}_k, \mathbf{w}_1, \cdots, \mathbf{w}_l | \mathbf{y}_1, \cdots, \mathbf{y}_l)$ is equivalent to maximizing $p(\mathbf{x}_0, \mathbf{w}_1, \cdots, \mathbf{w}_l | \mathbf{y}_1, \cdots, \mathbf{y}_l)$. Therefore, the 4–D VAR solution is optimal not only with respect to the model trajectory consisting of $(\mathbf{x}_0, \cdots, \mathbf{x}_l)$, but also with respect to $\mathbf{x}_k$ at a single observation time even though the model error is taken into consideration; this property is called *"consistent optimality."*

c) **Optimal transferability property**

Another optimal property of 4–D VAR is the so-called *"transferable optimality property,"* which refers to a certain equivalence between simultaneous and batch processing, in that information from one batch can be fully transferred to the next one via a background term defined by the 4–D VAR solution at the end of the time window of the previous batch. This property is intimately related to certain equivalences between 4–D VAR and Kalman smoothers (to be discussed in the next section).

For a *perfect linear model, it is possible to prove that the 4–D VAR solution for the batch processing of two sequential batches of observations,* $\{\mathbf{y}_1, \cdots, \mathbf{y}_l\}$ *and* $\{\mathbf{y}_{l+1}, \cdots, \mathbf{y}_{l+m}\}$, *using the optimal 4–D VAR solution of the first batch as the background at the observation time corresponding to the processing of the second batch, is identical to the 4–D VAR solution for the simultaneous processing of the entire set of observations* $\{\mathbf{y}_1, \cdots, \mathbf{y}_l, \mathbf{y}_{l+1}, \cdots, \mathbf{y}_{l+m}\}$. To prove this property, consider the cost function associated with the second batch, namely

$$
J(\mathbf{x}_l) = \frac{1}{2}\left(\mathbf{x}_l - \mathbf{x}_l^{(1)}\right)^T (\mathbf{P}_l^a)^{-1}\left(\mathbf{x}_l - \mathbf{x}_l^{(1)}\right)
$$
$$
+ \frac{1}{2}\sum_{k=l+1}^{l+m}\left[\mathbf{h}_k\mathbf{L}\,(k,l)\,\mathbf{x}_l - \mathbf{y}_k\right]^T \mathbf{R}_k^{-1}\left[\mathbf{h}_k\mathbf{L}\,(k,l)\,\mathbf{x}_l - \mathbf{y}_k\right], \qquad (4.75)
$$

where $\mathbf{x}_l^{(1)}$ denotes the 4–D VAR solution associated with the (first) batch $\{\mathbf{y}_1, \cdots, \mathbf{y}_l\}$, namely

$$
\mathbf{x}_l^{(1)} = \mathbf{L}\,(l,0)\left[\mathbf{x}_0^b + (\mathbf{H}_{0,l})^{-1}\sum_{k=1}^{l}\mathbf{L}(k,0)^T\mathbf{h}_k^T\mathbf{R}_k^{-1}\left(\mathbf{y}_k - \mathbf{y}_k^b\right)\right], \qquad (4.76)
$$

where

$$
\mathbf{y}_k^b = \mathbf{h}_k\mathbf{L}\,(k,0)\,\mathbf{x}_0^b. \qquad (4.77)
$$

Note also that the inverse, $(\mathbf{P}_l^a)^{-1}$, of the analysis error covariance matrix $\mathbf{P}_l^a$ associated with $\mathbf{x}_l^{(1)}$, is given by

$$
(\mathbf{P}_l^a)^{-1} = \mathbf{L}^{-T}(l,0)\,\mathbf{H}_{0,l}\mathbf{L}^{-1}(l,0)\,(l,0). \qquad (4.78)
$$

The 4–D VAR solution, $\mathbf{x}_l^{(2)}$, corresponding to the functional $J(\mathbf{x}_l)$ defined in Eq. (4.75) is obtained following the procedure outlined in Section 4.4.1, to obtain

$$\mathbf{x}_l^{(2)} = \mathbf{x}_l^{(1)} + (\mathbf{H}_{l,l+m})^{-1} \sum_{k=l+1}^{l+m} \mathbf{L}^T(k,l)\,\mathbf{h}_k^T\mathbf{R}_k^{-1}\left[\mathbf{y}_k - \mathbf{h}_k\mathbf{L}(k,l)\,\mathbf{x}_l^{(1)}\right], \quad (4.79)$$

where

$$\mathbf{H}_{l,l+m} = (\mathbf{P}_l^a)^{-1} + \sum_{k=l+1}^{l+m} \mathbf{L}^T(k,l)\,\mathbf{h}_k^T\mathbf{R}_k^{-1}\mathbf{h}_k\mathbf{L}(k,l). \quad (4.80)$$

Substituting Eq. (4.78) into Eq. (4.80) yields

$$\mathbf{H}_{l,l+m} = \mathbf{L}^{-T}(l,0)\,\mathbf{H}_{0,l+m}\mathbf{L}^{-1}(l,0)\,(l,0), \quad (4.81)$$

where

$$\mathbf{H}_{0,l+m} = \mathbf{B}_0^{-1} + \sum_{k=1}^{l+m} \mathbf{L}^T(k,0)\,\mathbf{h}_k^T\mathbf{R}_k^{-1}\mathbf{h}_k\mathbf{L}(k,0). \quad (4.82)$$

is the Hessian associated with the 4–D VAR solution when all of the observation are processed simultaneously. Substituting the expression of $\mathbf{x}_l^{(1)}$ from Eq. (4.76) together with Eq. (4.81) into the expression of $\mathbf{x}_l^{(2)}$ in Eq. (4.79) yields

$$
\begin{aligned}
\mathbf{x}_l^{(2)} = \mathbf{L}(l,0) &\left\{ \mathbf{x}_0^b + \mathbf{H}_{0,l}^{-1}\sum_{k=1}^{l}\mathbf{L}^T(k,0)\,\mathbf{h}_k^T\mathbf{R}_k^{-1}(\mathbf{y}_k - \mathbf{y}_k^b) \right. \\
&\left. + \mathbf{H}_{0,l+m}^{-1}\mathbf{L}^T(0,l)\sum_{k=l+1}^{l+m}\mathbf{L}^T(k,l)\,\mathbf{h}_k^T\mathbf{R}_k^{-1}\left[\mathbf{y}_k - \mathbf{h}_k\mathbf{L}(k,l)\,\mathbf{x}_l^{(1)}\right] \right\} \\
= \mathbf{L}(l,0) &\left\{ \mathbf{x}_0^b + \mathbf{H}_{0,l}^{-1}\sum_{k=1}^{l}\mathbf{L}^T(k,0)\,\mathbf{h}_k^T\mathbf{R}_k^{-1}(\mathbf{y}_k - \mathbf{y}_k^b) \right. \\
&+ \mathbf{H}_{0,l+m}^{-1}\left[\sum_{k=1}^{l+m}\mathbf{L}^T(k,0)\,\mathbf{h}_k^T\mathbf{R}_k^{-1}(\mathbf{y}_k - \mathbf{y}_k^b) - \sum_{k=1}^{l}\mathbf{L}^T(k,0)\,\mathbf{h}_k^T\mathbf{R}_k^{-1}(\mathbf{y}_k - \mathbf{y}_k^b) \right. \\
&\left.\left. - \sum_{k=l+1}^{l+m}\mathbf{L}^T(k,l)\,\mathbf{h}_k^T\mathbf{R}_k^{-1}\mathbf{h}_k\mathbf{L}(k,l)\,(\mathbf{H}_{0,l})^{-1}\sum_{k=1}^{l}\mathbf{L}^T(k,0)\,\mathbf{h}_k^T\mathbf{R}_k^{-1}(\mathbf{y}_k - \mathbf{y}_k^b)\right]\right\}.
\end{aligned}
$$
$$(4.83)$$

On the other hand, the following relation holds

$$\mathbf{H}_{0,l+m} = \mathbf{H}_{0,l} + \sum_{k=l+1}^{l+m} \mathbf{L}^T (k,0) \, \mathbf{h}_k^T R_k^{-1} \mathbf{h}_k \mathbf{L} (k,0) . \qquad (4.84)$$

This equation is now substituted in the second equality in Eq. (4.83) to obtain, after some algebraic simplifications, the expression

$$\mathbf{x}_l^{(2)} = \mathbf{L} (l,0) \left\{ \mathbf{x}_0^b + \mathbf{H}_{0,l+m}^{-1} \sum_{k=1}^{l+m} \mathbf{L}^T (k,0) \, \mathbf{h}_k^T \mathbf{R}_k^{-1} \left(\mathbf{y}_k - \mathbf{y}_k^b \right) \right\} . \qquad (4.85)$$

In view of Eqs. (4.39) and (4.41), the above expression indicates that $\mathbf{x}_l^{(2)}$ is identical to the 4–D VAR solution that would be obtained at the observation time t_l when all $l + m$ observations are processed simultaneously. Due to the optimality property of the 4–D VAR procedure, this conclusion holds for every observation time level t_k, for $k = l, \ldots, l + m$.

The *"optimal transferability"* property discussed in the foregoing for a perfect *linear model* also holds when the *linear model* is not perfect, but is *affected by normally distributed uncorrelated modeling errors*. This property can be shown by considering a time window with observations $\{\mathbf{y}_1, \cdots, \mathbf{y}_l, \mathbf{y}_{l+1}, \cdots, \mathbf{y}_{l+m}\}$, which is separated into two batches, $\{\mathbf{y}_1, \cdots, \mathbf{y}_l\}$ and $\{\mathbf{y}_{l+1}, \cdots, \mathbf{y}_{l+m}\}$, respectively. The *"optimal transferability"* property can be proven by solving a 4–D VAR problem for the first batch, $\mathbf{Y}_l = \{\mathbf{y}_1, \cdots, \mathbf{y}_l\}$, then for the second one, with observations $\mathbf{Y}_{l+m} = \{\mathbf{y}_{l+1}, \cdots, \mathbf{y}_{l+m}\}$, and finally showing that the end result of this procedure is the same as would be obtained by solving a single 4–D VAR problem which takes all observations simultaneously into account. The starting point for proving this property is provided by the repeated use of Bayes' theorem, to express $p(\mathbf{x}_l, \cdots, \mathbf{x}_{l+m} | \mathbf{Y}_{l+m})$ in the form

$$p(\mathbf{x}_l, \cdots, \mathbf{x}_{l+m} | \mathbf{Y}_{l+m}) = \frac{p(\mathbf{x}_l, \cdots, \mathbf{x}_{l+m}, \mathbf{Y}_l, \mathbf{y}_{l+1}, \cdots, \mathbf{y}_{l+m})}{p(\mathbf{Y}_{l+m})}$$

$$= \frac{p(\mathbf{Y}_l)}{p(\mathbf{Y}_{l+m})} p(\mathbf{x}_l | \mathbf{Y}_l, \mathbf{x}_{l+1}, \cdots, \mathbf{x}_{l+m}, \mathbf{y}_{l+1}, \cdots, \mathbf{y}_{l+m})$$

$$\times p(\mathbf{y}_{l+1}, \cdots, \mathbf{y}_{l+m} | \mathbf{x}_{l+1}, \cdots, \mathbf{x}_{l+m}, \mathbf{x}_l, \mathbf{Y}_l) \, p(\mathbf{x}_{l+1}, \cdots, \mathbf{x}_{l+m} | \mathbf{x}_l, \mathbf{Y}_l) .$$

$$(4.86)$$

Since the sequence $\mathbf{x}_k$ is Markov for all indices $k = 1, \cdots, l + m$, it follows that

$$p\left(\mathbf{x}_l | \mathbf{Y}_l, \mathbf{x}_{l+1}, \cdots, \mathbf{x}_{l+m}, \mathbf{y}_{l+1}, \cdots, \mathbf{y}_{l+m}\right) = p\left(\mathbf{x}_l | \mathbf{Y}_l\right). \qquad (4.87)$$

Using Eq. (4.21) indicates that

$$\begin{aligned}
& p\left(\mathbf{y}_{l+1}, \cdots, \mathbf{y}_{l+m} | \mathbf{x}_{l+1}, \cdots, \mathbf{x}_{l+m}, \mathbf{x}_l, \mathbf{Y}_l\right) \\
& = p\left(\mathbf{y}_{l+1}, \cdots, \mathbf{y}_{l+m} | \mathbf{x}_{l+1}, \cdots, \mathbf{x}_{l+m}\right) \\
& = p_{v_{l+1}}\left(\mathbf{y}_{l+1} - \mathbf{h}_{l+1}\mathbf{x}_{l+1}\right) \cdots p_{v_{l+m}}\left(\mathbf{y}_{l+m} - \mathbf{h}_{l+m}\mathbf{x}_{l+m}\right). \qquad (4.88)
\end{aligned}$$

Furthermore, since $\mathbf{w}_k$ is independent of $\mathbf{Y}_l$ for $k > l$, and since the sequence $\mathbf{x}_k$ is Markov, it follows that

$$\begin{aligned}
p\left(\mathbf{x}_{l+1}, \cdots, \mathbf{x}_{l+m} | \mathbf{x}_l, \mathbf{Y}_l\right) & = p\left(\mathbf{x}_{l+1}, \cdots, \mathbf{x}_{l+m} | \mathbf{x}_l\right) \\
& = p\left(\mathbf{x}_{l+m} | \mathbf{x}_{l+m-1}\right) \cdots p\left(\mathbf{x}_{l+1} | \mathbf{x}_l\right) \\
& = p_{w_{l+1}}\left[\mathbf{x}_{l+1} - \mathbf{L}\left(l+1, l\right)\mathbf{x}_l\right] \times \cdots \\
& \times p_{w_{l+m}}\left[\mathbf{x}_{l+m} - \mathbf{L}\left(l+m, l+m-1\right)\mathbf{x}_{l+m-1}\right]. \\
& \qquad (4.89)
\end{aligned}$$

Substituting Eq. (4.87), Eq. (4.88), and Eq. (4.89) into Eq. (4.86) shows that maximizing $p\left(\mathbf{x}_l, \cdots, \mathbf{x}_{l+m} | \mathbf{Y}_{l+m}\right)$ is equivalent to minimizing the cost function

$$\begin{aligned}
J = & \frac{1}{2} \sum_{k=l+1}^{l+m} \mathbf{w}_k^T \mathbf{Q}_k^{-1} \mathbf{w}_k + \frac{1}{2}\left(\mathbf{x}_l - \mathbf{x}_l^{(1)}\right)^T \left(\mathbf{P}_l^a\right)^{-1}\left(\mathbf{x}_l - \mathbf{x}_l^{(1)}\right) \\
& + \frac{1}{2} \sum_{k=l+1}^{l+m}\left[\mathbf{h}_k\mathbf{L}\left(k, l\right)\mathbf{x}_l - \mathbf{y}_k\right]^T \mathbf{R}_k^{-1}\left[\mathbf{h}_k\mathbf{L}\left(k, l\right)\mathbf{x}_l - \mathbf{y}_k\right], \qquad (4.90)
\end{aligned}$$

subject to the constraints expressed in Eq. (4.19), where $\mathbf{x}_l^{(1)}$ and $\mathbf{P}_l^a$ are provided by the 4–D VAR process over the first batch. This end result proves the assertion that the "*optimal transferability*" property also holds, as previously anticipated, for a linear model affected by normally distributed uncorrelated modeling errors.

4.5 Numerical Experience with Unconstrained Minimization Methods for 4–D VAR Using the Shallow Water Equations

The two-dimensional limited-area shallow water equations (SWE) are widely used in meteorology and oceanography for testing new algorithms since they contain most of the physical degrees of freedom (including gravity waves) present in the more sophisticated 3D primitive equation models. The 2D SWE equations may be written as

$$\frac{\partial u}{\partial t} + u\frac{\partial u}{\partial x} + v\frac{\partial u}{\partial y} - fv + \frac{\partial \phi}{\partial x} = 0, \tag{4.91}$$

$$\frac{\partial v}{\partial t} + u\frac{\partial v}{\partial x} + v\frac{\partial v}{\partial y} + fu + \frac{\partial \phi}{\partial y} = 0, \tag{4.92}$$

$$\frac{\partial \phi}{\partial t} + u\frac{\partial \phi}{\partial x} + v\frac{\partial \phi}{\partial y} + \phi\left(\frac{\partial u}{\partial x} + \frac{\partial v}{\partial y}\right) = 0, \tag{4.93}$$

where $f = 10^{-4}$ sec^{-1} is the Coriolis parameter, while u, v, and ϕ are the two components of the velocity field and the geopotential field, respectively. The cost functional is defined as a weighted sum of squared differences between the observations and the corresponding prediction model values:

$$J = W_\phi \sum_{n=1}^{N_\phi} \left(\phi_n - \phi_n^{obs}\right)^2 + W_V \sum_{n=1}^{N_V} \left[\left(u_n - u_n^{obs}\right)^2 + \left(v_n - v_n^{obs}\right)^2\right], \tag{4.94}$$

where N_ϕ is the total number of geopotential observations available over the assimilation window (t_0, t_R), and N_V is the total number of wind vector observations. The quantities u_n^{obs}, v_n^{obs}, and ϕ_n^{obs} are the observed values for the northward wind component, the eastward wind component, and the geopotential field respectively, while the quantities u_n, v_n, and ϕ_n are the corresponding computed model values. The weighting factors W_ϕ and W_V are taken to be the inverse of estimates of the statistical root-mean-square observational errors for the geopotential and wind components respectively; the actual values used in the numerical experiments reported in this section are $W_\phi = 10^{-4}m^{-4}s^4$ and $W_V = 10^{-2}m^{-2}s^2$.

The SWE are discretized with a centered difference scheme in space, and an explicit leapfrog integration scheme in time. A rectangular domain of size L= 6000 km, D= 4400 km is used along with discretization parameters $\Delta x = 300$ km, $\Delta y = 220$ km, and $\Delta t = 600$ s, thus providing a grid of 21×21 points in space and 60 time steps in the assimilation window (10 hours). Hence, the vector of control variables for the initial control problem has 1 323 components. If the boundary conditions of a limited-area model are also included as components of the control vector, all three field variables on the boundary perimeter must be stored in memory, for all the time steps, thus increasing the dimension of the vector of control variables to 14 763.

For the numerical experiments to be discussed in the next section, the observational data consisted of the model-integrated values for wind and geopotential at each time step starting from the Grammeltvedt [80] initial conditions:

$$h = H_0 + H_1 \tanh \frac{9(y - y_0)}{2D} + H_2 \text{sech} \frac{9(y - y_0)}{D} \sin \frac{2\pi x}{L}, \qquad (4.95)$$

where L is the length of the channel on the β plane, D is the width of the channel, and $y_0 = D/2$ is the middle of the channel. The numerical values used for these quantities are as follows: $H_0 = 2000$ m, $H_1 = -220$ m, $H_2 = 133$ m, $g = 10$ msec^{-2}, $L = 6000$ km , $D = 4400$ km, $f = 10^{-4}$ sec^{-1}, $\beta = 1.5 \times 10^{-11}$ sec^{-1}m^{-1}. The initial velocity fields were derived from the initial height field via the geostrophic relationship

$$u = -\frac{g}{f} \frac{\partial h}{\partial y}, \qquad (4.96)$$

$$v = \frac{g}{f} \frac{\partial h}{\partial x}. \qquad (4.97)$$

Random perturbations of these fields were then used as initial guesses for the solution.

4.5.1   Performance of LMQN Methods

The four LMQN methods (CONMIN-CG, E04DGF, L-BFGS, and BBVSCG) were separately applied to minimize the functional J defined in Eq. (4.94) using the SWE model. The CONMIN-CG and BBVSCG algorithms failed after the first iteration, even when both the gradient scaling and the nondimensional scaling were applied. The L-BFGS algorithm was successful only in concert with gradient scaling. On the other hand, the E04DGF algorithm worked only with the nondimensional shallow water equations model. This indicates that using additional scaling is essential for the success of the LMQN minimization algorithms when applied to large-scale minimization problems.

The comparative performances of the E04DGF and L-BFGS algorithms are presented in Table 4.1, for two objective functionals, as follows: for the first objective functional, $J_1(\mathbf{x_0})$, only the initial conditions $\mathbf{x_0} \equiv [u(t_0), v(t_0), \phi(t_0)]$ are considered as control variables, while for the second functional, $J_2(\mathbf{x_0}, \mathbf{v})$, both the initial conditions and a time-dependent function $\mathbf{v}$, defined on the boundary, are considered as control variables. Although different scaling procedures were employed in the two algorithms, the minimization computations were comparably converged within an error less than 10^{-4}.

Table 4.1 indicates that most of the Central Processing Unit (CPU) time is spent on function calls ("Nfun") rather than in the minimization iteration ("Iter") loop. The CPU time ("MTM") needed for converging the respective computations, as well as the CPU time required by the respective algorithms for function calls ("FTM"), are also presented in Table 4.1. Comparing the number of function calls and CPU time, it is apparent that the computational cost of L-BFGS is much lower than that of E04DGF. In 66 iterations with 89 function calls, the L-BFGS algorithm converged. On the other hand, the E04DGF algorithm required 72 iterations and 203 function calls to reach the same convergence criterion. This difference in the CPU time spent in minimization indicates that the L-BFGS algorithm uses less than half of the total CPU time required by the E04DGF algorithm.

As intuitively expected and as the results in Tables 4.1 confirm, controlling both the boundary and initial conditions is a much more difficult problem than controlling only the initial conditions. Adding the boundary conditions as control variables introduces the following additional difficulties:

1. The dimensionality of the Hessian of the objective function increases by about one order of magnitude (from 10^3 to 10^4); hence, the condition number of the Hessian will increase like $O\left(N^{\frac{d}{2}}\right)$, where

d is the dimensionality of the space variables and N is the number of components of the vector of control variables.

2. The perturbation of the boundary conditions creates locally an ill-posed problem, which is indicated by an increase of high frequency noise near the boundary. In turn, the condition number of the Hessian of the objective function increases.

TABLE 4.1
SWE problem solved with Limited Memory Quasi-Newton (LMQN) methods

Control Variable	Algorithm	Iter	Nfun	MTM total CPU time	FTM (function calls CPU time)
Initial	EO4DGF	72	203	36.89	33.56
	L-BGFS	66	89	15.53	14.76
Initial+	EO4DGF	160	481	87.31	79.98
Boundary	L-BGFS	179	468	80.70	77.81

4.5.2   Performance of Truncated Newton (T-N) Methods

This sections illustrates the performances of two distinct truncated-Newton "packages" called for convenience *TN1* and *TN2*, respectively, characterized by the following features:

1. *TN1* is a package developed by Nash ([140], [142], [144]). It uses a modified Lanczos algorithm with an automatically-supplied diagonal preconditioner.

2. *TN2* ("TNPACK") has been developed by Schlick and Fogelson ([179], [180],Schlick and Overton [181]). It is designed for structured separable problems for which the user provides a sparse preconditioner for a preconditioned Conjugate-Gradient (CG) method. In *TN2*, a sparse modified Cholesky factorization based on the Yale Sparse Matrix Package is used to factorize the preconditioner, which may not necessarily be positive definite. Schlick [178] has also implemented two modified Cholesky factorizations in *TN2*.

Table 4.2 presents a comparison of the performances of the algorithms *TN1* and *TN2* for the SWE model described in Eqs. (4.91) through (4.97). The specified maximum number of permitted inner iterations per outer iteration (indicated in Table 4.2 by the value of the descriptor Maximum Inner Iterations Conjugate Gradient (MXITCG)) has a small impact on the performance of *TN1* but a significant impact on that of *TN2*. For both *TN1* and *TN2*, the cost for large MXITCG is much lower than that for small MXITCG. Furthermore, *TN2* with MXITCG = 50 performs much better than *TN2* with MXITCG = 50 in terms of Newton iterations, CG iterations, function evaluations, and CPU time. These results suggest that *TN2* would perform best for large-scale minimization problems, especially in conjunction with a suitable preconditioner.

For comparison, results for a version of *TN1* without the diagonal preconditioning are also included in Table 4.2. These results indicate that the use of preconditioning in *TN1* accelerates its performance, as expected. Without preconditioning, *TN1* requires more function evaluations. The importance of preconditioning increases with increasing dimensionality of the minimization problem under consideration.

TABLE 4.2
Initial control SWE problem

Algorithm	MXITCG	ITER	NFUN	MTN	FTM
TN1	3	19	20	12.20	11.31
	50	20	26	13.79	12.89
TN1	3	63	64	38.82	37.15
(no prec.)	50	39	40	32.78	31.53
TN2	3	81	82	68.68	67.21
	50	4	5	16.41	16.30

Table 4.3 includes results for control variables comprising the initial and the boundary conditions. Comparison of the results in Tables 4.1 and 4.3 indicates that the T-N methods are competitive with L-BFGS. Although not shown in these tables, it is worth mentioning that the *TN2* algorithm was more accurate than the other algorithms. Overall, for the SWE paradigm meteorological problem, the T-N methods required far fewer iterations and function calls than the LMQN methods.

TABLE 4.3

SWE problem solved with Truncated Newton (T-N)
LMQN methods

Control Variable	Algorithm	Iter	Nfun	MTM	FTM
Initial	TN1	19	70	12.20	11.31
	TN2	4	96	16.41	16.30
Initial+	TN1	70	283	49.96	46.22
Boundary	TN2	12	520	87.22	86.30

4.6   Treatment of Model Errors in Variational Data Assimilation

The goal of *variational data assimilation* (VDA) is to find the trajectory that best fits (in a least-squared sense) the observational data over an assimilation time interval by adjusting the initial conditions supplied for the forward model integration. In the so-called *strong constraint VDA* (or *classical VDA*), it is assumed that the forecast model perfectly represents the evolution of the actual atmosphere. The "best fit" model trajectory is obtained by adjusting only the initial conditions via the minimization of a cost functional that is subject to the model equations as strong constrains.

However, numerical weather prediction (NWP) models are imperfect since subgrid processes are not included; furthermore numerical discretizations produce additional dissipative and dispersion errors. Modelling errors also arise from the incomplete mathematical modeling of the boundary conditions and forcing terms, and from the simplified representation of physical processes and their interactions in the atmosphere. Usually, all of these modeling imperfections are collectively called *model error* (ME). Model error is formally introduced as a correction to the time derivatives of model variables, so that the state of the atmosphere, represented by the vector $\mathbf{x}(t)$, evolves as follows:

$$\frac{d\mathbf{x}(t)}{dt} = \mathbf{M}[\mathbf{x}(t)] + \mathbf{T}[\boldsymbol{\eta}(t)], \qquad (4.98)$$

where the matrix $\mathbf{M}[.]$ denotes all the mathematical operations involved in the model, η represents the model error and $\mathbf{T}[.]$ is an operator that accounts for the fact that only certain components of the state vector have modeling errors. The ME usually varies both spatially and temporally, and contains both systematic and stochastic components. The operator $\mathbf{T}[\cdot]$ maps the space of the ME to the space of the model state, $\mathbf{x}$. If the model state has an associated error at every grid point, then $\mathbf{T}[\cdot]$ is identically equal to the unit matrix, $\mathbf{I}$ and, the dimension of η is equal to that of the model state, $\mathbf{x}$. On the other hand, if it is *a priori* known that the numerical model has some severe drawbacks (e.g., modeling of the atmosphere in certain regions of the globe like the poles) then the operator $\mathbf{T}[\cdot]$ should be specified in such a way that only those model grid points (e.g., at that pole) have modeling errors, while the rest of the model states are free of ME (see, e.g., ref. [85]).

In *variational continuous assimilation*, the evolution of ME is modeled as a continuous process, obeying an initial value problem of the form

$$\frac{d\eta}{dt} = \Phi[\eta(t), \mathbf{x}(t)] + \mathbf{q}(t). \tag{4.99}$$

Often, the forcing term $\mathbf{q}(t)$ is neglected or absorbed in the form of $\Phi(\eta)$. The function $\Phi(\eta)$ is obviously important: the ME monotonically decreases in time, if $\Phi(\eta) < 0$ for all η; the ME remains constant, if $\Phi(\eta) = 0$; the ME grows in time, if $\Phi(\eta) > 0$ for all η,

The initial condition η_0 is determined as part of minimizing the following cost functional:

$$J(\mathbf{x}_0, \eta_0) \equiv J_b + J_0 + J_\eta, \tag{4.100}$$

subject to the (hard) constraints

$$\frac{d\mathbf{x}(t)}{dt} = \mathbf{M}[\mathbf{x}(t)] + \mathbf{T}[\eta(t)], \quad \mathbf{x}(t_0) = \mathbf{x}_0;$$
$$\frac{d\eta}{dt} = \Phi[\eta(t), \mathbf{x}(t)], \quad \eta(t_0) = \eta_0. \tag{4.101}$$

The functionals J_b, J_0, and J_η are defined as follows

$$J_b \equiv \frac{1}{2}(\mathbf{x}_0 - \mathbf{x}^b)^T \mathbf{B}^{-1}(\mathbf{x}_0 - \mathbf{x}^b), \tag{4.102}$$

$$J_0 \equiv \frac{1}{2} \sum_{i=0}^{n} [\mathbf{H}(\mathbf{x}(t_i)) - \mathbf{y}^o(t_i)]^T \mathbf{R}^{-1} [\mathbf{H}(\mathbf{x}(t_i)) - \mathbf{y}^o(t_i)], \tag{4.103}$$

$$J_\eta \equiv \frac{1}{2} (\boldsymbol{\eta}_0 - \boldsymbol{\eta}^b)^T \mathbf{Q}^{-1} (\boldsymbol{\eta}_0 - \boldsymbol{\eta}^b), \tag{4.104}$$

where $\mathbf{Q}$ is the model error covariance matrix.

The constrained minimization of $J(\mathbf{x}_0, \boldsymbol{\eta}_0)$ can be performed by introducing the Lagrange multipliers $\mathbf{x}^*$ and $\boldsymbol{\eta}^*$ to construct the following augmented Lagrangian functional:

$$
\begin{aligned}
L(\mathbf{x}, \boldsymbol{\eta}, \mathbf{x}^*, \boldsymbol{\eta}^*) &= J(\mathbf{x}_0, \boldsymbol{\eta}) \\
&+ \int_{t_0}^{t_n} \left\langle \mathbf{x}^*, \frac{d\mathbf{x}(t)}{dt} - \mathbf{M}[\mathbf{x}(t)] - \boldsymbol{\eta}(t) \right\rangle dt \\
&+ \int_{t_0}^{t_n} \left\langle \boldsymbol{\eta}^*, \frac{d\boldsymbol{\eta}}{dt} - \boldsymbol{\Phi}[\boldsymbol{\eta}(t), \mathbf{x}(t)] \right\rangle dt
\end{aligned}
\tag{4.105}
$$

where $\langle \cdot, \cdot \rangle$ denotes the customary inner product of two vectors (i.e., $\langle \mathbf{a}, \mathbf{b} \rangle \equiv \mathbf{a} \cdot \mathbf{b}$).

The extrema of L are the same as the extrema of $J(\mathbf{x}_0, \boldsymbol{\eta}_0)$, and are obtained by requiring that the following equations ("optimality conditions") be satisfied:

$$\frac{\partial L}{\partial \mathbf{x}} = \mathbf{0}, \quad \frac{\partial L}{\partial \boldsymbol{\eta}} = \mathbf{0}, \tag{4.106}$$

$$\frac{\partial L}{\partial \mathbf{x}^*} = \mathbf{0}, \quad \frac{\partial L}{\partial \boldsymbol{\eta}^*} = \mathbf{0}. \tag{4.107}$$

Equations (4.107) yield the equations describing the evolution of model state and ME, while Eqs. (4.106) yield the adjoint equations that must be satisfied by the Lagrange multipliers ("adjoint functions") $\mathbf{x}^*, \boldsymbol{\eta}*$, which are:

$$-\frac{d\mathbf{x}^*(t)}{dt} = [\frac{\partial \mathbf{M}}{\partial \mathbf{x}}]^T \mathbf{x}^* + [\frac{\partial \boldsymbol{\Phi}}{\partial \mathbf{x}}]^T \boldsymbol{\eta}^*$$

$$+ \delta(t - t_i) \sum_{i=0}^{n} [\frac{\partial \mathbf{H}}{\partial \mathbf{x}}]^T \mathbf{R}^{-1} [\mathbf{H}(\mathbf{x}(t_i)) - \mathbf{y}^o(t_i)], \quad \mathbf{x}^*(t_n) = \mathbf{0}, \tag{4.108}$$

$$-\frac{d\boldsymbol{\eta}^*(t)}{dt} = [\frac{\partial \boldsymbol{\Phi}}{\partial \boldsymbol{\eta}}]^T \boldsymbol{\eta}^* + \mathbf{x}^*, \quad \boldsymbol{\eta}^*(t_n) = \mathbf{0}. \tag{4.109}$$

The optimality conditions (4.108) and (4.109) also yield the gradient of the cost functional $J(\mathbf{x}_0, \boldsymbol{\eta}_0)$ with respect to the model state $\mathbf{x}_0$ and the ME state

η_0 namely,

$$\nabla_{x_0} J = \nabla_{x_0} J_b + \nabla_{x_0} J_0 = \mathbf{B}^{-1}[\mathbf{x}_0 - \mathbf{x}^b] + \mathbf{x}^*(t_0), \qquad (4.110)$$

$$\nabla_{\eta_0} J = \nabla_{\eta_0} J_\eta + \nabla_{\eta_0} J_0 = \mathbf{Q}^{-1}[\boldsymbol{\eta}_0 - \boldsymbol{\eta}^b] + \boldsymbol{\eta}^*(t_0). \qquad (4.111)$$

Note that the evolution of $\mathbf{x}^*$ and $\boldsymbol{\eta}^*$ is coupled via the operator $\boldsymbol{\Phi}[.]$, and the adjoint equations evolve backwards in time. As usual, the backward integration of the adjoint models (4.110) and (4.111) from time $t_n \to t_0$ yields the values of initial adjoint states $\mathbf{x}^*(t_0)$ and $\boldsymbol{\eta}^*(t_0)$. Note that Eq. (4.108) couples the evolution of the adjoint variables corresponding to the model states and ME. This mapping also increases the complexity involved in the backward integration of the adjoint models, cf. Eqs. (4.110) and (4.111). The computation of both $\nabla_{x_0} J$ and $\nabla_{\eta_0} J$ doubles the size of the system to be optimized by comparison to computing only $\nabla_{x_0} J$, when the model errors are neglected.

5

4–D VAR in Numerical Weather Prediction Models

CONTENTS

5.1 The Objective of 4–D VAR

The objective of four-dimensional variational data assimilation (4–D VAR) is to find a model solution which best fits observational data distributed over some space and time intervals. A popular measure of the lack of fit between model forecast and observations is a cost function, J, mathematically describing a weighted least-square norm. Assuming that the observations are given by analyzed fields, such a cost function can be represented in the form

$$J(\vec{\mathbf{X}}(t_0)) = \frac{1}{2} \sum_{r=0}^{R} \left[\mathbf{X}(t_r) - \mathbf{X}^{obs}(t_r)\right]^T \mathbf{W}(t_r) \left[\mathbf{X}(t_r) - \mathbf{X}^{obs}(t_r)\right], \quad (5.1)$$

where: $\mathbf{X}(t_r)$ is a vector of dimension N containing all model variables over all grid points at time t_r; R is the number of time levels for the analyzed fields

227

in the assimilation window; $\mathbf{X}^{obs}(t_r)$ is the observational counterpart of the model variable $\mathbf{X}(t_r)$; $\mathbf{W}(\mathbf{t_r})$ is an $N \times N$ diagonal matrix of weighting coefficients, usually taken to be the inverse covariance matrix of the observations errors.

The objective of 4–D VAR is to find the optimal set of control variables, usually the initial conditions and/or boundary conditions, such that the cost function J is minimized subject to the constraints of satisfying the geophysical model. In order to minimize the cost function, we need to know the gradient of this function with respect to the control variables. A straightforward way of computing this gradient is to perturb each control variable in turn and estimate the change in the cost function. But this method is impractical when the number of control variables is large (as is the case for a typical meteorological model, e.g., 10^7). This means that one estimation of the gradient requires 10^7 model integrations. Furthermore, the iterative minimization of the cost function requires several gradient estimations on the way to finding a local minimum. Often, these gradient estimations are not sufficiently accurate to guarantee achieving convergence of the minimization process. Several studies from Zou (e.g., [226], [228], [233]) showed that strong nonlinearities and on-off processes may negatively impact the convergence of the minimization process in 4–D VAR.

In contradistinction to using forward model computations, the adjoint method yields the exact gradient of the cost function with respect to the control variables by integrating the adjoint model only once backwards in time. Such a backward integration of the adjoint model is of similar complexity to a single integration of the forward model. Another key advantage of adjoint variational data assimilation is the possibility to minimize the cost function using standard unconstrained minimization algorithms (usually iterative descent methods see, e.g., Gill et al. [76] and Navon and Legler [151] [230]).

In the geosciences, implementing 4–D VAR data assimilation for various models using finite element methods was probably first started in the work of Zhu et al.[224]. Significant computational challenges arise in the derivation of the adjoint of such a model, including the coding of iterative Successive Over Relaxation (SOR) and Gauss Seidel methods (as detailed, e.g., in Chen et al.

[32]), and the on-off discontinuities that appear with inclusion of radiation and precipitation physical processes in data assimilation.

Though the assimilation of precipitation data can effectively improve the forecast, it is difficult to reproduce precipitation in the analysis field. The model does not produce precipitation when all model levels are unsaturated (e.g., the precipitation process is switched off) but the variational method cannot turn a switch "on" since it can only make a continuous modification of the atmospheric state and cannot deal with discontinuous on/off changes. For precipitation observations, highly nonlinear parameterization schemes must be linearized for developing the adjoint version of the model required by the minimization procedure for the cost function. Alternatively, simpler physics schemes, which are not a direct linearization of the full model physics (i.e., the "linear model" is not tangent linear to the nonlinear full model), can be coded for the linear model. It is not clear what physical phenomena may be lost in this simplification. Several works (e.g., refs. [9], [219], [220], [226]), describe technical issues involved in the minimization of the cost function which arise from these characteristics of the physics. In particular, the work of Xu ([219], [220], [221]) used so–called "generalized adjoint formulations" to deal with on/off switches triggered at discrete time levels by threshold conditions. However, as shown in Xu [220], the discrete resulting solution is not continuously dependent on the initial state, so the cost function contains zigzag discontinuities which preclude the use of traditional time discretization with a switch time determined by interpolation as a continuous function of the initial state. Consequently, such difficulties have led to simplified physics models at all the operational Numerical Weather Prediction (NWP) centers implementing 4–D VAR assimilations procedures (see, e.g., Janisková et al. [95], [96]).

Of course, it is desirable to have a linearized model that approximates as closely as possible the sensitivity of the full nonlinear model; otherwise the forecast model may not be in balance with its own analysis, producing so–called "model spin–up" (e.g., Tompkins and Janisková [198]). Furthermore, multi-incremental approaches can exhibit discrete transitions affecting the stability of the overall minimization process Mahfouf and Rabier [133]. On the one hand, nonlinear models have steadily evolved in complexity in order to improve forecast skill. For example, the prognostic cloud scheme introduced into the European Centre for Medium-Range Weather Forecasts (ECMWF) model according to Tiedtke [196] and Jakob [94] includes many highly non-linear processes that are often controlled by threshold "switches." On the other hand, even if it were possible to construct the tangent linear and, respectively, adjoint models corresponding to this complex cloud scheme, the validity of these models would be restricted due to these thresholds and their value would be questionable. Specific methods to remove nonlinearities in the diffusion schemes were needed in order to create a functional assimilation system. The practical compromise is to generate a smooth tangent linear model containing as few "discrete on/off" processes as possible while still retaining the important aspects of the full nonlinear scheme. Issues related to using non-smooth optimization methods to address discontinuities is an ongoing research topic (see, e.g., Homescu and Navon [92] and Zhang et al. [223]).

The process of initialization of operational models requires filtering of spurious gravity waves; this process is referred to as *nonlinear normal mode initialization* (NNMI, see Machenhauer [132]). Additional penalty terms were introduced into the formulation of cost functional to perform NNMI efficient by Machenhauer [132]. Typically, operational centers include in their respective cost functional a penalty term J_C of the form

$$J_C = a \left(|\mathbf{N_G} \Delta \mathbf{x}_0|^2 + \sum_{i=1}^{\text{maxslot}} |\mathbf{N_G} \Delta \mathbf{x}_i|^2 \right),$$

where $\mathbf{N_G}$ denotes the operator needed to calculate the tendency of the gravity wave modes based on Machenhauer [132] in an incremental 4–D VAR procedure, $\Delta \mathbf{x}_0$ denotes the increment at the initial time before the initialization, and $\Delta \mathbf{x}_i$ denotes the increment evolved, according to the tangent linear model from the initial time to the representative time of the ith time slot after the initialization. The summation is from the first time interval (i = 1) to the last time interval (i = maxslot), while a is a constant (typically 3.0×10^{-2} s^4/m^2) determined empirically. Although this penalty term is introduced to suppress the gravity wave in the increment $\Delta \mathbf{x}_i$, it is also effective to stabilize the overall computation. The imbalance between the mass and wind fields causes large amplitude inertia-gravity oscillations to occur in numerical prediction models based upon primitive equations. Many initialization procedures have been developed to control these oscillations, e.g., static initialization (see, e.g., Phillips [159]) dynamic initialization (see, e.g., Miyakoda and Moyer [136]) nonlinear normal-mode initialization (see, e.g., Machenhauer [132] and Baer [6]) and bounded derivative initialization (see, e.g., Semazzi and Navon [184]).

An alternative approach to these initialization methods is the use of time filters. A digital filter has been used to initialize data for a limited-area shallow-water model in ref. [131], and for a baroclinic limited-area model in ref. [130]. In both studies, the numerical models were integrated forward and backward from the initial time. The time series of the model variables produced by the integrations are processed by a digital filter which removes the high-frequency components from the initial data. The forward and backward integrations are performed adiabatically, with all diabatic processes and horizontal diffusion disabled. The adiabatic digital-filtering initialization effectively removes the high frequencies while inducing very small changes to the initial fields as shown, e.g., in ref. [162] for 4–D VAR data assimilation with a shallow water equations model.

Although effective, the NNMI is difficult to implement as either a strong or a weak constraint; furthermore, the "adjoint NNMI" is also required. Consequently, simpler methods of gravity wave control were sought. In particular, Zou et al. ([231], [232]) demonstrated that augmented Lagrangian and penalty methods could control high-frequency noise well in the context of a shallow water equations model. However, these methods require the solution of several 4–D VAR problems, starting from the unconstrained problem and increasing the penalty parameter with each solution (see ref. [148] for a description of exterior penalty methods), which makes the implementation of penalty methods impractical for 4–D VAR operational forecast models. The search for more practical methods led to the consideration of simple penalty terms that are solved only once with a fixed value of the penalty parameter; Zou et al. [233] demonstrated that simple penalty terms involving the surface pressure tendency and the divergence were able to suppress much of the high frequency noise. However, penalty methods in which the respective equations are solved only once depend heavily on the arbitrary choice of penalty parameter, and their well-known ill-conditioning (see, e.g., ref. [76]) may cause slow convergence.

5.2　Computation of Cost Functional Gradient Using the Adjoint Model

This section illustrates the use of the adjoint model for computing the gradient, ∇J, of the cost functional identified in Eq. (5.1). The forward model employed for this illustration uses a two–level time integration scheme, which can be written as:

$$\mathbf{X}(t + \Delta t) = \mathbf{F}\left[\mathbf{X}(t)\right], \text{ for } t \geq t_0, \tag{5.2}$$

where $\mathbf{F}[\mathbf{X}(t)]$ represents operators that perform model integration for a given time step. To calculate the gradient $\nabla J[\mathbf{X}(t_0)]$ of the cost function with respect to the initial condition $\mathbf{X}(t_0)$, we define the quantity J' to denote the change in the cost function resulting from a small perturbation $\mathbf{X}'(t_0)$ about the initial conditions $\mathbf{X}(t_0)$, i.e.,

$$J'\left[\mathbf{X}(t_0)\right] = J\left[\mathbf{X}(t_0) + \vec{X}'(t_0)\right] - J\left[\mathbf{X}(t_0)\right]. \tag{5.3}$$

In limit as $\|\mathbf{X}'\| \to 0$, J' becomes the directional derivative of J in the direction $\mathbf{X}'(t_0)$, namely

$$J'\left[\mathbf{X}(t_0)\right] = \left\{\nabla J\left[\mathbf{X}(t_0)\right]\right\}^T \mathbf{X}'(t_0). \tag{5.4}$$

The left side of the above expression is computed from Eqs. (5.1) and (5.3) to obtain

$$J'\left[\mathbf{X}(t_0)\right] = \sum_{r=0}^{R} \left\{\mathbf{W}(t_r)\left[\mathbf{X}(t_r) - \mathbf{X}^{obs}(t_r)\right]\right\}^T \mathbf{X}'(t_r), \tag{5.5}$$

where $\mathbf{X}'(t_r)$ is the perturbation in the forecast resulting from the initial perturbation $\mathbf{X}'(t_0)$. Equating Eqs. (5.4) and (5.5) yields

$$\left\{\nabla J\left[\mathbf{X}(t_0)\right]\right\}^T \mathbf{X}'(t_0) = \sum_{r=0}^{R} \left\{\mathbf{W}(t_r)\left[\mathbf{X}(t_r) - \mathbf{X}^{obs}(t_r)\right]\right\}^T \mathbf{X}'(t_r), \tag{5.6}$$

which indicates that the gradient of the cost function with respect to the initial conditions can be found provided that $\mathbf{X}'(t_r)$ can be expressed as a function of $\mathbf{X}'(t_0)$. To express $\mathbf{X}'(t_r)$ in terms of $\mathbf{X}'(t_0)$, we linearize the forecast model

Eq. (5.2) about current model solution $\mathbf{X}(t)$. The linearized version of the forecast model (i.e., the tangent linear model) can be written as

$$\mathbf{X}'(t + \Delta t) = \mathbf{D}(t)\mathbf{X}'(t), \tag{5.7}$$

where the matrix $\mathbf{D}$ is defined as

$$\mathbf{D}(t) = \frac{\partial \mathbf{F}[\mathbf{X}(t)]}{\partial \mathbf{X}(t)}, \tag{5.8}$$

and comprises operator matrices that depend on the state $\mathbf{X}(t)$. Applying Eq. (5.7) repeatedly backwards in time starting at $t = t_r$ yields

$$
\begin{aligned}
\mathbf{X}'(t_r) &= \mathbf{D}(t_r - \Delta t)\mathbf{X}'(t_r - \Delta t) \\
&= \mathbf{D}(t_r - \Delta t)\mathbf{D}(t_r - 2\Delta t)\mathbf{X}'(t_r - 2\Delta t) \\
&= \cdots \\
&= \mathbf{D}(t_r - \Delta t)\mathbf{D}(t_r - 2\Delta t)\ldots\mathbf{D}(t_0)\mathbf{X}'(t_0) \\
&= \mathbf{P}_r\mathbf{X}'(t_0), \tag{5.9}
\end{aligned}
$$

where $\mathbf{P}_r$ represents the result of applying all the operator matrices in the linear model to obtain $\mathbf{X}'(t_r)$ from $\mathbf{X}'(t_0)$, i.e.,

$$\mathbf{P}_r = \mathbf{D}(t_r - \Delta t)\mathbf{D}(t_r - 2\Delta t)\ldots\mathbf{D}(t_0). \tag{5.10}$$

Replacing Eqs. (5.9) and (5.10) transforms the latter into the following expression:

$$
\begin{aligned}
\left\{\nabla J\left[\mathbf{X}(t_0)\right]\right\}^T \mathbf{X}'(t_0) &= \sum_{r=0}^{R}\left\{\mathbf{W}(t_r)\left[\mathbf{X}(t_r) - \mathbf{X}^{obs}(t_r)\right]\right\}^T \left[\mathbf{P}_r\mathbf{X}'(t_0)\right] \\
&= \sum_{r=0}^{R}(\mathbf{P}_r^*)^T \mathbf{W}(t_r)\left[\mathbf{X}(t_r) - \mathbf{X}^{obs}(t_r)\right]^T \mathbf{X}'(t_0), \tag{5.11}
\end{aligned}
$$

which implies that

$$\nabla J\left[\mathbf{X}(t_0)\right] = \sum_{r=0}^{R}(\mathbf{P}_r^*)^T \mathbf{W}(t_r)\left[\mathbf{X}(t_r) - \mathbf{X}^{obs}(t_r)\right], \tag{5.12}$$

where $(\mathbf{P}^*_r)^T$ is the conjugate transpose of the operator matrix $\mathbf{P}_r$, namely:

$$(\mathbf{P}^*_r)^T = [\mathbf{D}^*(t_0)]^T [\mathbf{D}^*(t_0 + \Delta t)]^T \dots [\mathbf{D}^*(t_r - \Delta t)]^T. \tag{5.13}$$

We now define the adjoint equations for the adjoint functions $\mathbf{Y}_r(t)$ as

$$\mathbf{Y}_r(t_0) = (\mathbf{P}^*_r)^T \mathbf{Y}(t_r), \quad r = 1, \cdots, R, \tag{5.14}$$

using the "final time" condition

$$\mathbf{Y}_r(t_r) = \mathbf{W}(t_r) \left[\mathbf{Y}(t_r) - \mathbf{Y}^{obs}(t_r) \right]. \tag{5.15}$$

Replacing Eq. (5.15) into Eq. (5.14) yields

$$\mathbf{Y}_r(t_0) = (\mathbf{P}^*_r)^T \mathbf{W}(t_r) \left[\mathbf{Y}(t_r) - \mathbf{Y}^{obs}(t_r) \right]. \tag{5.16}$$

Replacing Eq. (5.16) into Eq. (5.12) yields the following expression for calculating the gradient of the cost function $\nabla J(\vec{\mathbf{X}}(t_0))$ with respect to the initial condition $\vec{\mathbf{X}}(t_0)$:

$$\nabla J \left[\vec{\mathbf{X}}(t_0) \right] = \sum_{r=0}^{R} \mathbf{Y}_r(t_0). \tag{5.17}$$

The above expression indicates that $\nabla J[\vec{\mathbf{X}}(t_0)]$ can be obtained by integrating the adjoint model defined in Eq. (5.14) from the final time t_R, at $r = R$, to t_0, with zero initial conditions for the adjoint variables at time t_R, and with the weighted differences $\mathbf{Y}_r(t_r)$ imposed on the right hand side of the adjoint equations whenever an analysis time t_r ($r = 0, 1, \cdots, R$) is reached. Thus, a single integration of the adjoint model can yield the value $\nabla J \left[\vec{X}(t_0) \right]$ of the gradient of the cost function with respect to the initial conditions. A complete discussion related to the special problems of managing adjoint calculations for the grid interpolations in the semi-Lagrangian model is provided in ref. [124].

Since the 4–D VAR models comprise many thousands of lines of code, it is necessary to verify the correctness of the linearization and adjoint coding. The correctness of the adjoint of each operator can be checked by applying the following identity due to Navon et al. [153]:

$$[(\mathbf{AQ})^*]^T(\mathbf{AQ}) = (\mathbf{Q}^*)^T[(\mathbf{A}^*)^T(\mathbf{AQ})], \tag{5.18}$$

where $\mathbf{Q}$ represents the input of the original code and $\mathbf{A}$ represents either a single DO loop or a subroutine. The left side of Eq (5.18) involves the tangent linear code, while the right side also involves the adjoint code $(\mathbf{A}^*)^T$. The adjoint code is correct (compared with the tangent linear model) when the identity Eq. (5.18) is satisfied within machine accuracy. When the test expressed by Eq. (5.18) is fulfilled, the adjoint of each matrix operator may still contain errors caused by the linearization of the nonlinear models. Therefore, verification of the gradient obtained through the backwards integration of the adjoint model in time is also mandatory. The Taylor expansion of $J(\mathbf{X} + \Delta\mathbf{X})$ can be written as:

$$J(\mathbf{X}+\Delta\mathbf{X}) = J(\mathbf{X})+\nabla J(\mathbf{X})^T(\Delta\mathbf{X})+\frac{1}{2}(\Delta\mathbf{X})^T\nabla^2 J(\mathbf{X})(\Delta\mathbf{X})+HOT. \tag{5.19}$$

where the Higher-Order-Term (HOT) denotes terms involving third and higher–order derivatives, and where $\nabla^2 J(\mathbf{X})$ denotes the $N \times N$ Hessian matrix of the cost with respect to the control variable vector. Defining now $\Delta\mathbf{X}$ to be the particular vector

$$\Delta\vec{\mathbf{X}} = \alpha\vec{\mathbf{h}}, \quad \text{with} \quad \vec{\mathbf{h}} = \frac{\nabla J\left(\vec{\mathbf{X}}\right)}{\|\nabla J\left(\vec{\mathbf{X}}\right)\|}, \tag{5.20}$$

in Eq. (5.19) leads to the following formula

$$\Phi(\alpha) = 1.0 + \frac{1}{2}\frac{\nabla J\left(\vec{\mathbf{X}}\right)^T\nabla^2 J\left(\vec{\mathbf{X}}\right)\nabla J\left(\vec{\mathbf{X}}\right)}{\|\nabla J\left(\vec{\mathbf{X}}\right)\|^3}\alpha + O\left(\alpha^2\right). \tag{5.21}$$

If the cost function J is viewed as a quadratic function of the control variables $\mathbf{X}$, the quantity Φ is unity plus a linear term in α. Furthermore, if $\nabla^2 J\left(\vec{\mathbf{X}}\right)$ is positive definite, which is always the case if J has a unique minimum, the term $\frac{1}{2}\frac{\nabla J(\vec{\mathbf{X}})^T\nabla^2 J(\vec{\mathbf{X}})\nabla J(\vec{\mathbf{X}})}{\|\nabla J(\vec{\mathbf{X}})\|^3}$ is positive. In this case, the value of Φ approaches unity monotonically from some value greater than 1.0 as α

decreases to zero. If the cost function J is exactly quadratic in $\mathbf{X}$, the higher–order terms are zero and $\nabla^2 J(\mathbf{X})$ is a semipositive definite diagonal matrix. In this case, Φ approaches unity not only monotonically but also linearly and this holds even for large values of α, as long as the perturbation in the control variables does not generate a computational instability that would prevent the model from being integrated. Because of these properties, Eq. (5.21) provides a very valuable tool for verifying the accuracy of the computed gradient of the cost functional J.

As described in the previous section, adjoint coding for each segment of code in the forward model is achieved by finding the conjugate transpose of the operator matrix in the corresponding tangent linear model. Since the sequence of the conjugate transpose of the multiplication of many matrices reverses the original order in the tangent linear model code, writing the adjoint code proceeds from the bottom to the top in the linear model. In the following section, we will present examples illustrating the adjoint coding of the Fast Fourier Transform (FFT) and inverse FFT operators (which are used in the direct solver for elliptic equations), and the adjoint coding for "on/off" discontinuous operations.

5.3   Adjoint Coding of the FFT and of the Inverse FFT

Standard solvers of elliptic equation often perform "fast Fourier transform" (FFT) and inverse FFT operations. The FFT is defined as follows: given N discretized quantities h_k $(k = 0, 1, 2, \ldots, N)$, the corresponding N discretized FFT transformation coefficients $H_n(n - 0, \ldots, N - 1)$ are given by the formula

$$H_n = \sum_{k=0}^{N-1} h_k e^{2\pi i k n/N}, \quad (n = 0, 1, 2, \ldots, N - 1). \tag{5.22}$$

The corresponding inverse FFT is provided by the following formula:

$$h_k = \frac{1}{N} \sum_{n=0}^{N-1} H_n e^{-2\pi i k n/N}, \quad (k = 0, 1, 2, \ldots, N - 1). \tag{5.23}$$

Using the notation $E \equiv e^{i2\pi/N}$, we can express the FFT operator in matrix form as follows:

$$
\begin{pmatrix}
H_0 \\
H_1 \\
H_2 \\
\cdot \\
\cdot \\
\cdot \\
H_{N-1}
\end{pmatrix}
= \mathbf{A}
\begin{pmatrix}
h_0 \\
h_1 \\
h_2 \\
\cdot \\
\cdot \\
\cdot \\
h_{N-1}
\end{pmatrix}, \tag{5.24}
$$

where $\mathbf{A}$ denotes the matrix:

$$
\mathbf{A} =
\begin{pmatrix}
1 & 1 & 1 & \cdots & 1 & 1 \\
1 & E^{1\times 1} & E^{1\times 2} & \cdots & E^{1\times(N-2)} & E^{1\times(N-1)} \\
1 & E^{2\times 1} & E^{2\times 2} & \cdots & E^{2\times(N-2)} & E^{2\times(N-1)} \\
\cdot & \cdot & \cdot & \cdots & \cdot & \cdot \\
\cdot & \cdot & \cdot & \cdots & \cdot & \cdot \\
\cdot & \cdot & \cdot & \cdots & \cdot & \cdot \\
1 & E^{(N-1)\times 1} & E^{(N-1)\times 2} & \cdots & E^{(N-1)\times(N-2)} & E^{(N-1)\times(N-1)}
\end{pmatrix} \tag{5.25}
$$

Writing the adjoint to FFT is equivalent to writing the conjugate transpose of matrix $\mathbf{A}$ into algebraic coding. Denoting the conjugate of E as $E^* = e^{-2\pi i/N}$ and observing that $E^N = E^0 = 1$, the conjugate transpose of $\mathbf{A}$ is obtained from (5.25) as

$$
(\mathbf{A}^*)^T =
\begin{pmatrix}
1 & 1 & 1 & \cdots & 1 & 1 \\
1 & E^{*1\times 1} & E^{*1\times 2} & \cdots & E^{*1\times(N-2)} & E^{*1\times(N-1)} \\
1 & E^{*2\times 1} & E^{*2\times 2} & \cdots & E^{*2\times(N-2)} & E^{*2\times(N-1)} \\
\cdot & \cdot & \cdot & \cdots & \cdot & \cdot \\
\cdot & \cdot & \cdot & \cdots & \cdot & \cdot \\
\cdot & \cdot & \cdot & \cdots & \cdot & \cdot \\
1 & E^{*(N-1)\times 1} & E^{*(N-1)\times 2} & \cdots & E^{*(N-1)\times(N-2)} & E^{*(N-1)\times(N-1)}
\end{pmatrix} \tag{5.26}
$$

Therefore, the operation adjoint to Eq. (5.22) becomes:

$$
\begin{pmatrix} \hat{h}_0 \\ \hat{h}_1 \\ \hat{h}_2 \\ \cdot \\ \cdot \\ \cdot \\ \vec{h}_{N-1} \end{pmatrix} = (\mathbf{A}^*)^T \begin{pmatrix} \hat{H}_0 \\ \hat{H}_1 \\ \hat{H}_2 \\ \cdot \\ \cdot \\ \cdot \\ \hat{H}_{N-1} \end{pmatrix} , \quad \text{or}
$$

$$
\hat{h}_k = \sum_{n=0}^{N-1} \hat{H}_n e^{-2\pi i k n / N}, \quad (k = 0, 1, 2, \ldots, N-1). \tag{5.27}
$$

Comparing Eq. (5.27) to Eq. (5.23) shows that

$$
(\mathbf{A}^*)^T = N \mathbf{A}^{-1}. \tag{5.28}
$$

Therefore, to obtain the adjoint of the FFT, one simply calls the inverse FFT routine and multiplies the output with the factor N (i.e., with the data sample number). Inverting Eq. (5.28) yields the relations

$$
[(\mathbf{A}^{-1})^*]^T = \frac{1}{N} \mathbf{A}, \tag{5.29}
$$

which indicates that the adjoint of the inverse FFT is obtained by calling the FFT routine and multiplying the output by a factor of $\frac{1}{N}$.

5.4   Developing Adjoint Programs for Interpolations and "On/Off" Processes

Consider the following interpolation operation for a variable Y, with grid point values $Y(I)$ given at grid point locations $X(I)$, $I = 1, 2, 3, \ldots, N$:

$$X = C \,(some\,value)$$
$$J = integer\,(X)$$
$$JP1 = J + 1$$
$$\alpha = X - J$$
$$\beta = JP1 - X$$
$$Y = \alpha * Y(JP1) + \beta * Y(J)$$

The purpose of taking the integer J of a model variable X and using this integer as the index $(JP1)$ of some array Y whose elements are also model variables, is to keep track of those model variables $(X$ and $Y)$, which appear in the interpolation, in order to use them correspondingly in the linearized (tangent) and adjoint code segments. Denoting the original state variable with overbars, the linear tangent coding becomes:

$$\bar{X} = C$$
$$J = integer\,(\bar{X})$$
$$JP1 = J + 1$$
$$\bar{\alpha} = \bar{X} - J$$
$$\bar{\beta} = JP1 - \bar{X}$$
$$\alpha' = X'$$
$$\beta' = -X'$$
$$Y' = \bar{\alpha} * Y'\,(JP1) + \alpha' * \bar{Y}\,(JP1)$$
$$+ \bar{\beta} * Y'\,(J) + \beta' * \bar{Y}\,(J)$$

The adjoint program corresponding to the above (linearized tangent) program is

$$\bar{X} = C$$
$$J = integer\left(\bar{X}\right)$$
$$JP1 = J + 1$$
$$\bar{\alpha} = \bar{X} - J$$
$$\bar{\beta} = JP1 - \bar{X}$$
$$\beta' = Y' * \bar{Y}\left(J\right)$$
$$Y'\left(J\right) = \bar{\beta} * Y' + Y'\left(J\right)$$
$$\alpha' = Y' * \bar{Y}\left(JP1\right)$$
$$Y'\left(JP1\right) = \bar{\beta} * Y' + Y'\left(JP1\right)$$
$$X' = -\beta' + X'$$
$$X' = \alpha' + X'$$

The "on-off" processes appear in the code used for solving the original nonlinear problem (henceforth called the "nonlinear code") as IF statements that depend on the model variables. For example, in an operational large-scale code that models precipitation and evaporation, a typical IF statement assumes the form:

---

if $q > q_s$ **then**
    instructions for precipitation
end if
if $q > 0.8q_s$ **then**
    instructions for evaporation
end if

In order to develop the adjoint code, a bit vector is used in each IF statement to record the route of the forward integration of the nonlinear model. Therefore, the modified nonlinear code assumes the form:

---

if $q > q_s$ **then**

    IPRE=1

    instructions for precipitation

else

    $IPRE = -1$

end if

if $q > 0.8q_s$ **then**

    $IEVP = 1$

    instructions for evaporation

else

    $IEVP = -1$

end if

The same bit vector is used for the IF statement in the corresponding programming for the tangent linear model, namely:

---

if $IPRE > 0$ **then**

    linearized instructions for precipitation

end if

if $IEVP > 0$ **then**

    linearized instructions for evaporation

end if

The adjoint operation is based on the tangent linear code and follows the same route backwards, namely

---

if $IEVP > 0$ **then**

    adjoint code of instructions for evaporation

end if

if $IPRE > 0$ **then**

    adjoint code of instructions for precipitation

end if

As the foregoing illustrative example indicates, the correctness of the adjoint code can be verified by performing an additional integration of the nonlinear model with bit vectors added in order to determine the routes for the IF statements included in the physical processes. This additional integration of the nonlinear model with added bit vectors is needed for the verification of both the tangent linear model and the adjoint model.

5.5 Construction of Background Covariance Matrices

Modelling and specification of the covariance matrix of background error constitute important components of any data assimilation system. The main attributes of the background error covariance matrix $\mathbf{B}$ are:

1. To spread out the information from the observations; correlations in the background covariance matrix will perform spatial spreading of information from observation points to a finite domain surrounding them;

2. To provide statistically consistent increments at the neighboring grid points and levels of the model;

3. To ensure that observations of one model variable (e.g., temperature) produce dynamically consistent increments in the other model variables (e.g., vorticity and divergence).

For operational models, a typical background covariance matrix contains $10^7 \times 10^7$ elements. Therefore, non-essential components of this important covariance matrix may need to be neglected in order to produce a computationally feasible algorithm.

Construction of background error covariances has been addressed in ref. [90] by the so–called "innovation method," in which the background errors are assumed to be independent of observation errors. Writing the innovation for the ith observation in the form $\mathbf{d}_i = \mathbf{y}_i - \mathbf{H}_i(\mathbf{x}_b)$ denoting the background error by ε, the observation error by η and neglecting the representativeness error, it follows that $Var(\mathbf{d}_i) = Var(\eta_i) + Var(\mathbf{H}_i(\varepsilon))$ and $Cov(\mathbf{d}_i, \mathbf{d}_k) = Cov(\mathbf{H}_i(\varepsilon), \mathbf{H}_k(\varepsilon))$.

The so-called National Meteorological Center (NMC) method was introduced in ref. [57] as a surrogate for samples of background error using differences between forecasts of different length that verify at the same time. The ensemble method for constructing background covariances was proposed in ref. [62], while ref. [93] proposed using statistical structures of forecast errors. One can attempt to disentangle information about the statistics of background error from the available information (innovation statistics), or one can try to find a surrogate quantity whose error statistics can be argued to be similar to those of the unknown background errors.

Customarily, the background covariance matrix (denoted here as $\mathbf{B}$) is used to construct a functional J_b defined as

$$J_b(\mathbf{X}) = \frac{1}{2}(\mathbf{X}_0 - \mathbf{X}_b)^T \mathbf{B}^{-1}(\mathbf{X}_0 - \mathbf{X}_b), \tag{5.30}$$

where $\mathbf{X}_0 - \mathbf{X_b}$ represents the departures of the model variables at start of the analysis from the background field $\mathbf{X_b}$. The term J_b is added to the cost functional that measures the deviations between the model computations and the corresponding observations, in order to construct the following overall cost functional:

$$J(\mathbf{X}_0) = \frac{1}{2}(\mathbf{X}_0 - \mathbf{X_b})^T \mathbf{B}^{-1}(\mathbf{X}_0 - \mathbf{X_b})$$
$$+ \frac{1}{2} \sum_{r=0}^{R} [\mathbf{HX}(t_r) - \mathbf{X}^{obs}(t_r)]^T \mathbf{R}^{-1}[\mathbf{HX}(t_r) - \mathbf{X}^{obs}(t_r)]. \tag{5.31}$$

Spatial variations of background covariances can be modeled by various means, such as:

1. wavelets expansions (see, e.g., ref. [60] and [67]);

2. diffusion operators using numerical integration of the diffusion equation to perform convolutions for covariance modelling (see, e.g., Vukićević and Errico [210]);

3. digital filters to synthesize inhomogeneous covariances, by allowing the filter coefficients to vary with location and using combinations of several filters to do so.

In practice, digital filters are closely related to diffusion operators since one time step integration of the diffusion equation is equivalent to application of the digital filter. Since it is difficult to determine filter coefficients from data, diffusion operators may be preferable to use in practice.

5.6   Characterization of Model Errors in 4–D VAR

The 4–D VAR data assimilation procedure is affected by errors attributable to: (i) the dynamical model (e.g., poor representation of processes, omissions or incorrect formulations of key processes, numerical approximations); and (ii) observations or measurements (e.g., sensor design, performance, noise, sample averaging, aliasing).

For dynamically evolving systems the model errors are expected to depend on time and, possibly, on the model state variables. Typically, at any time step, t_k, the evolution of a model error $\boldsymbol{\eta}_k$ is assumed (see, e.g., ref. [86]) to evolve according to the following discrete equation:

$$\boldsymbol{\eta}_k = \mathbf{T}_k(\mathbf{e}_k) + \mathbf{q}_k, \tag{5.32}$$

where $\mathbf{e}_k$ represents time-varying systematic components of the model errors, $\mathbf{T}_k$ describes the distribution of systematic errors in the model equations, and $\mathbf{q}_k$ (stochastic component) is an unbiased, serially correlated, normally distributed random vector with a known covariance matrix. The evolution of $\mathbf{e}_k$ is, in turn, modeled by assuming that it depends on the state vector $\mathbf{x}_k$, namely:

$$\mathbf{e}_{k+1} = \mathbf{g}_k(\mathbf{x}_k, \mathbf{e}_k). \tag{5.33}$$

Three forms were proposed for the evolution of the above systematic component model errors, as follows:

1. constant in time:

$$\mathbf{e}_{k+1} = \mathbf{e}_k, \quad \mathbf{T}_k = \mathbf{I}, \tag{5.34}$$

which is suitable for modeling errors in source terms and boundary conditions;

2. evolving in time:

$$\mathbf{e}_{k+1} = \mathbf{F}_k \mathbf{e}_k \ , \quad \mathbf{T}_k \equiv \mathbf{I}, \tag{5.35}$$

where $\mathbf{F}_k$ is a linear model, appropriate for representing discretization errors:

3. spectral form:

$$\mathbf{e}_{k+1} = \mathbf{T}_k \, \mathbf{e}_k, \tag{5.36}$$

where $\mathbf{T}_k$ is a block diagonal matrix with diagonal entries given by

$$I, \quad I\sin(\frac{\kappa}{N\tau}), \quad I\cos(\frac{\kappa}{N\tau}),$$

where τ is a constant time scale.

While Eq. (5.32) describes the evolution of model errors has been considered as a *discrete process*, Vidard et al. [208] considered a continuous time evolution of model errors, consistent with the fact that model equations are first written as differential equations and then discretized (in space, time, etc.). Considering that the initial vector of model errors at the initial time t_0 is

$$\boldsymbol{\eta}(t_0) = \boldsymbol{\eta}_0, \tag{5.37}$$

the evolution of model errors in time was modeled in ref. [148] using the evolution equation

$$\frac{d\eta}{dt} = \boldsymbol{\Phi}[\boldsymbol{\eta}(t), \mathbf{x}(t)] + \mathbf{q}(t), \tag{5.38}$$

where $\mathbf{q}(t)$ denotes the stochastic component. The simplest form of Eq. (5.38) is the exponential growth equation

$$\frac{d\eta}{dt} = \eta(t). \tag{5.39}$$

In ref. [44], the model errors were considered to be correlated in time and their evolution was considered to follow a Markov process in a simple Kalman filtering framework. The Markovian assumption is based on the observation that as the numerical model is integrated in time, errors show a trend of serial correlation in both time and space.

The most important property of a Markov process is that the state at any future time depends only on its present value, but not on its value in the past. Considering the model error as a Markov process, its evolution between any two successive time steps, t_k and t_{k+1}, is governed by the relation

$$\boldsymbol{\eta}_{k+1} = \mu \mathbf{G}_k[\boldsymbol{\eta}_k] + (1 - \mu)\mathbf{q_k}, \tag{5.40}$$

where μ is a scalar, $\mathbf{q}_k$ is the random component of model errors, and $\mathbf{G}_k$ is a linear operator. Reference [44] discusses the implementation of two different forms of this operator in a Kalman filter setting. Using $\mathbf{G}_k \equiv \mathbf{I}$ in Eq. (5.40), references [236], [238] and [240] provided results obtained using the National Center for Environmental Prediction (NCEP) regional weather prediction system.

In the "weak constraint" 4–D VAR, the following cost functional $J(\mathbf{x_0}, \boldsymbol{\eta_0})$ is minimized with respect to the initial state variables $\mathbf{x}_0$ and initial model errors $\boldsymbol{\eta}_0$:

$$
\begin{aligned}
J(\mathbf{x_0}, \boldsymbol{\eta_0}) = {}& \frac{1}{2}[\mathbf{x}_0 - \mathbf{x}^b]^T \mathbf{B}^{-1}[\mathbf{x}_0 - \mathbf{x}^b] \\
& + \frac{1}{2}\sum_{i=0}^{n}[\mathbf{H}(\mathbf{x}(t_i)) - \mathbf{y}^o(t_i)]^T \mathbf{R}^{-1}[\mathbf{H}(\mathbf{x}(t_i)) - \mathbf{y}^o(t_i)] \\
& + \frac{1}{2}[\boldsymbol{\eta}_o - \boldsymbol{\eta}^b]^T \mathbf{Q}^{-1}[\boldsymbol{\eta}_o - \boldsymbol{\eta}^b],
\end{aligned} \tag{5.41}
$$

where $\mathbf{Q}$ denotes the model error covariance matrix. The minimization of $J(\mathbf{x_0}, \boldsymbol{\eta_0})$ is subject to the following equations as constraints:

$$\frac{d\mathbf{x}(t)}{dt} = \mathbf{M}[\mathbf{x}(t)] + \boldsymbol{\eta}(t); \quad \mathbf{x}(t_0) = \mathbf{x}_0, \tag{5.42}$$

$$\frac{d\boldsymbol{\eta}}{dt} = \boldsymbol{\Phi}[\boldsymbol{\eta}(t), \mathbf{x}(t)]; \quad \boldsymbol{\eta}(t_0) = \boldsymbol{\eta}_0. \tag{5.43}$$

The above constrained minimization problem becomes an unconstrained problem by introducing the following augmented Lagrangian functional

$$L(\mathbf{x}, \boldsymbol{\eta}, \mathbf{x}^*, \boldsymbol{\eta}^*) = J(\mathbf{x}_0, \boldsymbol{\eta}_0)$$

$$+ \int_{t_0}^{t_n} \langle\, \mathbf{x}^*, \{\frac{d\mathbf{x}(t)}{dt} - \mathbf{M}[\mathbf{x}(t)] - \boldsymbol{\eta}(t)\} \,\rangle \, dt$$

$$+ \int_{t_0}^{t_n} \langle\, \boldsymbol{\eta}^*, \{\frac{d\boldsymbol{\eta}}{dt} - \boldsymbol{\Phi}[\boldsymbol{\eta}(t), \mathbf{x}(t)]\} \,\rangle \, dt, \qquad (5.44)$$

where $\mathbf{x}^*$, $\boldsymbol{\eta}^*$ are the Lagrange multiplier vectors corresponding to $\mathbf{x}$ and $\boldsymbol{\eta}$, respectively, while the angular brackets $\langle,\rangle$ denote the Euclidean inner product. The necessary conditions for the minimization of L defined in Eq. (5.44) are

$$\frac{\partial L}{\partial \mathbf{x}} = \mathbf{0}, \quad \frac{\partial L}{\partial \boldsymbol{\eta}} = \mathbf{0}, \quad \frac{\partial L}{\partial \mathbf{x}^*} = \mathbf{0}, \quad \frac{\partial L}{\partial \boldsymbol{\eta}^*} = \mathbf{0}. \qquad (5.45)$$

Alternatively, model errors can be introduced as "model biases" into the cost functional to be minimized within the 4–D VAR data assimilation framework, as proposed, e.g., in ref. [51]. This alternative viewpoint is usually called the "strong constraint" 4–D VAR, and is based on using the augmented control vector

$$\mathbf{z}^T = [\mathbf{x}^T, \boldsymbol{\beta}^T], \qquad (5.46)$$

which includes the parameters $\boldsymbol{\beta}$ representing model biases in addition to the model state vector $\mathbf{x}$. The analysis is then obtained by minimizing the cost functional

$$J(\mathbf{z}) = (\mathbf{z}^b - \mathbf{z})^T \mathbf{Z}^{-1} (\mathbf{z}^b - \mathbf{z}) + [\mathbf{y} - \mathbf{h}(\mathbf{z})]^T \mathbf{R}^{-1} [\mathbf{y} - \mathbf{h}(\mathbf{z})]. \qquad (5.47)$$

with respect to the new control vector $\mathbf{z}$, whose background estimate $\mathbf{z}^b$ must now include a prior estimate $\boldsymbol{\beta}^b$ of the bias parameters. The observation operator $\mathbf{h}$ may depend on (some of) the newly introduced bias parameters. The matrix $\mathbf{Z}$ represents an augmented background-error covariance operator, which, in principle, should include cross-covariances among parameters and state vector components.

Controlling modeling errors in addition to the model's initial conditions in the weak constraint 4–D VAR doubles the size of the optimization problem by comparison to the strong constraint 4–D VAR. Furthermore, if the stochastic component is included in the model error formulation, then the random realization would need to be saved at each model time step. Consequently, the size of the optimization problem would be tripled. The size of the model error control vector can be reduced by projecting it onto the subspace of eigenvectors corresponding to the leading eigenvalues of the adjoint-tangent linear operators (see, e.g., ref. [207]).

Most data assimilation systems are not equipped to handle large, systematic corrections; they were designed to make small adjustments to the background fields that are consistent with the presumed multivariate and spatial structures of random errors. Statistics of "observed-minus-background" residuals provide a different, sometimes more informative, view on systematic errors afflicting the model or observations. Operational NWP centers routinely monitor time and space averaged background residuals associated with different components of the observing system, providing information on the quality of the input data as well as on the performance of the assimilation system. In general, small root-mean-square residuals imply that the system is able to accurately predict future observations. Nonzero mean residuals, however, indicate the presence of biases in the observations and/or their model-predicted equivalents. The crucial problem linked to parametrizing model error is that of developing physically meaningful representations of model errors that can be clearly distinguished from possible observation errors. This issue is the subject of intensive ongoing research.

5.7 The Incremental 4–D VAR Algorithm

5.7.1 Introduction

During the 1990s, the 4–D VAR methodology matured and was adopted at several important international Numerical Weather Prediction centers. However, although 4–D VAR cost function and gradient can be evaluated at the cost of one integration of the forecast model followed by one integration of the adjoint model, the computational cost to implement it was still prohibitive since a typical minimization requires between 10 and 100 evaluations of the gradient. The cost of the adjoint model is typically 3 times that of the forward model ref. [84]. The analysis window in a typical operational model such as the ECMWF system is 12 hours. Thus, the cost of the analysis is roughly equivalent to between 20 and 200 days of model integration with 10^8 variables, making it computationally prohibitive for NWP centers that have to deliver timely forecasts to the public. Amongst the methods that greatly facilitated the adoption, application, and implementation of 4–D VAR data assimilation at major operational centers and contributed to advance of the technique, the *incremental* 4–D VAR method is of paramount importance. Courtier et al. [38] (based on an idea of Derber) introduced the incremental formulation of the 4–D VAR. The incremental algorithm reduces the cost of 4–D VAR mainly by reducing the resolution of the model, thus allowing the 4–D VAR method to become computationally feasible. Other simplifications introduced by the incremental 4–D VAR method will be briefly described below. The nonlinearity of the model and/or of the observation operator can produce multiple minima in the cost function, which will impact the convergence of the minimization algorithm. The incremental 4–D VAR algorithm removes the nonlinearities in the cost minimization by using a forward integration of the linear model instead of a nonlinear one. It also uses a coarser resolution model and eliminates most of the time-consuming physical packages. In this section we will address several algorithmic aspects of incremental 4–D VAR that are used in present-day implementations of 4–D VAR data assimilation.

Some aspects related to the incremental method versus the full 4–D VAR were addressed by Li et al. [125]. They conducted a set of four-dimensional variational assimilation (4–D VAR) experiments using both a standard method and an incremental method and compared the corresponding performances.

5.7.2   The 4–D VAR Incremental Method

Courtier et al. [38] devised an incremental 4–D VAR algorithm which removes nonlinearities in the minimization by using a forward integration of a linear model instead of a nonlinear one. The minimization of the cost functional is carried out at a reduced model resolution which leads to an effective reduction of computational cost and memory requirements.

The 4–D VAR problem consists in finding the state at time t_0 that minimizes the cost function:

$$\mathcal{J}(\mathbf{x}_0) = \frac{1}{2}(\mathbf{x}_0 - \mathbf{x}^b)^T \mathbf{B}^{-1}(\mathbf{x}_0 - \mathbf{x}^b) + \frac{1}{2}\sum_{i=0}^{N}[(\mathbf{H}_i(\mathbf{x}_i) - \mathbf{y}_i)^T \mathbf{R}_i]^{-1}[\mathbf{H}_i(\mathbf{x}_i) - \mathbf{y}_i]$$

(5.48)

subject to the states $\mathbf{x}_i$ satisfying the NWP model as a strong constraint. In optimal control language this is referred to as Partial Differential Equations (PDE) constrained optimization. We consider a discrete nonlinear dynamical system given by the equation

$$\mathbf{x}_{i+1} = \mathbf{M}_i(\mathbf{x}_i),$$

(5.49)

where $\mathbf{x}_i \in \mathbb{R}^n$ is the state vector at time t_i and $\mathbf{M}_i$ is the nonlinear model operator that propagates the state at time t_i to time t_{i+1} for $i = 0, 1, \ldots, N-1$. We assume that we have imperfect observations $\mathbf{y}_i \in \mathbb{R}^{p_i}$ at times $t_i, i = 0, \ldots, N$.

Here $\mathbf{H}_i : \mathbb{R}^{p_i} \to \mathbb{R}^n$ is known as the **observation operator** and maps the state vector to the observation space. $\mathbf{B}$ is the background error covariance matrix and $\mathbf{R}_i$ are the observation error covariance matrices. In the incremental formulation the solution to the nonlinear minimization problem is approximated by a sequence of minimizations of linear quadratic cost functions.

We define $\mathbf{x}_0^{(k)}$ to be the kth estimate to the solution and linearize the cost function Eq. (5.48) around the model trajectory forecast from this estimate. The following estimate is defined by

$$\mathbf{x}_0^{k+1} = \mathbf{x}_0^k + \delta\mathbf{x}_0^k,$$

(5.50)

where the perturbation $\delta x_0^k \in \mathbb{R}^n$ is a solution of the linearized cost function

$$\tilde{\mathcal{J}}^{(k)}(\delta \mathbf{x}_0^{(k)}) = \frac{1}{2}[\delta \mathbf{x}_0^{(k)} - (\mathbf{x}^b - \mathbf{x}_0^{(k)})]^T \mathbf{B}^{-1}[\delta \mathbf{x}_0^{(k)} - (\mathbf{x}^b - \mathbf{x}_0^{(k)})]$$

$$+ \frac{1}{2}\sum_{i=1}^{N}(\mathbf{H}_i \delta \mathbf{x}_i^{(k)} - \mathbf{d}_i^{(k)})^T \mathbf{R}^{-1}(\mathbf{H}_i \delta \mathbf{x}_i^{(k)} - \mathbf{d}_i^{(k)}). \qquad (5.51)$$

The cost function is written in terms of increments with respect to the background state. The increment satisfies the linear dynamical equation $\partial \mathbf{x}_{i+1} = \mathbf{M}_i \partial \mathbf{x}_i$ and $\mathbf{d}_i^k = \mathbf{y}_i^0 - \mathbf{H}_i \mathbf{x}_i^{(k)}$ is the innovation departure or observation increment at time t_i. The quantity $\mathbf{H}_i$ is the linearized observation operator, and $\mathbf{M}_i$ depicts the tangent linear model operator evaluated at the current estimate of the nonlinear trajectory, usually referred to as the linearization state.

The process of minimization is identical to the usual 4–D VAR algorithm except that the control variable is the increment at time t_0 and the increment trajectory is obtained by integration of the linear model. The reference trajectory required by the linear and adjoint models comes from the background integration and is not updated at every iteration. Correspondingly, the iterative procedure of minimizing the incremental cost function is called the *inner loop*, which is much cheaper computationally to implement due to the aforementioned simplifications.

When the quadratic cost function is approximated in this way, the 4–D VAR algorithm no longer converges to the solution of the original problem. The analysis increments are calculated at a reduced resolution and must be interpolated to the high-resolution model's grid. This drawback is partially overcome by executing after a number of inner-loops, one *outer loop* which is updating the high-resolution reference trajectory and the observation departures. Correspondingly, the iterative procedure of minimizing the incremental cost function is called the *outer loop*. After each *outer loop* update, it is possible to use a progressively higher resolution for the inner-loops. Such a procedure was carried out in a multi-incremental algorithm proposed by Veerse and Thépaut [206].

The incremental method was shown by Lawless [113] to be equivalent to an inexact Gauss-Newton method applied to the original nonlinear cost function. The outer loop iterations can be shown to be locally convergent under certain conditions, provided that the inner loop minimization is solved to sufficient accuracy (see, e.g., ref. [81]). In practice, however, very few outer loop steps are performed, typically three. The inclusion of full physics in the adjoint model requires the 4–D VAR algorithm to overcome the negative effect of strong nonlinearities present in physics parameterization packages while being able to take advantage of the positive aspects resulting from consistency between the forecasting nonlinear model and adjoint model.

Several approaches have been proposed for mitigating the negative effect of strong nonlinearities in physical processes included in the adjoint model. These approaches involved either direct modifications or simplifications to physical parameterizations. Zupanski [237], and Tsuyuki [204] showed beneficial effects when smoothing formulas are used to replace those with discontinuities. The ECMWF system uses simplified physics in the adjoint model, although modifications or simplifications may lead to inconsistencies between the nonlinear forecasting model and the adjoint model.

In a further, *multi-incremental*, extension of the incremental 4–D VAR method, the inner loop resolution is increased after each iteration of the *outer loop*. In particular, the information about the shape of the cost function obtained during the early low-resolution iterations provides a very effective preconditioner for subsequent iterations at higher resolution, thus reducing the number of costly iterations. The inner loops can be efficiently minimized using the conjugate gradient method, provided the cost function is purely quadratic, i.e. when the operators involved in the definition of the cost function (the model and the observation operators) are linear. For this reason, the inner loops have been completely linearized; the nonlinear effects are all gathered at the outer loop level.

Figure 5.1 illustrates the interplay between the *inner* and the *outer loops* in the implementation of incremental 4–D VAR method. The operator **S** projects from the full model to the coarse mesh model and the pseudo-inverse operator, $\mathbf{S}^{-1}$, projects from the inner-loop coarse mesh model to the nonlinear fine mesh full-physics model.

Outer loops are performed at high resolution using the full nonlinear model. Inner loop iterations are performed at lower resolution using the tangent-linear forecast model, linearized around a 12-hour succession of model states ("the trajectory") obtained through interpolation from high resolution

(S denotes the truncation operator, J the cost function and x the atmospheric state vector). Both the "incremental" and the "multi-incremental" algorithms use the same high resolution (T511, currently increased to T799) at the outer loop level. However, the inner loops use a lower resolution (T95) in the "incremental" algorithm, and an increasingly higher (up to T159) in the "multi-incremental" algorithm.

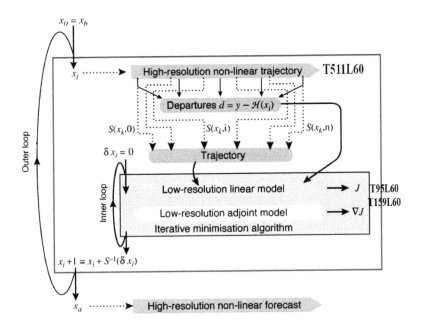

FIGURE 5.1
Incremental and multi-incremental 4–D VAR algorithms

5.7.3   Preconditioning of Incremental 4–D VAR

Recall that the Hessian of the 4–D VAR cost function is given by

$$\mathbf{B}^{-1} = \mathbf{H}^T \mathbf{R}^{-1} \mathbf{H}. \tag{5.52}$$

Defining an operator $\mathbf{L}$ such that

$$\mathbf{B} = \mathbf{L}\mathbf{L}^T, \tag{5.53}$$

and defining the change of variables

$$\boldsymbol{\xi} = \mathbf{L}^{-1}\partial\mathbf{x},$$

(5.54)

transforms the cost function in Eq. (5.51) into the form:

$$\hat{J} = \frac{1}{2}\boldsymbol{\xi}^T\boldsymbol{\xi} + \frac{1}{2}(\mathbf{HL}\boldsymbol{\xi} - \mathbf{d})^T\mathbf{R}^{-1}(\mathbf{HL}\boldsymbol{\xi} - \mathbf{d}).$$

(5.55)

Using the operator $\mathbf{L}$ enables us to write the Hessian of $\hat{J}$ in the form

$$(\nabla^2\hat{J}) = \mathbf{I} + \mathbf{L}^T\mathbf{H}^T\mathbf{R}^{-1}\mathbf{HL},$$

(5.56)

where its smallest eigenvalue is $\lambda_N = 1$. Furthermore, the Hessian of the cost function can be written in the spectral form

$$(\nabla^2\hat{J}) = \sum_{k=1}^{N}\lambda_k\boldsymbol{\nu}_k\boldsymbol{\nu}_k^T,$$

(5.57)

where λ_k and ν_k are its eigenvalues and eigenvectors, respectively. Choosing a preconditioning based on the K leading eigenvectors of the Hessian $(\nabla^2\hat{J})$ makes it possible to separate the leading contributions in the form

$$(\nabla^2\hat{J}) = \sum_{k=1}^{K}\mu_k\lambda_k\boldsymbol{\nu}_k\boldsymbol{\nu}_k^T + \sum_{k=K+1}^{N}\lambda_k\boldsymbol{\nu}_k\boldsymbol{\nu}_k^T.$$

(5.58)

where the scalars μ_k are related to the specific form of the preconditioning. Choosing, for example, the scalars μ_k such that $\mu_k\lambda_k < \lambda_{k+1}$, causes the spectral radius, $\kappa(\nabla^2\hat{J})$, of the Hessian in Eq. (5.57) to become

$$\kappa(\nabla^2\hat{J}) = \lambda_{K+1}/\lambda_N.$$

(5.59)

Additional preconditioning can be performed if additional information about the Hessian of the transformed cost function is available. One way to obtain such additional information is to use the Lanczos algorithm to perform the inner loop minimization. The Lanczos method produces estimates of the leading eigenvectors and eigenvalues of the Hessian of the function being minimized. This methodology is employed presently in the Weather Research Forecast community model (see e.g., [203]).

In a comparison between conjugate gradient and quasi-Newton algorithms, the number of iterations for each inner loop has been kept fixed to 25 for both algorithms. Laroche and Gauthier [112] and Lawless et al. [114] have shown that the best results are obtained when inner loop iterations are stopped early enough to avoid overfitting observational noise, as this would introduce spurious gradients in the following trajectory integration.

5.7.4   Summary and Discussion

There are still many other open problems to consider in incremental 4–D VAR. Part of the computational issues related to the incremental method were addressed in the work of Trémolet ([199], [200], [201]), and skillfully summarized by Fisher [61].

Robert et al. ([170], [171]) demonstrated how the incremental 4–D VAR data assimilation method can be applied with an efficient preconditioning applied to an oceanographic problem. The approach consists in performing a few iterations of the reduced-order (using EOF/POD) 4–D VAR prior to the incremental 4–D VAR in the full space in order to achieve faster convergence. It allows the global cost of the assimilation to be reduced by a factor of 2 without affecting the quality of the solution.

Lawless et al. [114] investigated the convergence of incremental four-dimensional variational data assimilation (4–D VAR) when an approximation to the tangent linear model is used within the inner loop. Using a model of the one-dimensional shallow water equations, they perform data assimilation experiments using an exact tangent linear model and using an inexact linear model (based on a perturbation forecast model). They find that the two assimilations converge at a similar rate and the analyses are also similar, with the difference between them depending on the amount of noise in the observations. They then present the incremental 4–D VAR algorithm as a Gauss-Newton iteration for solving a least-squares problem and consider its fixed points.

Lawless and Nichols [115] derived a new stopping criterion for the inner loop minimizations of incremental 4–D VAR, which guarantees convergence of the outer loops. This new criterion gives improved convergence, as compared to other commonly used inner loop stopping criteria.

Gratton et al. [81] continued this investigative line and examined:

1. "truncated" Gauss-Newton methods, in which the inner linear least-squares problem is not solved exactly;

2. "perturbed" Gauss-Newton methods, in which the true linearized inner problem is approximated by a simplified, or perturbed, linear least-squares problem.

They present conditions ensuring that the truncated and perturbed Gauss-Newton methods converge and derive rates of convergence for the iterations. A practical application to the problem of data assimilation in a 1–D shallow water equations system is finally presented.

Trémolet [199] finds that linearization errors in incremental 4–D VAR are larger than expected and that large errors appear very early in the assimilation window. He also shows that higher resolution 4–D VAR will require more accurate linear physics than currently available. A modification of the computation of the trajectory around which the problem is linearized is shown to improve the accuracy of the linearization.

Trémolet [200] analyzed the operational implementation at ECMWF of the incremental four-dimensional data assimilation which is using coarser resolution in the inner loop and the variational data assimilation system (4–D VAR) is run with two outer loop iterations. He re-evaluates the convergence of 4–D VAR at outer loop level with the current ECMWF system. Experimental results show that 4–D VAR in its current implementation does diverge after four outer loop iterations.

Further investigations have shown that convergence can be obtained when the inner and outer loops are executed at the same resolution, or at least with the same time step. This is explained by the presence of gravity waves which propagate at different speeds in the linear and nonlinear models. These gravity waves are related to the shape of the leading eigenvector of the Hessian of the 4–D VAR cost function: this eigenvector is determined by surface pressure observation and controls the behavior of the minimization algorithm.

5.8   Open Research Issues

The construction of covariances for model errors, on the one hand, and back-ground errors, on the other hand, are closely connected. Consequently, the geosciences community focuses on both model error parameterizations along with the linked problem of estimation of model bias using 4–D VAR data assimilation. There is a consensus that background error statistics are crucial for any 4–D VAR data assimilation system in order to provide statistically consistent increments at the neighboring grid points and all vertical levels of the operational model. Implementation of multivariate statistics and various geophysical balances are crucial for realistic flow-dependent construction of the background error covariance matrix, so research in this domain is still intensive. References [71] and [97] highlight the role of the representativeness error in data assimilation. Extensions to allow for anisotropic and inhomoge-neous background error correlations are now state-of-the-art procedures for background error covariance matrix formulation.

Constructing covariance matrices for model errors is particularly difficult for underobserved systems, such as oceans. More generally, significant research efforts are currently devoted to developing efficient algorithms for optimiza-tion of nonsmooth functionals in the presence of "on/off" physical processes. For such functionals, the adjoint model integration produces values of subgra-dients (instead of gradients) of these cost functions with respect to the model's control variables at discontinuous points. Minimization of these cost functions using conventional differentiable optimization algorithms may encounter dif-ficulties. To illustrate the performances of differentiable and nondifferentiable optimization algorithms, the work in ref. [223] considered an idealized dis-continuous model and an actual shallow convection parameterization, both including on–off switches. The conclusions of this work indicated that:

1. The differentiable optimization, such as the limited memory quasi-Newton (L-BFGS) algorithm, works well for minimizing a nondif-ferentiable cost function, especially when the changes made in the forecast model at switching points to the model state are not too large;

2. For a nondifferentiable optimization algorithm: introducing a local smoothing that removes discontinuities may insert artificial stationary points, which will increase significantly the difficulty of finding the true minimum of a nonsmooth cost function;

3. A nondifferentiable optimization bundle method (modified to use only one subgradient that can be calculated by the conventional adjoint model) was able to find the true minima in cases where the differentiable minimization failed. Recent advances in this topic show potential of large-scale limited memory bundle algorithms (see e.g., Steward et al. [192]).

4. A set-valued adjoint model can be used to develop a bundle method that yields higher convergence scores for a vertical diffusive convective adjustment model used in ocean general circulation models. This is because the set-valued adjoint model can produce all supporting subgradients at singular points.

5. Perhaps the best evaluation of different methods should be based on a detailed cost-benefit analysis for a particular data assimilation problem at hand.

In cases of meso-scale data assimilation an examination of small scale features may reveal problems with a particular minimization method that have not been noticed in larger scales. Work in ref. [239] showed that the maximum likelihood ensemble filter acts as a very efficient minimizer for nondifferentiable cost functions and observation operators.

6

Appendix A

CONTENTS

6.1   Frequently Encountered Probability Distributions

This appendix is intended to provide a quick reference for selected properties of distributions commonly used for data analysis, evaluation, and assimilation.

For further reference, note that the *moment generating function (MGF)*, denoted in this appendix as $M_x(t)$, is defined as the expectation of e^{tx}, namely $M_x(t) = E(e^{tx}) = \int_{-\infty}^{\infty} e^{tx} p(x) \, dx$, where x is a random variable and t is a real number. The multivariate *MGF* is defined analogously as

$$M_x(\mathbf{t}) = E\left(\exp\left(\sum_{i=1}^{n} x_i t_i\right)\right) = \int_{\delta x_x} \exp\left(\sum_{i=1}^{n} x_i t_i\right) p(\mathbf{x}) \, d\mathbf{x}, \qquad (6.1)$$

where the vector $(t_1, t_2, \ldots, t_n)$ has components t_i defined symmetrically around the origin $(0, 0, \ldots, 0)$, namely $-t_{oi} < t_i < t_{oi}$, with $t_{oi} > 0$, $(i = 1, \ldots, n)$.

Degenerate Distribution: Consider that x is a random variable which can assume only a single value, namely, $x = c$, where c is a real number. The distribution for x is called the *degenerate distribution*, and the corresponding probability density is given by

$$p(x) = \delta(x - c), \qquad (6.2)$$

where $-\infty \leq x \leq \infty$, and δ denotes the Dirac delta functional. The *MGF* for this distribution is

$$M_x(t) = e^{tc}, \quad (t \in \mathcal{R}), \tag{6.3}$$

while the mean value is $m_o = c$, and the variance is $\mu_2 = 0$.

Discrete Uniform Distribution: Consider that x is a random variable, which can assume only the integer values $x = 1, \ldots, n$. Each of these values carries equal probability. The distribution for x is called the *discrete uniform distribution*, and the corresponding probability is given by

$$p(x) = \begin{cases} 1/n, & (\text{x=1,2,}\ldots,\text{n}) \\ 0, & \text{otherwise.} \end{cases} \tag{6.4}$$

Using the Dirac delta functional, Eq. (6.4) can also be written in the alternative form $p(x) = (1/n) \sum_{i=1,n} \delta(x-i)$. The *MGF* for this distribution is

$$M_x(t) = \left[e^t \left(1 - e^{nt}\right)\right] / \left[n \left(1 - e^t\right)\right], \quad (t \in \mathcal{R}). \tag{6.5}$$

Using either the above *MGF* or by direct calculations, the mean value of x is obtained as $m_o = (n+1)/2$, while the variance of x is obtained as $\mu_2 = \left(n^2 - 1\right)/12$.

Continuous Uniform Distribution: Consider that x is a random variable, which can assume any real value in the nondegenerate (i.e., $a < b$) interval $I(a, b)$. The distribution for x is called the *continuous uniform distribution*, and the corresponding probability density function is given by

$$p(x) = \begin{cases} 1/(b-a), & a \leq x < b; \ a, b \in \mathcal{R}, \\ 0, & \text{otherwise.} \end{cases} \tag{6.6}$$

The *MGF* for this distribution is

$$M_x(t) = \begin{cases} \left[(e^{tb} - e^{ta}]/[(b-a)t], & t \in \mathcal{R}, \quad t \neq 0, \\ 1, & t = 0. \end{cases} \tag{6.7}$$

From the above MGF, the mean value is obtained as $m_o = (a+b)/2$, while the variance is obtained as $\mu_2 = (b-a)^2/12$. This distribution is employed wherever the range of a finite random variable is bounded and there is no *a priori* reason for favoring one value over another within that range. In practical applications, the continuous uniform distribution is often used in Monte Carlo analysis or computational methods, and as a prior for applying Bayes' theorem [10] in the extreme situation when no information is available prior to performing an experiment.

Bernoulli Distribution: *Bernoulli trials* are defined as random trials in which the outcomes can be represented by a random variable having only two values, say $x = 0$ and $x = 1$. Such a model can be applied to a random trial whose outcomes (events) are described by "yes or no," "on or off," "black or white," "success or failure," etc. Suppose that only one Bernoulli trial is performed, for which the probability of "success" ($x = 1$) is denoted by s, for $0 < s < 1$, while the probability of "failure" ($x = 0$) is ($1-s$). The probability distribution that describes this trial is called the *Bernoulli distribution*, and has the form

$$P(x) = \begin{cases} s^x(1-s)^{1-x}, & \text{x=0, 1} \\ 0, & \text{otherwise .} \end{cases} \tag{6.8}$$

The MGF for this distribution is given by

$$M_x(t) = se^t + 1 - s, \quad (t \in \mathscr{R}) . \tag{6.9}$$

From the above MGF, the mean value is obtained as $m_o = s$, while the variance is obtained as $\mu_2 = s(1-s)$. The Bernoulli distribution provides a basis for the binomial and related distributions.

Binomial and multinomial distributions: Consider a series of n independent trials or observations, each having two possible outcomes, usually referred to as "success" and "failure," respectively. Consider further that the probability for success takes on a constant value, s, and consider that the quantity of interest is the accumulated result of n such trials (as opposed to the outcome of just one trial). The set of n trials can thus be regarded as a single measurement characterized by a *discrete* random variable x, *defined to be the total number of successes.* Thus, the sample space is defined to be the set of possible values of x successes given n observations. If the measurement were repeated many times with n trials each time, the resulting values of x would occur with relative frequencies given by the *binomial distribution,* which is defined as

$$P(x) = \frac{n!}{x!\,(n-x)!} s^x (1-s)^{n-x}, \quad (x = 0, 1, \ldots, n), \tag{6.10}$$

where x is the random variable, while n and s are parameters characterizing the binomial distribution. Note that the binomial distribution is symmetric for $s = 1/2$. The MGF is obtained as

$$M_x(t) = \left(se^t + 1 - s\right)^n, \quad (t \in \mathscr{R}) . \tag{6.11}$$

The mean value of x is given by

$$m_o = \sum_{n=0}^{\infty} x \frac{n!}{x!\,(n-x)!} s^x (1-s)^{n-x} = ns, \tag{6.12}$$

while the variance is given by $\mu_2 = ns(1-s)$.

If the space $\mathscr{E}$ of all possible simple events is partitioned in $m+1$ (instead of just two, as was the case for the binomial distribution) compound events A_i, $(i = 1, \ldots, m+1)$, the binomial distribution can be generalized to the *multinomial distribution* by considering a set of nonnegative integer variables $(x_1, x_2, \ldots, x_{m+1})$, satisfying the conditions $\sum_{i=1}^{m+1} x_i = n$, $\sum_{i=1}^{m+1} A_i = \mathscr{E}$, and $\sum_{i=1}^{m+1} s_i = 1$ (since one of the outcomes must ultimately be realized). Then, the multinomial distribution for the joint probability for x_1 outcomes of type 1, x_2 of type 2, etc. is given by

$$P(x_1, \ldots, x_{m+1}) = \frac{n!}{x_1! \ldots x_{m+1}!} s_1^{x_1} \ldots s_{m+1}^{x_{m+1}} . \tag{6.13}$$

The MGF for this distribution is

$$M_{x_1 \ldots x_{m+1}}(t_1, \ldots, t_m) = \left(\sum_{i=1}^{m} s_i e^{t_i} + s_{m+1}\right)^n, \quad (t_i \in \mathscr{R}, \ i = 1, \ldots, m) . \tag{6.14}$$

The variances and covariances for this distribution are, respectively:

$$\mu_{ii} = \sigma_i^2 = n\, s_i\, (1 - s_i), \quad \text{for } i = j, \text{ and}$$

$$\mu_{ij} = -n\, s_i s_j, \qquad \text{for } i \neq j \ (i, j = 1, \ldots, m+1). \qquad (6.15)$$

Since μ_{ij} is negative (anti-correlated variables), it follows that, if in n trials, bin i contains a larger than average number of entries ($x_i > n\, s_i$), then the probability is increased that bin j will contain a smaller than average number of entries.

Geometric Distribution: The geometric distribution is also based on the concept of a Bernoulli trial. Consider that s, $0 < s < 1$, is the probability that a particular Bernoulli trial is a success, while $1 - s$ is the corresponding probability of failure. Also, consider that x is a random variable that can assume the infinite set of integer values $(1, 2, \ldots)$. The *geometric distribution* gives the probability that the first $x - 1$ trials will be failures, while the xth trial is a success. Therefore, it is the distribution of the "waiting time" for a success. Thus, the probability function characterizing the geometric distribution is

$$P(x) = s\,(1 - s)^{x-1}, \quad (x = 1, 2, \ldots). \qquad (6.16)$$

The *MGF* for this distribution is

$$M_x(t) = s e^t / \left[1 - (1 - s)\, e^t\right], \quad (t < -\ln(1 - s)). \qquad (6.17)$$

From the above *MGF*, the mean value is obtained as $m_o = (1/s)$, while the variance is obtained as $\mu_2 = (1 - s)/s^2$.

Negative Binomial (Pascal) Distribution: The negative binomial (Pascal) distribution also employs the concept of a Bernoulli trial. Thus, consider that s, $0 < s < 1$, is the probability of success in any single trial and $1 - s$ is the corresponding probability of failure. This time, though, the result of interest is *the number of trials that are required in order for r successes to occur*, $(r = 1, 2, \ldots)$. Note that at least r trials are needed in order to have r successes. Consider, therefore, that x is a random variable that represents the number of additional trials required (beyond r) before obtaining r successes, so that $(x = 0, 1, 2 \ldots)$. Then, the form of the Pascal probability distribution is found to be

$$P(x) = C_{mx} \, s^r (1-s)^n, \quad (x = 0, 1, 2, \ldots), \tag{6.18}$$

where $m = x + r - 1$ and C_{mx} is the binomial coefficient. The MGF for the binomial distribution is found from Eq. (6.18) to be

$$M_n(t) = \left[s / \left(1 - (1-s) \, e^t \right) \right]^r, \quad (t < -\ln(1-s)). \tag{6.19}$$

It follows from Eq. (6.19) that the mean value for the Pascal distribution is $m_o = [r(1-s)/s]$, while the variance is $\mu_2 = [r(1-s)/s^2]$.

Poisson Distribution: In the limit of many trials, as n becomes very large and the probability of success s becomes very small, but such that the product ns (i.e., the expectation value of the number of successes) remains equal to some finite value ν, the binomial distribution takes on the form

$$P(x) = \frac{\nu^x}{x!} e^{-\nu}, \quad (x = 0, 1, \ldots), \tag{6.20}$$

which is called the *Poisson distribution* for the integer random variable x. The corresponding MGF is given by the expression

$$M_x(t) = \exp\left[\nu \left(e^t - 1 \right) \right], \quad (t \in \mathscr{R}). \tag{6.21}$$

From the above MGF, or from a direct calculation, the expectation value of the Poisson random variable x is obtained as

$$m_o = E(x) = \sum_{x=0}^{\infty} x \frac{\nu^x}{x!} e^{-\nu} = \nu, \tag{6.22}$$

and the variance is obtained as

$$\mu_2 = V(x) = \sum_{x=0}^{\infty} (x - \nu)^2 \frac{\nu^x}{x!} = \nu. \tag{6.23}$$

Although the Poisson variable x is discrete, it can be treated as a continuous variable if it is integrated over a range $\Delta x \gg 1$. An example of a Poisson random variable is the number of decays of a certain amount of radioactive material in a fixed time period, in the limit that the total number of possible decays (i.e., the total number of radioactive atoms) is very large and the probability for an individual decay within the time period is very small.

Exponential distribution: The exponential Probability Density Function (*PDF*), $p(x)$, of the continuous variable x ($0 \le x \le \infty$) is defined by

$$p(x) = \frac{1}{\xi} e^{-x/\xi} , \qquad (6.24)$$

where ξ a is real-valued parameter. The expectation value of x is given by

$$m_o = E(x) = \frac{1}{\xi} \int_0^\infty x e^{-x/\xi} \, dx = \xi , \qquad (6.25)$$

and the variance of x is given by

$$\mu_2 = V(x) = \frac{1}{\xi} \int_0^\infty (x - \xi)^2 e^{-x/\xi} \, dx = \xi^2 . \qquad (6.26)$$

The *MGF* is given by the expression

$$M_x(t) = \lambda/(\lambda - t) ; \quad \lambda \equiv 1/\xi, \quad (t \in \mathcal{R}, t < \lambda) . \qquad (6.27)$$

The exponential distribution is widely used in radioactivity applications and in equipment *failure rate analysis*. The failure rate is defined as the reciprocal of the mean time to failure, i.e., $\xi = 1/\lambda$. The quantity $R(t_o) = e^{-\lambda t_o}$ is usually denoted as the reliability (of the equipment) at time $t_o > 0$. It should not be surprising that the reliability of a piece of equipment should decline with age, and statistical experience has shown that this decline is well represented by the exponential distribution.

Gaussian Distribution: Perhaps the single most important distribution in theoretical as well as in applied statistics is the *Gaussian (or normal) PDF* of the continuous random variable x (with $-\infty < x < \infty$), defined as

$$p(x) = \frac{1}{\sqrt{2\pi\sigma^2}} \exp\left(\frac{-(x - \nu)^2}{2\sigma^2}\right) . \qquad (6.28)$$

Note that the Gaussian distribution has two parameters, ν and σ^2, which are, by design, the mean and variance of x, i.e.,

$$E\left(x\right) = \int_{-\infty}^{\infty} x \frac{1}{\sqrt{2\pi\sigma^2}} \exp\left(\frac{-(x-\nu)^2}{2\sigma^2}\right) dx = \nu, \qquad (6.29)$$

$$V\left(x\right) = \int_{-\infty}^{\infty} (x-\mu)^2 \frac{1}{\sqrt{2\pi\sigma^2}} \exp\left(\frac{-(x-\nu)^2}{2\sigma^2}\right) dx = \sigma^2. \qquad (6.30)$$

The *MGF* for the Gaussian distribution is

$$M_x\left(t\right) = \exp\left[\nu t + \left(\sigma^2 t^2/2\right)\right], \quad (t \in \mathscr{R}). \qquad (6.31)$$

The *standard normal distribution* is obtained by setting $\nu = 0$ and $\sigma = 1$ in Eq. (6.28), to obtain

$$p\left(x\right) = \exp\left(-x^2/2\right)/(2\pi)^{1/2}, \qquad (6.32)$$

with *MGF*

$$M_x\left(t\right) = \exp\left(t^2/2\right), \quad (t \in \mathscr{R}). \qquad (6.33)$$

The function

$$P\left(x\right) = \int_{-x}^{x} p\left(z\right) dz, \quad (x > 0), \qquad (6.34)$$

with $p\left(z\right)$ given by Eq. (6.32), represents the integrated probability of an event with $-x \le z < x$, for the standard normal distribution.

The first four derivatives of the *MGF*, $M_x\left(t\right)$, of the standard normal distribution are obtained as:

$$M_x^{(1)}\left(t\right) = t\exp\left(t^2/2\right),$$
$$M_x^{(2)}\left(t\right) = \left(1+t^2\right)\exp\left(t^2/2\right),$$
$$M_x^{(3)}\left(t\right) = \left(3t+t^3\right)\exp\left(t^2/2\right),$$
$$M_x^{(4)}\left(t\right) = \left(3+6t^2+t^4\right)\exp\left(t^2/2\right).$$

Evaluating the above expressions at $t = 0$ yields:

$$M_x^{(1)}(0) = \nu_1 = m_o = 0\,,$$

$$M_x^{(2)}(0) = \nu_2 = \mu_2 = \sigma^2 = 1\,,$$

$$M_x^{(3)}(0) = \nu_3 = \mu_3 = \alpha_3 = 0\,,$$

$$M_x^{(4)}(0) = \nu_4 = \mu_4 = \alpha_4 = 3\,.$$

Thus, a normal distribution is symmetric (since it has skewness $\alpha_3 = 0$) and is mesokurtic (since it has kurtosis $\alpha_4 = 3$). Furthermore, since $m_o = 0$ for all standard distributions, the respective raw and central moments are equal to each other. In particular, it can be shown that $\mu_{2k-1} = 0$, $(k = 1, 2, \ldots)$, which indicates that all of the odd central moments of the normal distribution vanish. Furthermore, it can be shown that the even central moments are given by $\mu_{2k} = (1)(3)\ldots(2k-1)\sigma^{2k}$, $(k = 1, 2, \ldots)$. These results highlight the very important feature of the normal distribution that all its nonzero higher-order central moments can be expressed in terms of a single parameter, namely the standard deviation. This is one of several reasons why the Gaussian distribution is arguably the single most important *PDF* in statistics.

Another prominent practical role played by the Gaussian distribution is as a replacement for either the binomial distribution or the Poisson distribution. The circumstances under which such a replacement is possible are given by the *DeMoivre-Laplace theorem* (given below without proof), which can be stated as follows: *Consider the binomial distribution, denoted as $p_b(k)$, $(k = 1, \ldots, n)$ (n and s denote the usual parameters of the binomial distribution, while the subscript "b" denotes "binomial"), and consider that a and c are two nonnegative integers satisfying $a < c < n$. Furthermore, consider the standard normal distribution, denoted as $p_{sn}(x)$ (the subscript "sn" denotes "standard normal"). Finally, define the quantities α and β as*

$$\alpha \equiv (a - ns - 0.5)/[ns(1-s)]^{1/2} \text{ and } \beta \equiv (c - ns + 0.5)/[ns(1-s)]^{1/2}\,.$$

Then, for large n, the following relation holds

$$\sum_{k=a}^{c} p_b(k) \approx \int_{\alpha}^{\beta} p_{sn}(x)\,dx. \tag{6.35}$$

Thus, the DeMoivre-Laplace theorem indicates that the sum of the areas of contiguous histogram segments, representing discrete binomial probabilities, approximately equals the area under the corresponding continuous Gaussian curve spanning the same region. The binomial becomes increasingly symmetrical as $s \to 0.5$, so the approximation provided by the DeMoivre-Laplace theorem is accurate even for relatively small n, since the Gaussian is intrinsically symmetric. When $s \ll 1$, and n is so large that $\sigma^2 = ns(1 - s) = \nu(1 - \nu/n) \approx \nu$, even though $\nu \gg 1$, then the normal distribution can be shown to be a reasonably good approximation to both the corresponding Poisson distribution and to the binomial distribution (particularly when $x \approx \nu$).

Another fundamental role played by the Gaussian distribution is highlighted by the *Central Limit Theorem*, which essentially states that *the sum of n independent continuous random variables x_i with means μ_i and variances σ_i^2 becomes a Gaussian random variable with mean $\mu = \sum_{i=1}^{n} \mu_i$ and variance $\sigma^2 = \sum_{i=1}^{n} \sigma_i^2$ in the limit that n approaches infinity.* This statement holds under fairly general conditions, regardless of the form of the individual *PDFs* of the respective random variables x_i. The central limit theorem provides the formal justification for treating measurement errors as Gaussian random variables, as long as the total error is the sum of a large number of small contributions.

The behavior of certain distributions for limiting cases of their parameters can be investigated more readily by using characteristic or moment generating functions, rather than their *PDFs*. For example, taking the limit $s \to 0$, $n \to \infty$, with $\nu = sn$ (constant), in the characteristic function for the binomial distribution yields the characteristic function of the Poisson distribution:

$$\phi(k) = \left[s\left(e^{ik} - 1\right) + 1 \right]^n = \left(\frac{\nu}{n}\left(e^{ik} - 1\right) + 1 \right)^n \to \exp\left[\nu\left(e^{ik} - 1\right) \right] .$$

Note also that a Poisson variable x with mean ν becomes a Gaussian variable in the limit as $\nu \to \infty$. This fact can be shown as follows: although the Poisson variable x is discrete, when it becomes large, it can be treated as a continuous variable as long as it is integrated over an interval that is large compared to unity. Next, the Poisson variable x is transformed to the variable

$$z = \frac{x - \nu}{\sqrt{\nu}} .$$

The characteristic function of z is

$$\varphi_z(k) = E\left(e^{ikz}\right) = E\left(e^{ikx/\sqrt{\nu}} e^{-ik\sqrt{\nu}}\right) = \varphi_x\left(k/\sqrt{\nu}\right) e^{-ik\sqrt{\nu}},$$

where φ_x is the characteristic function of the Poisson distribution. Expanding the exponential term and taking the limit as $\nu \to \infty$ yields

$$\varphi_z(k) = \exp\left[\nu\left(e^{ik/\sqrt{\nu}} - 1\right) - ik\sqrt{\nu}\right] \to \exp\left(-\frac{1}{2}k^2\right).$$

The last term on the right side of the above expression is the characteristic function for a Gaussian with zero mean and unit variance. Transforming back to the original Poisson variable x, one finds, therefore, that for large ν, x follows a Gaussian distribution with mean and variance both equal to ν.

Multivariate Normal Distribution: Consider that $\mathbf{x}$ is a (column) vector of dimension n, with components $x_i \in \mathscr{R}$, $(i = 1, \ldots, n)$. Consider further that $\mathbf{m}_o$ is a (column) vector of constants with components denoted as $m_{oi} \in \mathscr{R}$, $(i = 1, \ldots, n)$. Finally, consider that $\mathbf{V}$ is a real, symmetric, and positive definite $n \times n$ matrix with elements denoted by μ_{ij}, $(i, j = 1, \ldots, n)$. Then, the *multivariate normal distribution for n variables* is defined by the probability density function

$$p(\mathbf{x}) = (2\pi)^{-n/2} [\det(\mathbf{V})]^{-1/2} \exp\left[-(1/2)(\mathbf{x} - \mathbf{m}_o)^T \mathbf{V}^{-1} (\mathbf{x} - \mathbf{m}_o)\right], \tag{6.36}$$

where the superscript "T" denotes matrix transposition, and where $\mathbf{V}^{-1}$ is the inverse of $\mathbf{V}$. The quantity

$$Q \equiv (\mathbf{x} - \mathbf{m}_o)^T \mathbf{V}^{-1} (\mathbf{x} - \mathbf{m}_o), \tag{6.37}$$

is called a *multivariate quadratic form*. The *MGF* corresponding to Eq. (6.36) is given by

$$M_x(t) = \exp\left[\mathbf{t}^T \mathbf{m}_o + (1/2)\, \mathbf{t}^T \mathbf{V}^{-1} \mathbf{t}\right], \quad (t_i \in \mathscr{R};\ i = 1, \ldots, n). \tag{6.38}$$

Note that $\mathbf{m}_o$ is the *mean vector*, while $\mathbf{V}$ is actually the covariance matrix for this distribution; note also that

$$E\left(x_i\right) = \left[\partial M_x\left(t\right)/\partial t_i\right]_{t=0} = m_{oi},$$

$$E\left(x_i x_j\right) = \left[\partial^2 M_x\left(t\right)/\partial t_i \partial t_j\right]_{t=0} = m_{oi} m_{oj} + \mu_{ij}.$$

In particular, the Gaussian *PDF* for two random variables x_1 and x_2 becomes

$$p\left(x_1, x_2; m_{o1}, m_{o2}, \sigma_1, \sigma_2, \rho\right) = \frac{1}{2\pi\sigma_1\sigma_2\sqrt{1-\rho^2}}$$

$$\times \exp\left\{-\frac{1}{2\left(1-\rho^2\right)}\left[\left(\frac{x_1 - m_{o1}}{\sigma_1}\right)^2 + \left(\frac{x_2 - m_{o2}}{\sigma_2}\right)^2 - 2\rho\left(\frac{x_1 - m_{o1}}{\sigma_1}\right)\left(\frac{x_2 - m_{o2}}{\sigma_2}\right)\right]\right\}$$

where $\rho = \text{cov}\left(x_1, x_2\right)/\left(\sigma_1\sigma_2\right)$ is the correlation coefficient.

The following properties of the multivariate normal distribution are often used in applications to data evaluation and analysis:

1. *If $\mathbf{x}$ is a normally distributed random vector of dimension n with corresponding mean vector $\mathbf{m}_o$ and positive definite covariance matrix $\mathbf{V}$, and if $\mathbf{A}$ is an $m \times n$ matrix of rank m, with $m \le n$, then the m-dimensional random vector $\mathbf{y} = \mathbf{A}\mathbf{x}$ is also normally distributed, with a mean (vector) equal to $\mathbf{A}\mathbf{m}_o$ and a positive definite covariance matrix equal to $\mathbf{A}\mathbf{V}\mathbf{A}^T$, where the superscript "T" denotes "transposition."*

2. *If $\mathbf{x}$ is a normally distributed random vector of dimension n, with corresponding mean vector $\mathbf{m}_o$ and positive definite covariance matrix $\mathbf{V}$, and if $\mathbf{A}$ is a nonsingular $n \times n$ matrix such that $\mathbf{A}^T\mathbf{V}^{-1}\mathbf{A} = \mathbf{I}$ (the diagonal unit matrix), then $\mathbf{y} = \mathbf{A}^{-1}\left(\mathbf{x} - \mathbf{m}_o\right)$ is a random vector with a zero mean vector, $\mathbf{0}$, and unit covariance matrix, $\mathbf{I}$. This linear transformation is the multivariate equivalent of the standard-variable transformation for the univariate Gaussian distribution.*

3. *If $\mathbf{x}$ is a normally distributed random vector of dimension n with*

corresponding mean vector $\mathbf{m}_o$ *and positive definite covariance matrix* $\mathbf{V}$, *then the quadratic form*

$$Q = (\mathbf{x} - \mathbf{m}_o)^T \mathbf{V}^{-1} (\mathbf{x} - \mathbf{m}_o) \qquad (6.39)$$

is a statistic distributed according to the chi-square distribution with n *degrees of freedom.*

4. *If* $\mathbf{x}$ *is a normally distributed vector with mean vector* $\mathbf{m}_o$ *and covariance matrix* $\mathbf{V}$, *then the individual variables* $x_1, x_2, \ldots, x_n$ *are mutually independent if and only if all the elements* μ_{ij} *in* $\mathbf{V}$ *are equal to zero for* $i \neq j$.

The geometrical features of multivariate normal distributions can be most readily visualized by considering the bivariate distribution, for which the quadratic form of Eq. (6.39) becomes

$$Q = \frac{\left\{ \left[(x_1 - m_{o1})^2 / \sigma_1^2 \right] - \left[2\rho(x_1 - m_{o1})(x_2 - m_{o2}) / (\sigma_1 \sigma_2) \right] + \left[(x_2 - m_{o2})^2 / \sigma_2^2 \right] \right\}}{(1 - \rho^2)},$$

$$(6.40)$$

where σ_1 and σ_2 are the standard deviations for x_1 and x_2, respectively, and ρ is the correlation coefficient (i.e., $\mu_{11} = \sigma_1^2$, $\mu_{22} = \sigma_2^2$, and $\mu_{12} = \mu_{21} = \rho\sigma_1\sigma_2$ are the elements of the covariance matrix $\mathbf{V}$). Considering, in three dimensions, that x_1 and x_2 are two rectangular coordinates in a plane while x_3 is the coordinate of elevation above this plane, a plot of the bivariate Gaussian surface $x_3 = p(x_1, x_2)$ forms a smooth "hill" centered on the position $x_1 = m_{o1}$ and $x_2 = m_{o2}$. Regardless of their azimuthal orientation, planes $x_1 = const.$ or $x_2 = const.$ (parallel to x_3) that pass through the central point (m_{o1}, m_{o2}) will cut the surface $x_3 = p(x_1, x_2)$ into univariate Gaussian profiles. Planes perpendicular to x_3 (i.e., $x_3 = c = const.$) will cut the surface $x_3 = p(x_1, x_2)$ into ellipses, provided that $0 < c < \left[2\pi\sigma_1\sigma_2 (1 - \rho^2) \right]^{-1/2}$; note that $Q = -2\ln\left[2\pi\sigma_1\sigma_2 c(1 - \rho^2)^{1/2} \right]$ when $x_3 = p(x_1, x_2) = c$. The projections of these ellipses onto the $x_3 = 0$ plane will be imbedded in rectangles centered at (m_{o1}, m_{o2}), with sides $a_1 = 2\beta\sigma_1$ (along the x_1-coordinate axis) and $a_2 = 2\beta\sigma_2$ (along the x_2-coordinate axis). The correlation coefficient ρ determines the shape and orientation of the surface. Thus, when $\rho = \pm 1$, the surface degenerates and lies entirely in a plane perpendicular to the $x_3 = 0$

plane. More generally, when $|\rho| < 1$, the ellipses take on various orientations that depend upon the magnitude and sign of ρ. In addition, the shapes of the profiles produced by horizontal and vertical planes cutting through the surface $x_3 = p(x_1, x_2)$ also depend upon the relative magnitudes of σ_1 and σ_2.

Log-Normal Distribution: The random variable $x = e^y$, where y is a Gaussian random variable with mean μ and variance σ^2, is distributed according to the log-normal distribution

$$p(x) = \frac{1}{\sqrt{2\pi\sigma^2}} \frac{1}{x} \exp\left(-\frac{(\log x - \mu)^2}{2\sigma^2}\right). \tag{6.41}$$

Since the log-normal and Gaussian distributions are closely connected to one another, they share many common properties. For example, the moments of the log-normal distribution can be obtained directly from those of the normal distribution, by noting that

$$E\left(x^k\right) = E\left(e^{ky}\right) = M_y\left(t\right)|_{t=k}. \tag{6.42}$$

Thus, the expectation value and variance, respectively, for the log-normal distribution are

$$E\left(x\right) = \exp\left(\mu + \frac{1}{2}\sigma^2\right), V\left(x\right) = \exp\left(2\mu + \sigma^2\right)\left[\exp\left(\sigma^2\right) - 1\right].$$

Applying the central limit theorem to the variable $x = e^y$ shows that a variable x that stems from the product of n factors (i.e., $x = x_1 x_2 \ldots x_n$) will follow a log-normal distribution in the limit as $n \to \infty$. Therefore, the log-normal distribution is often used to model random errors which change a result by a multiplicative factor. Since the log-normal function is distributed over the range of positive real numbers, and has only two parameters, it is particularly useful for modeling nonnegative phenomena, such as analysis of incomes, classroom sizes, masses or sizes of biological organisms, evaluation of neutron cross sections, scattering of subatomic particles, etc.

Cauchy Distribution: The Cauchy distribution is defined by the probability density function

$$p(x) = (\lambda/\pi)\left[\lambda^2 + (x - \nu)^2\right]^{-1}, \tag{6.43}$$

for $x \in \mathcal{R}$, $\nu \in \mathcal{R}$, and $\lambda \in \mathcal{R}$ $(\lambda > 0)$. Although this distribution is normalized (i.e., the *zeroth* raw moment exists), the expectations that define the higher-order raw moments are divergent. Mathematical difficulties can be alleviated, however, by confining the analysis to the vicinity of $x = \nu$; if needed, this distribution can be arbitrarily truncated.

Gamma Distribution: Consider that x is a nonnegative real variable ($x \in \mathcal{R}$, $x > 0$), and α and β are positive real constants ($\alpha \in \mathcal{R}$, $\alpha > 0$; $\beta \in \mathcal{R}$, $\beta > 0$). Then the probability density function for the gamma distribution is defined as

$$p(x) = x^{\alpha-1}e^{-(x/\beta)}/[\beta^{\alpha}\Gamma(\alpha)], \quad \Gamma(\alpha) \equiv \int_0^{\infty} t^{\alpha-1}e^{-t}dt. \tag{6.44}$$

Recall that, for all $\alpha > 0$, $\Gamma(\alpha+1) = \alpha\Gamma(\alpha)$, $\Gamma(1) = 1$, $\Gamma(1/2) = \pi^{1/2}$, and

$$\Gamma[n+(1/2)] = [(2n-1)(2n-2)\ldots3\cdot2\cdot1]\,\Gamma(1/2)/2^n. \tag{6.45}$$

The *MGF* for the gamma distribution is

$$M_x(t) = 1/(1-\beta t)^{\alpha}, \quad (t \in \mathcal{R}, \; t < 1/\beta), \tag{6.46}$$

while the mean value and variance are $m_o = \alpha\beta$ and $\mu_2 = \alpha\beta^2$, respectively. Note that when $\alpha = 1$ and $\beta = 1/\lambda$, the gamma distribution reduces to the exponential distribution.

Beta Distribution: Consider that x is a real variable in the range $0 \le x \le 1$, and α and β are positive real parameters ($\alpha \in \mathcal{R}$, $\alpha > 0$; $\beta \in \mathcal{R}$, $\beta > 0$). The probability density function for the beta distribution is defined as

$$p(x) = \{\Gamma(\alpha+\beta)/[\Gamma(\alpha)\Gamma(\beta)]\}x^{\alpha-1}(1-x)^{\beta-1}. \tag{6.47}$$

Since the *MGF* for the beta distribution is inconvenient to use, it is easier to derive the mean value and the variance directly from Eq. (6.47); this yields $m_o = \alpha/(\alpha+\beta)$ and $\mu_2 = \alpha\beta/[(\alpha+\beta)^2(\alpha+\beta+1)]$. The beta distribution is often used for weighting probabilities along the unit interval.

Student's t$-$Distribution: The *t*-distribution was discovered by W. Gosset (who published it under the pseudonym "Student"), and arises when considering the quotient of two random variables. For a real variable x ($x \in \mathcal{R}$), the probability density for the t-distribution is defined as

$$p(x) = \Gamma\left[(n+1)/2\right]\left[1 + \left(x^2/n\right)\right]^{-(n+1)/2} / \left[(n\pi)^{1/2}\,\Gamma\left(n/2\right)\right],$$

$$(n = 1, 2, 3, \ldots). \qquad (6.48)$$

The t-distribution function does *not* have an *MGF*. However, certain moments do exist and can be calculated directly from Eq. (6.48). Thus, the mean value is $m_o = 0$, for $n > 1$, and the variance is $\mu_2 = n/(n-2)$, for $n > 2$. Note that the conditions $n > 1$ and $n > 2$ for the existence of the mean value and variance, respectively, arise from the fact that the t-distribution for $n = 1$ is equivalent to the Cauchy distribution for $\nu = 0$ and $\lambda = 1$. The other limiting case of the t-distribution is the Gaussian ($n = \infty$).

F-Distribution: Just as the t-distribution, the F-distribution also arises when considering the quotient of two random variables; its probability density is given by

$$p(x) = (n/m)^{n/2}\,\Gamma\left[(n+m)/2\right]x^{(n/2)-1}$$

$$/\left\{\Gamma\left(n/2\right)\Gamma\left(m/2\right)\left[1 + (nx/m)\right]^{(n+m)/2}\right\}, \qquad (6.49)$$

where $x \in \mathcal{R}$, $x > 0$, and the parameters m and n are positive integers called *degrees of freedom*. The *MGF* for the F-distribution does not exist, but the mean value is $m_o = m/(m-2)$, for $m > 2$, and the variance is

$$\mu_2 = m^2\,(2m + 2n - 4)/\left[n(m-2)^2\,(m-4)\right], \quad \text{for } m > 4.$$

Few-Parameter Distribution and Pearson's Equation: K. Pearson made the remarkable discovery that the differential equation

$$(dp/dx) = p(x)\,(d-x)/\left(a + bx + cx^2\right), \qquad (6.50)$$

yields several of the univariate probability density functions considered in the foregoing. In particular, Eq. (6.50) yields:

1. The normal distribution if $a > 0$, $b = c = 0$, while d is arbitrary,

2. The exponential distribution if $a = c = d = 0$, $b > 0$,

3. The gamma distribution if $a = c = 0$, $b > 0$, and $d > -b$,

4. The beta distribution if $a = 0$, $b = -c$, and $d > 1 - b$.

The solutions to Eq. (6.50) are known as Pearson's curves, and they underscore the close relationships among the probability distributions mentioned above.

Chi-square χ^2-Distribution: The χ^2 (chi-square) distribution of the continuous variable x $(0 \leq x < \infty)$ is defined as

$$p(x) = \frac{1}{2^{n/2}\,\Gamma(n/2)}x^{n/2-1}e^{-x/2}, \quad (n = 1, 2, \dots), \quad (6.51)$$

where the parameter n is the *number of degrees of freedom*. The mean value and variance of x are given by $m_o = n$ and $\mu_2 = 2n$, respectively.

The χ^2-distribution is related to the sum of squares of normally distributed variables: given N independent Gaussian random variables x_i with known mean μ_i and variance σ_i^2, the random variable

$$x = \sum_{i=1}^{N} \frac{(x_i - \mu_i)^2}{\sigma_i^2}, \quad (6.52)$$

is distributed according to the χ^2-distribution with N degrees of freedom. More generally, if the random variables x_i are not independent but are described by an N-dimensional Gaussian PDF, then the random variable

$$Q = (\mathbf{x} - \boldsymbol{\mu})^{+}\mathbf{V}^{-1}(\mathbf{x} - \boldsymbol{\mu}), \quad (6.53)$$

also obeys a χ^2-distribution with N degrees of freedom [cf. Eq. (6.39)]. Random variables following the χ^2-distribution play an important role in tests of goodness-of-fit, as highlighted, for example, by the method of least squares.

7

Appendix B

CONTENTS

7.1 Elements of Functional Analysis for Data Analysis and Assimilation

This appendix introduces and summarizes the most important properties of adjoint operators in conjunction with differential calculus in vector spaces, as used for data assimilation. Recall that a nonempty set $\mathscr{V}$ is called a *vector space* if any pair of elements $f, g \in \mathscr{V}$ can be: (i) added together by an operation called *addition* to give an element $f + g$ in $\mathscr{V}$, such that, for any f, g, $h \in \mathscr{V}$, the following properties hold:

$$f + g = g + f;$$
$$f + (g + h) = (f + g) + h;$$

There is a unique element 0 in $\mathscr{V}$ such that $f + 0 = f$ for all $f \in \mathscr{V}$; For each $f \in$ there is a unique element $(-f)$ in $\mathscr{V}$ such that $f + (-f) = 0$.

(ii) multiplied by any scalar α of a field $\mathscr{F}$ to give an element αf in $\mathscr{V}$; furthermore, for any scalars α, β, the following properties must hold:

$$\alpha \left(f + g \right) = \alpha f + \alpha g;$$
$$\left(\alpha + \beta \right) f = \alpha f + \beta f;$$
$$\left(\alpha \beta \right) f = \alpha \left(\beta f \right);$$
$$1 \times f = f.$$

The space $\mathscr{V}$ is called *a complex vector space* if the scalar field is $\mathscr{C}$, or a real vector space if the scalar field is $\mathscr{R}$. The members f, g, h of $\mathscr{V}$ are called *points*, elements, or *vectors* depending on the respective context.

A finite set $\mathscr{S} = \{f_j\}_1^n$ of vectors in $\mathscr{V}$ is called *linearly dependent* if and only if (henceforth abbreviated as *iff*) there are scalars $\alpha_1, \ldots, \alpha_n$, not all of which are zero, such that $\sum \alpha_j f_j = 0$; otherwise, $\mathscr{S}$ is called *linearly independent*. An arbitrary set $\mathscr{S}$ of vectors in $\mathscr{V}$ is linearly independent iff every finite nonempty subset of $\mathscr{S}$ is linearly independent; otherwise, the set $\mathscr{S}$ is linearly dependent. If there exists a positive integer n such that $\mathscr{V}$ contains n, but not $n + 1$ linearly independent vectors, then $\mathscr{V}$ is said to be *finite dimensional* with dimension n; $\mathscr{V}$ is *infinite dimensional* iff it is not finite dimensional. The finite set $\mathscr{S}$ of vectors in an n-dimensional space $\mathscr{V}$ is called a *basis* of $\mathscr{V}$ iff $\mathscr{S}$ is linearly independent, and each element of $\mathscr{V}$ may be written as $\sum_1^n \alpha_j f_j$ for some $\alpha_1, \ldots, \alpha_n \in \mathscr{C}$ and $f_1, \ldots, f_n \in \mathscr{S}$. An infinite-dimensional space is obtained by generalizing $\mathscr{R}^n$ or $\mathscr{C}^n$, and taking infinite sequences $f = (f_n)$ as the elements of the space. This space is known as a *sequence space* and is usually denoted by ℓ.

Consider that $\mathscr{S}$ is a subset of $\mathscr{V}$, and define its complement $\mathscr{V}/\mathscr{S}$ as the set of all elements in $\mathscr{V}$ that do not belong to $\mathscr{S}$. Next, define a new subset $\overline{\mathscr{S}} \subset \mathscr{V}$, called the *closure* of $\mathscr{S}$, by requiring that $f \in \overline{\mathscr{S}}$ iff there is a sequence of (not necessarily distinct) points of $\mathscr{S}$ converging to f. The set $\mathscr{S}$ is called *closed* iff $\mathscr{S} = \overline{\mathscr{S}}$. A subset $\mathscr{S}$ of $\mathscr{V}$ is said to be open iff its complement $\mathscr{V}/\mathscr{S}$ is closed. If $\mathscr{S}_1 \subset \mathscr{S}_2 \subset \mathscr{V}$ then $\mathscr{S}_1$ is said to be *open* in $\mathscr{S}_2$ iff it is the intersection of an open set with $\mathscr{S}_2$. A *neighborhood* of a point is any set that contains an open set that itself contains the point. A point f is an *interior point* of $\mathscr{S} \subset \mathscr{V}$ iff there is a neighborhood of f contained in $\mathscr{S}$. The *interior* $\mathscr{S}^0$ of $\mathscr{S}$ is the set of interior points of $\mathscr{S}$ (and is open). A point f is a *boundary point* of $\mathscr{S}$ iff every neighborhood of f contains points of both $\mathscr{S}$ and its complement $\mathscr{V}/\mathscr{S}$. The *boundary* $\partial\mathscr{S}$ of $\mathscr{S}$ is the set of boundary points of $\mathscr{S}$.

The *norm of a vector* $f \in \mathscr{V}$, denoted by $\|f\|$, is a nonnegative real scalar that satisfies the following relations, for all $f, g \in \mathscr{V}$:

$$\|f\| = 0 \text{ iff } f = 0;$$
$$\|\alpha f\| = |\alpha| \; \|f\| \text{ for any scalar } \alpha;$$
$$\|f + g\| \leq \|f\| + \|g\| \text{ (the triangle inequality).}$$

A vector space $\mathscr{V}$ endowed with a norm as defined above is called a *normed vector space*; the notation $\|\cdot\|$ highlights the role of the norm as a generalization of the customary distance in $\mathscr{R}^3$.

For finite dimensional spaces, the most frequently used vector norm is the *Hölder* or ℓ_p-norm, defined as

$$\|\mathbf{x}\|_p = \left(\sum_{i=1}^{n} |x_i|^p \right)^{1/p}, \quad p > 0. \tag{7.1}$$

The most frequently used values of the *Hölder norm* are

$$\|\mathbf{x}\|_1 = \sum_{i=1}^{n} |x_i|, \quad \|\mathbf{x}\|_2 = \sqrt{\sum_{i=1}^{n} |x_i|^2}, \quad \text{and} \quad \|\mathbf{x}\|_\infty = \max_i |x_i|.$$

The *norm of a square matrix* $\mathbf{A}$, denoted by $\|\mathbf{A}\|$, is a nonnegative real scalar that satisfies the following relations:

1. $\|\mathbf{A}\| > 0$, if $\mathbf{A} \neq \mathbf{0}$ (nonnegative);

2. $\|\alpha\mathbf{A}\| = |\alpha| \, \|\mathbf{A}\|$, for any $\alpha \in \mathscr{C}$ and any $\mathbf{A} \in \mathscr{R}^n$ (homogeneous);

3. $\|\mathbf{A} + \mathbf{B}\| \leq \|\mathbf{A}\| + \|\mathbf{B}\|$, for any $\mathbf{A}, \mathbf{B} \in \mathscr{R}^n$ (triangle inequality);

4. $\|\mathbf{AB}\| \leq \|\mathbf{A}\| \, \|\mathbf{B}\|$, for any $\mathbf{A}, \mathbf{B} \in \mathscr{R}^n$ (submultiplicative).

There are many matrix norms that satisfy the above relation; the most frequently used norm is the *matrix-Hölder norm induced by (or subordinate to) the Hölder vector norm*, defined as

$$\|\mathbf{A}\|_p \equiv \max_x \frac{\|\mathbf{Ax}\|_p}{\|\mathbf{x}\|_p} . \tag{7.2}$$

The most widely used matrix-Hölder norms are those obtained for the following values of p:

1. $p = 1$, *the maximum column sum matrix norm*, induced by the ℓ_1 vector norm: $\|\mathbf{A}\|_1 \equiv \max_x \frac{\|\mathbf{Ax}\|_1}{\|\mathbf{x}\|_1} = \max_j \sum_{i=1}^{n} |a_{ij}|$;

2. $p = 2$, the *spectral norm*, induced by the ℓ_2 vector norm:
$\|\mathbf{A}\|_2 \equiv \max_x \frac{\|\mathbf{Ax}\|_2}{\|\mathbf{x}\|_2} = \max_\lambda \{\sqrt{\lambda}\}$, where λ is an eigenvalue of $(\mathbf{A}^*)\,\mathbf{A}$;

3. $p = \infty$, the *maximum row sum matrix norm*, induced by the ℓ_∞ vector norm: $\|\mathbf{A}\|_\infty \equiv \max_x \frac{\|\mathbf{Ax}\|_\infty}{\|\mathbf{x}\|_\infty} = \max_i \sum_{j=1}^{n} |a_{ij}|$.

The *condition number* of a square matrix $\mathbf{A}$ is a nonnegative real scalar defined in connection with the Hölder matrix and vector norms as

$$\gamma_p\left(\mathbf{A}\right) \equiv \max_{u,v} \frac{\|\mathbf{Au}\|_p}{\|\mathbf{Av}\|_p}, \quad \|\mathbf{u}\|_p = \|\mathscr{V}\|_p = 1. \tag{7.3}$$

A geometrical interpretation of the condition number $\gamma_p\left(\mathbf{A}\right)$ can be envisaged by considering that the surface $\|\mathbf{x}\| = 1$ is mapped by the linear transformation $\mathbf{y} = \mathbf{Ax}$ onto some surface S. Then, the condition number $\gamma_p\left(\mathbf{A}\right)$ is the ratio of the largest to the smallest distances from the origin to points on S. This interpretation indicates that $\gamma_p\left(\mathbf{A}\right) \geq 1$, when $\mathbf{A}$ is nonsingular, and $\gamma_p\left(\mathbf{A}\right) = \infty$, when $\mathbf{A}$ is singular.

Rearranging the above definition leads to $\gamma_p\left(\mathbf{A}\right) = \left\|\mathbf{A}^{-1}\right\| \|\mathbf{A}\|$, which implies that $\gamma_p\left(\mathbf{A}\right)$ becomes large when $|\mathbf{A}|$ is small; in such cases, the matrix $\mathbf{A}$ is called *ill-conditioned*. Ill-conditioning may be regarded as an approach towards singularity, a situation that causes considerable difficulties when solving linear simultaneous equations, since a small change in the coefficients of the equations causes a large displacement in the solution, leading to loss of solution accuracy due to the loss of significant figures during the computation.

In the *sequence space* ℓ_p, the quantity $\|\cdot\|_p$, defined as

$$\|f\|_p = \left\{ \sum |f_n|^p \right\}^{1/p}, \quad (1 \leq p < \infty),$$

$$\|f\|_\infty = \sup |f_n| \tag{7.4}$$

is a norm on the subset of those f where $\|f\|_p$ is finite. Two numbers, p, q, with $1 \leq p, q \leq \infty$, are called *conjugate indices* iff $p^{-1} + q^{-1} = 1$; if $p = 1$, then $q = \infty$. When $1 \leq p \leq \infty$, and q is the *conjugate index*, the following inequalities hold for any $f, g \in {}_p$ (infinite values being allowed):

$$\|fg\|_1 \leq \|f\|_p \|g\|_q;$$

$$\|f + g\|_p \leq \|f\|_p + \|g\|_p.$$

A complex valued function f, defined on $\Omega \in \mathscr{R}^n$, is called *continuous at the point* $x_0 \in \Omega$ iff for each $\varepsilon > 0$ there exists $\delta > 0$ such that $|f(x) - f(x_0)| < \varepsilon$ whenever $x \in \Omega$ and $|x - x_0| < \delta$. The function f is said to be *uniformly continuous on* Ω iff for each $\varepsilon > 0$ there exists a $\delta > 0$ such that $|f(x) - f(x_0)| < \varepsilon$, whenever $x, x_0 \in \Omega$ and $|x - x_0| < \delta$.

Thus, when Ω is the finite interval $[a, b]$, continuity at a or b is to be interpreted as continuity from the right or left, respectively. If f is continuous on $[a, b]$, then it is bounded, but f need not be bounded if it is continuous only on (a, b). For functions of several variables, continuity in the sense of the above definition is sometimes referred to as "joint continuity" to distinguish it from "separate continuity"; the latter terminology means that the function is continuous in each variable in turn when the other variables are fixed. For example, if f is a function of two variables, separate continuity requires only that $f(x, \cdot)$ and $f(\cdot, y)$ should be continuous for fixed x and y, respectively. (The notation $f(x, \cdot)$ indicates that x is fixed, and the function is regarded as a function of its second argument only.) Uniform continuity is in general stronger than continuity, since the same δ must serve for every $x_0 \in \Omega$, but if $\Omega \subset \mathscr{R}^n$ is closed and bounded, then these concepts are equivalent.

The vector space of bounded continuous complex valued functions defined on $\Omega \subset \mathscr{R}^n$ is denoted by $\mathscr{C}(\Omega)$. The space $\mathscr{C}(\Omega)$ may be normed in several ways. The *sup norm* $\|\cdot\|$, defined as

$$\|f\| = \sup_{x \in \Omega} |f(x)|, \tag{7.5}$$

is most often used in practical applications. Generalizations of $\mathscr{C}(\Omega)$ are often used for treating differential equations. Thus, $\mathscr{C}(\Omega, \mathscr{R}^m)$ denotes the normed vector space of $\mathscr{R}^m$-valued functions equipped with the sup norm

$$\|f\| = \max_{1 \le j \le m} \sup_{x \in \Omega} |f_j(x)|. \tag{7.6}$$

The vector space denoted by $\mathscr{C}^k(\Omega, \mathscr{R}^m)$ consists of all $\mathscr{R}^m$-valued functions defined on Ω such that all partial derivatives up to and including those of order $k > 0$ of all components are bounded and continuous. The vector space $\mathscr{C}^\infty(\Omega, \mathscr{R}^m)$ consists of functions in $\mathscr{C}^k(\Omega, \mathscr{R}^m)$ such that $\mathscr{C}^\infty(\Omega, \mathscr{R}^m) = \bigcap_{k=1}^{\infty} \mathscr{C}^k(\Omega, \mathscr{R}^m)$. The space $\mathscr{C}^k(\overline{\Omega}, \mathscr{R}^m)$ consists of those continuous functions defined on $\overline{\Omega}$ which on Ω have bounded and *uniformly* continuous partial derivatives up to and including those of order k. (For $n > 1$, this convention avoids difficulties with the definition of derivatives on $\partial\Omega$, which is not necessarily a smooth set). It also follows that $\mathscr{C}^\infty(\overline{\Omega}, \mathscr{R}^m) = \bigcap_{k=1}^{\infty} \mathscr{C}^k(\overline{\Omega}, \mathscr{R}^m)$.

The spaces $\mathscr{C}_0^\infty(\Omega', \mathscr{R}^m)$, where $\Omega \subset \Omega' \subset \overline{\Omega}$, are used when it is convenient to exclude the boundary from consideration. These spaces consist of those functions in $\mathscr{C}^k(\Omega, \mathscr{R}^m)$ that have bounded support contained in the interior of Ω'. (Recall that the support of a function is the closure of the set on which the function is nonzero; the support may vary from function to function).

For example, for the finite interval $[a, b]$, a function is in $\mathscr{C}^1([a, b])$ iff it has a continuous derivative on (a, b) and has left and right derivatives at b and a, respectively, which are the limits of the derivatives in the interior. Another possibility is to take $\Omega = (a, b)$ and to set

$$\|f\|_{\mathscr{C}^k} = \sum_{j=0}^{k} \sup_{x \in [a,b]} \left| f^{(j)}(x) \right|, \tag{7.7}$$

where $f^{(j)}$ denotes the jth derivative of f. The above norm is often used as a basis for analysis in $\mathscr{C}^k\left(\overline{\Omega}\right)$. Corresponding norms may be defined when Ω is a subset of $\mathscr{R}^n$ for $n > 1$ by summing over the partial derivatives.

A comparison of the concept of "closeness" for two functions f and g in the sup norm and in $\|\cdot\|_1$ shows that the sup norm bounds the difference $|f(x) - g(x)|$ for every x, whereas $\|\cdot\|_1$ only restricts the average value of this difference. Convergence in sup norm is a very strong condition and implies convergence in $\|\cdot\|_1$; the converse implication is not true.

A sequence of functions (f_n) in a normed vector space $\mathscr{V}$ is called *Cauchy* iff $\lim\limits_{m,n\to\infty} \|f_n - f_m\| = 0$. A set $\mathscr{S}$ in a normed vector space $\mathscr{V}$ is said to be *complete* iff each Cauchy sequence in $\mathscr{S}$ converges to a point of $\mathscr{S}$. A complete normed vector space $\mathscr{V}$ is usually called a *Banach space*.

An *inner product*, denoted by $(\cdot,\cdot)$, on a normed vector space $\mathscr{V}$ is a complex (respectively, real) valued function on $\mathscr{V} \times \mathscr{V}$ such that for all f, g, $h \in \mathscr{V}$ and $\alpha \in \mathscr{C}$ (respectively, $\alpha \in \mathscr{R}$) the following properties hold:

1. $(f,f) \geq 0$; the equality $(f,f) = 0$ holds iff $f = 0$;

2. $(f, g+h) = (f,g) + (f,h)$;

3. $(f,g) = \overline{(g,f)}$, where the overbar denotes complex conjugation;

4. $(\alpha f, g) = \alpha\,(f,g)$.

A space $\mathscr{V}$ equipped with an inner product is called a *pre-Hilbert* or *inner product space*. If $\mathscr{V}$ is a real vector space, and the inner product is real-valued, then the respective space is called a *real pre-Hilbert space*.

A pre-Hilbert space that is *complete with respect to the norm* is called a *Hilbert space*, and is usually denoted as $\mathscr{H}$. The spaces $\mathscr{R}^n$ and $\mathscr{C}^n$ with the usual inner product are Hilbert spaces, and so is the infinite sequence space equipped with the inner product

$$(f,g) = \sum f_n \bar{g}_n , \qquad (7.8)$$

where $f = (f_1, f_2, ...)$, $g = (g_1, g_2, ...)$.

A set $\mathscr{K}$ of vectors in $\mathscr{H}$ is said to be *complete* iff $(f,\varphi) = 0$ for all $\varphi \in \mathscr{K}$ implies that $f = 0$. A countable set $\mathscr{K} = \{\varphi_n\}_{n=1}^{n=\infty}$ is called *orthonormal* iff $(\varphi_n, \varphi_m) = \delta_{nm}$ for all m, $n \geq 1$. The numbers (f, φ_n) are called the *Fourier coefficients* of f (with respect to $\mathscr{K}$), and the *Fourier series* of f is the formal series $\sum_n (f,\ \varphi_n)\,\varphi_n$. An orthonormal set $\mathscr{K} = \{\varphi_n\}$ is called an *orthonormal basis* of $\mathscr{H}$ iff every $f \in \mathscr{H}$ can be represented in the *Fourier series*

$$f = \sum_n (f,\ \varphi_n)\,\varphi_n\,. \qquad (7.9)$$

The mapping A is called an *operator* or a *function* from $\mathscr{V}$ into $\mathscr{W}$. The notation $A : \mathscr{S} \to \mathscr{W}$ indicates that A is an operator with domain $\mathscr{S}$ and range in $\mathscr{W}$, or, equivalently, that A maps $\mathscr{S}$ into $\mathscr{W}$. Note that an operator is always single-valued, in that it assigns exactly one element of its range to each element in its domain. Furthermore, although there is no strict distinction between "operator" and "function," it is customary to reserve "function" for the case when $\mathscr{V}$ and $\mathscr{W}$ are finite dimensional and to use "operator" otherwise. In view of its importance, one particular type of operator is given a name of its own: the operator from $\mathscr{V}$ into the field $\mathscr{F}$ of scalars (real or complex) is called a *functional*.

An operator A from $\mathscr{V}$ into $\mathscr{W}$ is called *injective* iff for each $g \in \mathbf{R}\,(A)$, there is exactly one $f \in \mathbf{D}\,(A)$ such that $Af = g$; A is called *surjective* iff $\mathbf{R}\,(A) = \mathscr{W}$, and A is called *bijective* iff it is both injective and surjective. The terms "one-to one," "onto," "one to one and onto," respectively, are common alternatives in the literature. An operator A from $\mathscr{V}$ into $\mathscr{W}$ is *continuous at the point* $f_0 = \mathbf{D}\,(A)$ iff for each $\varepsilon > 0$, there is a $\delta > 0$ such that $\|Af - Af_0\| < \varepsilon$ if $f \in \mathbf{D}\,(A)$ and $\|f - f_0\| < \delta$; A is said to be *continuous* iff it is continuous at every point of $\mathbf{D}\,(A)$.

An operator $\mathbf{L}$ from $\mathscr{V}$ into $\mathscr{W}$ with domain $\mathbf{D}\,(L)$ is called *linear* iff $L\,(\alpha f + \beta g) = \alpha L f + \beta L g$ for all $\alpha, \beta \in \mathscr{C}$ (or $\alpha, \beta \in \mathscr{R}$, if $\mathscr{V}$ and $\mathscr{W}$ are real spaces), and all $f, g \in \mathbf{D}\,(L)$. A linear operator is the vector space analogue of a function in one dimension represented by a straight line through the origin, that is, a function $\varphi : \mathscr{R} \to \mathscr{R}$ where $\varphi\,(x) = \lambda x$ for some $\lambda \in \mathscr{R}$. In particular, the identity operator, denoted by $\mathbf{I}$, is the operator from $\mathscr{V}$ onto itself such that $If = f$ for all $f \in \mathscr{R}$.

A wide variety of equations may be written in the form $Lf = g$, L a linear operator. For example, the simultaneous algebraic equations

$$\sum_{j=1}^{n} \alpha_{ij} f_j = g_i,\quad (i = 1, ..., m)\,,$$

define the operator L via the relation

$$(Lf)_i = \sum_{j=1}^{n} \alpha_{ij} f_j, \quad (i = 1, ..., m).$$

Then $L : \mathscr{C}^n \to \mathscr{C}^m$ is a linear operator and the above equation can be written as $Lf = g$. Conversely, every linear operator $\mathscr{C}^n \to \mathscr{C}^m$ may be expressed in the above form by choosing bases for $\mathscr{C}^n$ and $\mathscr{C}^m$. The above equations may also be put in matrix form, but note that there is a distinction between the matrix, which depends on the bases chosen, and the operator L, which does not. As a further example, the Fredholm integral equation $f(x) - \int_0^1 k(x, y) f(y) \, dy = g(x), 0 \le x \le 1$, where k and g are given and f is the unknown function, can also be written in operator form as $f - Kf = g$, where K is a linear operator. Similarly, the differential equation $a_0 f''(x) + a_1 f'(x) + a_2 f(x) = g(x), 0 \le x \le 1$, where a_0, a_1, $a_2 \in \mathscr{C}$, and g is a given continuous function, can be written in the form $Lf = g$, with $L : \mathscr{C}^2([0, 1]) \to \mathscr{C}([0, 1])$ being a linear operator.

The operator equation $Lf = g$, where L is a linear operator from $\mathscr{V}$ into $\mathscr{W}$ with domain $\mathbf{D}(L)$, may or may not have a solution f for every $g \in \mathbf{R}(L)$, depending on the following possibilities:

1. L *is not injective*, in which case a reasonable interpretation of the inverse operator L^{-1} is not possible. The equation $Lf = g$ always has more than one solution if $g \in \mathbf{R}(L)$.

2. L *is injective but not surjective*. In this case, the equation $Lf = g$ has exactly one solution if $g \in \mathbf{R}(L)$, but no solution otherwise. The inverse L^{-1} is the operator with domain $\mathbf{R}(L)$ and range $\mathbf{D}(L)$ defined by $f = L^{-1}g$. The set $\mathbf{N}(L) \subset \mathbf{D}(L)$ of solutions of the equation $Lf = 0$ is called the *null space* of L. Note that $\mathbf{N}(L)$ is a linear subspace, and $\mathbf{N}(L) = 0$ iff L is injective.

3. L is *bijective*. In this case, L^{-1} is a linear operator with domain $\mathscr{W}$, and $Lf = g$ has exactly one solution for each $g \in \mathscr{W}$.

If a linear operator L from $\mathscr{V}$ into $\mathscr{W}$ is continuous at some point $f \in \mathbf{D}(L)$, then L is continuous everywhere. A linear operator L is *bounded* on $\mathbf{D}(L)$ iff there is a finite number m such that $\|Lf\| \le m \|f\|$, with $f \in \mathbf{D}(L)$. If

L is not bounded on $\mathbf{D}(L)$, it is said to be *unbounded*. The infimum of all constants m such that this inequality holds is denoted by $\|L\|$, and is called the *operator norm* of L. Note that $\|Lf\| \leq \|L\|\,\|f\|$; this relationship may be compared to the relation $|\varphi(x)| = |\lambda x| = |\lambda|\,|x|$ for a linear operator $\varphi(x) = \lambda x$, $\varphi : \mathscr{R} \to \mathscr{R}$. Since $|\lambda|$ is a measure of the gradient of φ, the norm of the operator L may therefore be thought of as its maximum gradient.

Suppose that L is a (possibly unbounded) linear operator from a Banach space $\mathbf{B}$ into $\mathbf{B}$. The set $\rho(L)$ of complex numbers for which $(\lambda I - L)^{-1}$ belongs to the space of linear operators on $\mathbf{B}$ is called the *resolvent* set of L. For $\lambda \in \rho(L)$, the operator $R(\lambda; L) \equiv (\lambda I - L)^{-1}$ is known as the *resolvent* of L. The complement $\sigma(L)$ in $\mathscr{C}$ of $\rho(L)$ is the *spectrum* of L. A complex number λ is called an *eigenvalue (characteristic value)* of L iff the equation $\lambda f - Lf = 0$ has a nonzero solution. The corresponding nonzero solutions are called *eigenfunctions (characteristic functions)*, and the linear subspace spanned by these is called the *eigenspace (characteristic space)* corresponding to λ. The set $\sigma_p(L)$ of eigenvalues is known as the *point spectrum* of L. The set consisting of those $\lambda \in \sigma(L)$ for which $(\lambda I - L)$ is injective and $R(\lambda I - L)$ is dense (respectively, not dense) in $\mathbf{B}$ is called the *continuous spectrum* (respectively, the *residual spectrum*). Thus, $\lambda \in \rho(L)$ iff $\lambda I - L$ is *bijective*; in this case $\sigma(L)$ *is the union of the point, continuous and residual spectra, which are disjoint sets. If* $\mathbf{B}$ *is finite dimensional, then* $\sigma(L) = \sigma_p(L)$. The operator

$$l = \sum_{r=0}^{n} p_r(x) \left(\frac{d}{dx}\right)^r, \tag{7.10}$$

where p_r $(r = 0, 1, ..., n)$ are given functions on $\mathscr{R}$, is called a *formal ordinary differential operator* of order n. In a Hilbert space, it is convenient to refer loosely to any operator L obtained from l by setting $Lf = lf(x) = \sum_{r=0}^{n} p_r(x) f^{(r)}(x)$, for f in some specified domain, as a *differential operator*. For a general *partial differential equation* in n dimensions, the notational complexity can be reduced considerably by the use of multi-indices, which are defined as follows: a *multi-index* α is an n-tuple $(\alpha_1, \ldots, \alpha_n)$ of nonnegative integers. It is also convenient to use the notation $|\alpha| = \alpha_1 + \ldots + \alpha_n$ for a multi-index; even though this notation conflicts with the notation for the Euclidean distance $\mathscr{R}^n$, the meaning will always be clear from the context.

In the following, multi-indices will be denoted by α and β, and a point in $\mathscr{R}^n$ will be denoted as $x = (x_1, \ldots, x_n)$, with $|x|^2 = \sum x_j^2$ and $x^\alpha = x_1^\alpha \ldots x_n^\alpha$. The notation used for derivatives is $D_j = \partial/\partial x_j$ and $D^\alpha = D_1^{\alpha_1} \ldots D_n^{\alpha_n}$. Consider that $p_{\alpha\beta} \neq 0$ are complex-valued variable coefficients such that $p_{\alpha\beta} \in \mathscr{C}^\infty(\bar{\Omega})$, for multi-indices α and β, with $|\alpha| = |\beta| = m$. Consider, in addition, a function $\phi \in \mathscr{C}^{2m}$; then, the *formal partial differential operator* l of order $2m$ is defined as

$$l\phi \equiv \sum_{|\alpha|, |\beta| \leq m} (-1)^{|\alpha|} D^\alpha \left(p_{\alpha\beta} D^\beta \phi \right). \tag{7.11}$$

The operator l_P, defined as

$$l_P \phi \equiv (-1)^m \sum_{|\alpha| = |\beta| = m} D^\alpha \left(p_{\alpha\beta} D^\beta \phi \right), \tag{7.12}$$

is called the *principal part* of the formal partial differential operator l. Furthermore, the operator l^+ defined as

$$l^+ \phi \equiv \sum_{|\alpha|, |\beta| \leq m} (-1)^{|\alpha|} D^\alpha \left(\bar{p}_{\beta\alpha} D^\beta \phi \right), \quad (\phi \in \mathscr{C}^{2m}) \tag{7.13}$$

is called the *formal adjoint* of l. Iff $l = l^+$, then l is called *formally self-adjoint*.

Define $\mathbf{L}_p^{loc}$ to be the set of all functions that lie in $\mathbf{L}_p(\mathscr{S})$ for every set $\mathscr{S}$ bounded and closed in $\mathscr{R}^n$. Then, a function f in $\mathbf{L}_2^{loc}$ has a αth *weak derivative* iff there exists a function $g \in \mathbf{L}_0^{loc}$ such that

$$\int_\Omega g\phi \, dx = (-1)^{|\alpha|} \int_\Omega f \cdot (D^\alpha \phi) \, dx, \tag{7.14}$$

for all $\phi \in \mathscr{C}_0^\infty$; the function g is called the αth *weak derivative* of f, and we write $D^\alpha f = g$. Weak derivatives are unique in the context of $\mathbf{L}_2$ spaces, in the sense that if g_1 and g_2 are both weak α^{th} derivatives of f, then

$$\int_\Omega (g_1 - g_2) \, \phi \, dx = 0 \quad (\phi \in \mathscr{C}_0^\infty), \tag{7.15}$$

which implies that $g_1 = g_2$ almost everywhere in Ω. The above relation also implies that if a function has an αth derivative g in the ordinary sense in $\mathbf{L}_2^{loc}$, then g is the weak α^{th} derivative of f. The weak derivatives may be thought of as averaging out the discontinuities in f. Consequently, it is permissible to exchange the order of differentiation: $D_i D_j f = D_j D_i f$. In one dimension, a function has a weak first derivative iff it is absolutely continuous and has a first derivative in $\mathbf{L}_2^{loc}$. However, note that f may have an ordinary derivative almost everywhere without having a weak derivative. For example, if $f(x) = 1$ for $x > 0$ and $f(x) = 0$ for $x < 0$, then $\int_{-1}^1 f \phi' \, dx = \int_0^1 \phi' \, dx = -\phi(0)$, but f does not have a weak derivative since there is no $g \in \mathbf{L}_0^{loc}$ such that $\phi(0) = \int_{-1}^1 g \phi \, dx$ for all $\phi \in \mathscr{C}_0^\infty$.

The analysis of differential operators is carried out in special Hilbert spaces, called *Sobolev spaces*. The Sobolev space $\mathscr{H}^m$ [or $\mathscr{H}^m(\Omega)$ if the domain requires emphasis] of order m, where m is a nonnegative integer, consists of the set of functions f such that for $0 \le |\alpha| \le m$, all the weak derivatives $D^\alpha f$ exist and are in $\mathbf{L}_2$; furthermore, $\mathscr{H}^m$ is equipped with an inner product and a norm defined, respectively, as:

$$(f, g)_m = \sum_{|\alpha| \le m} \int_\Omega D^\alpha f \cdot \overline{D^\alpha g} \, dx, \qquad (7.16)$$

$$\|f\|_m^2 = (f, f)_m = \sum_{|\alpha| \le m} \int_\Omega |D^\alpha f|^2 \, dx. \qquad (7.17)$$

The above norm may be regarded as measuring the average value of the weak derivatives. Note that $\mathscr{H}^m$ is a proper subset of the set of functions with m^{th} weak derivatives, since the derivatives $D^\alpha f$ are required to be in $\mathbf{L}_2$, and not merely in $\mathbf{L}_2^{loc}$. The closure of $\mathscr{C}^\infty$ in $\| \cdot \|_m$ is $\mathscr{H}^m$, and the closure in $\mathscr{H}^m$ of $\mathscr{C}_0^\infty$ is the *Sobolev space* $\mathscr{H}_0^m$, of order m. The following chains of inclusions hold for the higher order Sobolev spaces:

$$\mathscr{C}^\infty \subset \ldots \subset \mathscr{H}^{m+1} \subset \mathscr{H}^m \subset \ldots \subset \mathscr{H}^0 = \mathbf{L}_2, \qquad (7.18)$$

$$\mathscr{C}_0^\infty \subset \ldots \subset \mathscr{H}_0^{m+1} \subset \mathscr{H}_0^m \subset \ldots \subset \mathscr{H}_0^0 = \mathbf{L}_2 \qquad (7.19)$$

The *bilinear form*, $B[f, \phi]$, associated with the formal differential operator l is defined for all functions $f, \phi \in \mathscr{H}_0^m$ as

$$B[f, \phi] = \sum_{|\alpha|, |\beta| \le m} (p_{\alpha\beta} D^\alpha f, D^\beta \phi)_o. \qquad (7.20)$$

Note that $B[f, \phi]$ is a bounded bilinear form on $\mathscr{H}_0^m \times \mathscr{H}_0^m$. The problem of finding $f \in \mathscr{H}_0^m$ such that $B[f, \phi] = (g, \phi)_0$ for all $\varphi \in \mathscr{H}_0^m$ and $g \in \mathbf{L}_2$ is called the *generalized Dirichlet problem*.

The space of continuous linear functionals on a Banach space $\mathbf{B}$ is called the *dual* of $\mathbf{B}$, and is denoted here by $\mathbf{B}^+$. For a bounded linear operator $L : \mathbf{B} \to \mathscr{C}$, the relation

$$g^+ (Lf) = L^+ g^+ (f), \tag{7.21}$$

required to hold for all $f \in \mathbf{B}$ and all $g^+ \in \mathscr{C}^+$, defines an operator L^+ from $\mathscr{C}^+$ into $\mathbf{B}^+$, called the *adjoint* of L. For example, the adjoint of an integral operator of the form $Kf(x) = \int_0^1 k(x, y) f(y) \, dy$, for a continuous kernel $k : [0, 1] \times [0, 1] \to \mathscr{R}$, is the operator $K^+ : \mathbf{L}_\infty (0, 1) \to \mathbf{L}_\infty (0, 1)$ defined as $K^+ g^+ (y) = \int_0^1 k(x, y) g^+ (x) \, dx$.

As the next example, consider the finite dimensional operator $L : \ell_1^{(n)} \to \ell_1^{(n)}$ corresponding to the matrix (α_{ij}), namely

$$(Lf)_i = \sum_{j=1}^{n} \alpha_{ij} f_j , \qquad (i = 1, ..., n) .$$

The corresponding adjoint operator, $L^+ : \ell_\infty^{(n)} \to \ell_\infty^{(n)}$, is represented by the transposed matrix $(\alpha_{ij})^T$, as can be seen by applying the definition of the adjoint operator:

$$g^+ (Lf) = \sum_{i=1}^{n} (Lf)_i g_i^+ = \sum_{i=1}^{n} g_i^+ \sum_{j=1}^{n} \alpha_{ij} f_j = \sum_{j=1}^{n} f_j \sum_{i=1}^{n} \alpha_{ij} g_i^+ = (L^+ g^+) (f) .$$

Thus, even in finite dimensions, the adjoint operator depends on the norm of the space and has more significance than the algebraic "transpose." For example, the relation $\|L\| = \|L^+\|$ will only hold if the dual space is correctly selected.

In an n-dimensional unitary space $\mathscr{R}$, the linear operator A^+ is called *adjoint* to the operator A iff, for any two vectors $\mathbf{x}, \mathbf{y}$ of $\mathscr{R}$, the following relationship holds

$$(A\mathbf{x}, \mathbf{y}) = (\mathbf{x}, A^+ \mathbf{y}) . \tag{7.22}$$

In an orthonormal basis $\mathbf{e}_1, \mathbf{e}_2, \ldots, \mathbf{e}_n$ in $\mathscr{R}$, the adjoint operator A^+ can be represented uniquely in the form

$$A^+\mathbf{y} = \sum_{k=1}^{n} \langle \mathbf{y}, A\mathbf{e}_k \rangle \, \mathbf{e}_k \,, \tag{7.23}$$

for any vector $\mathbf{y}$ of $\mathcal{R}$. Consider that A is a linear operator in a unitary space and that $\mathbf{A} = \|a_{ik}\|_1^n$ is the corresponding matrix that represents A in an orthonormal basis $\mathbf{e}_1, \mathbf{e}_2, \ldots, \mathbf{e}_n$. Then, the matrix $\mathbf{A}^+$ corresponding to the representation of the adjoint operator A^+ in the same basis is the complex conjugate of the transpose of $\mathbf{A}$, i.e.,

$$\mathbf{A}^+ = \overline{\mathbf{A}}^T . \tag{7.24}$$

The matrix $\mathbf{A}^+$ given by (7.24) is called the *adjoint of* $\mathbf{A}$. Thus, *in an orthonormal basis, adjoint matrices correspond to adjoint operators.*

The adjoint operator has the following properties:

1. $(A^+)^+ = A$,

2. $(A + B)^+ = A^+ + B^+$,

3. $(\alpha A)^+ = \bar{\alpha} A^+$ (α a scalar),

4. $(AB)^+ = B^+ A^+$.

5. If an arbitrary subspace $\mathscr{S}$ of $\mathcal{R}$ is invariant with respect to A, then the orthogonal complement $\mathscr{T}$ of the subspace $\mathscr{S}$ is invariant with respect to A^+.

Two systems of vectors $\mathbf{x}_1, \mathbf{x}_2, \ldots, \mathbf{x}_n$ and $\mathbf{y}_1, \mathbf{y}_2, \ldots, \mathbf{y}_n$ are by definition *bi-orthogonal* if

$$\langle \mathbf{x}_i, \mathbf{y}_k \rangle = \delta_{ik} \,, \quad (i, k = 1, 2, \ldots, m) \,, \tag{7.25}$$

where δ_{ik} is the Kronecker symbol.

If A is a linear operator of simple structure, then the adjoint operator A^+ is also of simple structure; therefore, complete systems of characteristic vectors $\mathbf{x}_1, \mathbf{x}_2, \ldots, \mathbf{x}_n$ and $\mathbf{y}_1, \mathbf{y}_2, \ldots, \mathbf{y}_n$ of A and A^+, respectively, can be chosen such that they are bi-orthogonal:

$$A\mathbf{x}_i = \lambda_i \mathbf{x}_i, \quad A^+\mathbf{y}_i = \lambda_i \mathbf{y}_i, \quad \langle \mathbf{x}_i, \mathbf{y}_k \rangle = \delta_{ik}, \quad (i, k = 1, 2, \ldots, n) \,. \tag{7.26}$$

Furthermore, if the operators A and A^+ have a common characteristic vector, then the corresponding characteristic values are complex conjugates. A linear operator A is called *normal* if it commutes with its adjoint:

$$AA^+ = A^+A. \tag{7.27}$$

A linear operator H is called *hermitian* if it is equal to its adjoint:

$$H^+ = H. \tag{7.28}$$

A linear operator U is called *unitary* if it is inverse to its adjoint:

$$UU^+ = I. \tag{7.29}$$

Just as in the case of operators, a matrix is called *normal* if it commutes with its adjoint, *hermitian* if it is equal to its adjoint, and *unitary* if it is inverse to its adjoint. Thus, in an orthonormal basis, a normal (hermitian, unitary) operator corresponds to a normal (hermitian, unitary) matrix. Note that every characteristic vector of a normal operator A is a characteristic vector of the adjoint operator A^+, i.e., if A is a normal operator, then A and A^+ have the same characteristic vectors. Furthermore, a linear operator is normal if and only if it has a complete orthonormal system of characteristic vectors. If A is a normal operator, then each of the operators A and A^+ can be represented as a polynomial in the other; these two polynomials are determined by the characteristic values of A.

In a Hilbert space $\mathscr{H}$, Eq. (7.21) becomes

$$(Lf, g) = (f, L^+g), \tag{7.30}$$

for all f and $g \in \mathscr{H}$; thus, Eq. (7.30) defines the bounded linear operator L^+, called the *(Hilbert space) adjoint* of L. The *Riesz Representation Theorem* ensures that for every element g^+ of the dual $\mathscr{H}^+$ of a Hilbert space $\mathscr{H}$, there is a unique element g of $\mathscr{H}$ such that $g^+(f) = (f, g)$ for all $f \in \mathscr{H}$. The equality $\|g^+\| = \|g\|$ also holds.

The existence of solutions for operator equations involving linear compact operators is elucidated by the *Fredholm Alternative Theorem* (in a Hilbert space $\mathscr{H}$), which can be formulated as follows: consider that $L : \mathscr{H} \to \mathscr{H}$ is a linear compact operator, and consider the equation

$$(\lambda I - L) = g, \quad g \in, \quad \lambda \neq 0. \tag{7.31}$$

Then, one of the following alternatives hold:

1. The homogeneous equation has only the zero solution; in this case, $\lambda \in \rho(L)$, where $\rho(L)$ denotes the *resolvent set* of L (thus, λ cannot be an eigenvalue of L); furthermore, $(\lambda I - L)^{-1}$ is bounded, and the inhomogeneous equation has exactly one solution $f = (\lambda I - L)^{-1}g$, for each $g \in \mathscr{H}$;

2. The homogeneous equation has a nonzero solution; in this case, the inhomogeneous equation has a solution, necessarily nonunique, iff $\langle g, \varphi^+ \rangle = 0$, for every solution φ^+ of the adjoint equation $\lambda \varphi^+ = L^+ \varphi^+$, where L^+ denotes the operator adjoint to L, and $\langle , \rangle$ denotes the inner product in the respective Hilbert space $\mathscr{H}$.

Consider now that L denotes an *unbounded* linear operator from $\mathscr{H}$ into $\mathscr{H}$, with domain $\mathbf{D}(L)$ dense in $\mathscr{H}$. Recall that the specification of a domain is an essential part of the definition of an unbounded operator. Define $\mathbf{D}(L^+)$ to be the set of elements g such that there is an h with $(Lf, g) = (f, h)$ for all $f \in \mathbf{D}(L)$. Let L^+ be the operator with domain $\mathbf{D}(L^+)$ and with $L^+ g = h$ on $\mathbf{D}(L^+)$, or, equivalently, consider that L^+ satisfies the relation

$$(Lf, g) = (f, L^+ g), \quad f \in \mathbf{D}(L), \quad g \in \mathbf{D}(L^+). \tag{7.32}$$

Then the operator L^+ is called the *adjoint* of L. Furthermore, a densely defined linear operator L from a Hilbert space into itself is called *self-adjoint* iff $L = L^+$. Note that necessarily $\mathbf{D}(L) = \mathbf{D}(L^+)$ for self-adjoint operators.

It is important to note that the operator theoretic concept of "adjoint" defined in Eq. (7.32) involves the boundary conditions in an essential manner, and is therefore a more comprehensive concept than the *formal adjoint operator* defined in Eq. (7.13), which merely described the coefficients of a certain differential operator. To illustrate the difference between the (operator theoretic) adjoint of an unbounded operator and the underlying formal adjoint operator, consider the formal differential operator $l = id/dx$ on the interval $[0, 1]$, and consider the operator $Lf = if$. Furthermore, denote by **A** the linear subspace of $\mathscr{H} = \mathbf{L}_2 (0, 1)$ consisting of absolutely continuous functions with derivatives in $\mathbf{L}_2 (0, 1)$. Suppose first that no boundary conditions are specified for L, so that $\mathbf{D} (L) = \mathbf{A}$; in this case, an application of Eq. (7.32) shows that the proper adjoint boundary conditions are $\mathbf{D} (L^+) = \{g : g \in \mathbf{A}, \ g (0) = g (1) = 0\}$. On the other hand, suppose that the boundary conditions imposed on L are $\mathbf{D} (L) = \{f : f \in \mathbf{A}, \ f (0) = f (1) = 0\}$; in this case, an application of Eq. (7.32) shows that the proper adjoint boundary conditions are $\mathbf{D} (L^+) = \mathbf{A}$. Note that L is not equal to L^+, in either of these two examples, although ℓ is "self-adjoint" in the sense of classical differential equation theory. To construct a self-adjoint operator from the formal operator $l = id/dx$ on $[0, 1]$, it is necessary to choose boundary conditions so as to ensure that $\mathbf{D} (L) = \mathbf{D} (L^+)$. Applying again Eq. (7.32) yields the proper domain as $\mathbf{D} (L) = \{f : f \in \mathbf{A}, \ f (1) = e^{i\theta} f (0), \ \theta \in \mathscr{R}\}$, which means that there are infinitely many self-adjoint operators based on the formal operator id/dx on the interval $[0, 1]$.

Consider that I is an open interval of the real line and $\mathscr{V}$ is a normed real space. The *derivative* of a mapping (or operator) $\phi : I \rightarrow \mathscr{V}$ at $t_0 \in I$ is defined as $\lim\limits_{t \rightarrow t_0} \frac{\phi(t) - \phi(t_0)}{t - t_0}$ if this limit exists, and is denoted by $\phi' (t_0)$. The limit is to be understood in the sense of the norm in $\mathscr{V}$, namely $\left\| \frac{\phi(t) - \phi(t_0)}{t - t_0} - \phi' (t_0) \right\| \rightarrow 0$ as $t \rightarrow 0$. Consider next that $\mathscr{V}$ and $\mathscr{W}$ are normed real spaces and $\mathbf{D}$ is an open subset of $\mathscr{V}$. Consider that $x_0 \in \mathbf{D}$ and that h is a fixed nonzero element in $\mathscr{V}$. Since $\mathbf{D}$ is open, there exists an interval $I = (-\tau, \tau)$ for some $\tau > 0$ such that if $t \in I$, then $x_0 + th \in \mathbf{D}$. If the mapping $\phi : I \rightarrow \mathscr{V}$ defined by $\phi (t) = F (x_0 + th)$ has a derivative at $t = 0$, then $\phi' (0)$ is called the *Gâteaux variation* of F at x_0 with increment h, and is denoted by $\delta F (x_0; h)$, i.e.,

$$\delta F (x_0; h) \equiv \frac{d}{dt} F (x_0 + th) \bigg|_{t=0} = \lim_{t \rightarrow 0} \frac{1}{t} \{F (x_0 + th) - F (x_0)\} . \qquad (7.33)$$

Note that Eq. (7.33) may be used to define $\delta F (x_0; h)$ when $\mathscr{V}$ is any linear space, not necessarily normed. When $\delta F (x_0; h)$ exists, it is homogeneous in h of degree one, i.e., for each real number λ, $\delta F (x_0; \lambda h)$ exists and is equal to $\lambda \delta F (x_0; h)$. The Gâteaux variation is a generalization of the notion of the directional derivative in calculus and of the notion of the first variation arising in the calculus of variations. The existence of the Gâteaux variation at $x_0 \in \mathbf{D}$ provides a local approximation property in the following sense:

$$F(x_0 + h) - F(x_0) = \delta F(x_0; h) + r(x_0; h), \tag{7.34}$$

where $\lim\limits_{t \to 0} \frac{r(x_0; th)}{t} = 0$.

The existence of $\delta F(x_0; h)$ implies the directional continuity of F at x_0, i.e.,

$$\| F(x_0 + th) - F(x_0) \| \to 0 \text{ as } t \to 0 \text{ for fixed } h, \tag{7.35}$$

but does not imply that F is continuous at x_0. This is equivalent to saying that, in general, Eq. (7.35) does not hold uniformly with respect to h on the bounded set $\{h : \|h\| = 1\}$. Note also that the operator $h \to \delta F(x_0; h)$ is not necessarily linear or continuous in h.

If F and G have a Gâteaux variation at x_0, then so does the operator $T = \alpha F + \beta G$, where α, β are real numbers, and the relation $\delta T(x_0; h) = \alpha \delta F(x_0; h) + \beta \delta G(x_0; h)$ holds. However, the chain rule for the differentiation of a composite function does not hold in general.

An operator F has a *Gâteaux differential* at x_0 if $\delta F(x_0; \cdot)$ is linear and continuous; in this case, $\delta F(x_0; \cdot)$ is denoted by $DF(x_0)$ and is called the *Gâteaux derivative*. The necessary and sufficient condition for $\delta F(x_0; h)$ to be linear and continuous in h is that F satisfies the following two relations:

1. To each h, there corresponds $\delta(h)$ such that $|t| \leq \delta$ implies

$$\| F(x_0 + th) - F(x_0) \| \leq M \| th \|, \tag{7.36}$$

   where M does not depend on h;

2. and

$$F(x_0 + th_1 + th_2) - F(x_0 + th_1) - F(x_0 + th_2) + F(x_0) = O(t). \tag{7.37}$$

Note that the chain rule does not necessarily hold for Gâteaux derivatives. Note also that if $\delta F(x_0; \cdot)$ is additive, then $\delta F(x_0; h)$ is directionally continuous in h, i.e.,

$$\lim_{\tau \to 0} \delta F (x_0; h + \tau k) = \delta F (x_0; h) . \tag{7.38}$$

An operator $F : \mathbf{D} \to \mathscr{W}$, where $\mathbf{D}$ is an open subset of $\mathscr{V}$, and $\mathscr{V}$ and $\mathscr{W}$ are normed real linear spaces, is called *Fréchet differentiable* at $x_0 \in \mathbf{D}$ if there exists a continuous linear operator $L (x_0) : \mathscr{V} \to \mathscr{W}$ such that the following representation holds for every $h \in \mathscr{V}$ with $x_0 + h \in \mathbf{D}$:

$$F (x_0 + h) - F (x_0) = L (x_0) h + r (x_0; h) , \quad \text{with} \quad \lim_{h \to 0} \frac{\| r (x_0; h) \|}{\| h \|} = 0. \tag{7.39}$$

The unique operator $L (x_0) h$ in Eq. (7.39) is called the *Fréchet differential* of F at x_0 and is usually denoted by $dF (x_0; h)$. The linear operator $F' (x_0) :$ $\mathscr{V} \to \mathscr{W}$ defined by $h \to dF (x_0; h)$ is called the *Fréchet derivative* of F at x_0, and $dF (x_0; h) = F' (x_0) h$. An operator F is Fréchet differentiable at x_0 iff (a) F is Gâteaux differentiable at x_0 and (b) Eq. (7.34) holds uniformly with respect to h on the set $\mathscr{S} = \{h : \|h\| = 1\}$. In this case the Gâteaux and Fréchet differentials coincide.

The Féchet differential has the usual properties of the classical differential of a function of one or several variables. In particular, the chain rule holds for $F \cdot G$ if F has a Fréchet differential and G has a Gâteaux differential. Note that the chain rule may not hold for $F \cdot G$ if F has a Gâteaux differential and G has a Fréchet differential. Fréchet differentiability of F at x_0 implies continuity of F at x_0.

Consider an operator $F : \mathscr{V} \to \mathscr{W}$, where $\mathscr{V}$ is an open subset of the product space $\mathbf{P} = \mathscr{E}_1 \times \ldots \times \mathscr{E}_n$. The *Gâteaux partial differential* at $u \equiv (u_1, \ldots, u_n)$ of F with respect to u_i is the bounded linear operator $D_i F (u_1, \ldots, u_n; h_i) : \mathscr{E}_i \to \mathscr{W}$ defined such that the following relation holds:

$$F (u_1, \ldots, u_{i-1}, u_i + h_i, u_{i+1}, \ldots, u_n) - F (u_1, \ldots, u_n)$$
$$= D_i F (u_1, \ldots, u_n; h_i) + R (u_1, \ldots, u_n; h_i) \tag{7.40}$$

where $\lim_{t \to 0} \frac{R(u_1, \ldots, u_n; th_i)}{t} = 0.$

The operator F is said to be *totally Gâteaux differentiable* at u_0 if F, considered as a mapping on $\mathscr{V} \subset \mathbf{P}$ into $\mathscr{W}$, is Gâteaux differentiable at u_0. This means that

$$F(u_1 + h_1, \ldots, u_n + h_n)$$
$$- F(u_1 \ldots u_n) = L(u_1, \ldots, u_n; h_1, \ldots, h_n) + R(u_1, \ldots, u_n; h_1, \ldots, h_n),$$

$$(7.41)$$

where the total Gâteaux differential L is a continuous linear operator in $h = (h_1 \ldots h_n)$ and where

$$\lim_{t \to 0} \left(t^{-1}\right) R(u_1, \ldots, u_n; th_1, \ldots, th_n) = 0.$$

The *Fréchet partial differential* at $(u_1, \ldots, u_n)$ of F with respect to u_i is the bounded linear operator $d_i F(u_1, \ldots, u_n; h_i)$ defined such that, for all $h_i \in \mathscr{E}_i$, with $(u_1, \ldots, u_{i-1}, u_i + h_i, u_{i+1}, \ldots, u_n) \in \mathscr{V}$, the following relation holds:

$$F(u_1, \ldots, u_{i-1}, u_i + h_i, \ldots, u_n) - F(u_i, \ldots, u_n) = d_i F(u_1, \ldots, u_n; h_i)$$
$$+ R(u_1, \ldots, u_n; h_i),$$

$$(7.42)$$

where $\frac{\|R(u_1, \ldots, u_n; h_i)\|}{\|h_i\|} \to 0$ as $h_i \to 0$.

The *total Fréchet differential* of F is denoted by $dF(u_1, \ldots, u_n; h_1, \ldots, h_n)$ and is defined as the linear mapping on $\mathscr{V} \subset \mathscr{E}_1 \times \ldots \times \mathscr{E}_n$ into $\mathscr{W}$, which is continuous in $h = (h_1, \ldots, h_n)$, such that the following relation holds:

$$\lim_{h \to 0} \frac{\|F(u_1 + h_1, \ldots, u_n + h_n) - F(u_1 \ldots u_n) - dF(u_1, \ldots, u_n; h_1, \ldots, h_n)\|}{\|h_1\| + \ldots + \|h_n\|}$$
$$= 0.$$

$$(7.43)$$

An operator $F : \mathscr{V} \subset \mathbf{P} \to \mathscr{W}$ that is totally differentiable at $(u_1, \ldots, u_n)$, is partially differentiable with respect to each variable, and its total differential is the sum of the differentials with respect to each of the variables. If F is totally differentiable at each point of $\mathscr{V}$, then a necessary and sufficient condition for $F' : \mathcal{V} \to L(\mathcal{E}_1 \times \ldots \times \mathcal{E}_n; \mathcal{W})$ to be continuous is that the partial derivatives $F'_i : \mathcal{V} \to L(\mathcal{E}_i; \mathcal{W})$, $(i = 1, \ldots, n)$, be continuous.

Higher order partial derivatives are defined by induction. Note, in particular, that if $F : \mathcal{V} \to \mathcal{W}$ is twice Fréchet (totally) differentiable at x_0, then the second-order partial derivatives $\frac{\partial^2 F(x_0)}{\partial x_i \partial x_j} \in L(\mathcal{E}_i, \mathcal{E}_j, \mathcal{W})$, $(i, j = 1, \ldots, n)$ exist, and

$$d^2 F(x_0; k_1, \ldots, k_n; h_1, \ldots, h_n) = \sum_{i,j=1}^{n} \frac{\partial^2 F(x_0)}{\partial x_i \partial x_j} k_i h_j. \qquad (7.44)$$

Thus, the second-order derivative $F''(x_0)$ may be represented by the array $\{\partial^2 F(x_0)/\partial x_i \partial x_j; (i, j = 1, \ldots, n)\}$. Note that $d^2 F(x_0; h, k)$ is symmetric in h and k; consequently, the mixed partial derivatives are also symmetric, i.e.,

$$\frac{\partial^2 F(x_0)}{\partial x_i \partial x_j} = \frac{\partial^2 F(x_0)}{\partial x_j \partial x_i}, \quad (i, j = 1, \ldots, n) . \qquad (7.45)$$

If the operator $F : \mathbf{D} \to \mathcal{W}$ has a nth-variation on $\mathbf{D}$, $\delta^n F(x_0 + th; h)$, which is continuous in t on $[0, 1]$, then the following Taylor expansion with an integral remainder holds for x_0, $x_0 + h \in \mathbf{D}$:

$$F(x_0 + h) = F(x_0) + \delta F(x_0; h) + \frac{1}{2} \delta^2 F(x_0; h) + \cdots$$

$$+ \frac{1}{(n-1)!} \delta^{n-1} F(x_0; h) + \int_0^1 \frac{(1-t)^{n-1}}{(n-1)!} \delta^n F(x_0 + th; h) \, dt.$$

$$(7.46)$$

Note that the integral in Eq. (7.46) exists and is a Banach space-valued integral in the Riemann sense; note also that $\delta^k F(x_0; h) = \delta^k F(x_0; h_1, \ldots, h_k)$ for $h_1 = \ldots = h_k = h$. Furthermore, if an operator $F : \mathbf{D} \to \mathcal{W}$ has a nth-order Fréchet differential, and if the map $\phi : [0, 1] \to F^{(n)}(x_0 + th)$ is bounded and its set of discontinuities is of measure zero, then the Taylor expansion in Eq. (7.46) becomes

$$F\left(x_0 + h\right) = F\left(x_0\right) + F'\left(x_0\right) h + \frac{1}{2}F''\left(x_0\right) h^2 + \ldots$$

$$+ \frac{1}{(n-1)\,!}F^{(n-1)}\left(x_0\right)\,h^{n-1} + \int_0^1 \frac{(1-t)^{n-1}}{(n-1)\,!}F^{(n)}\left(x_0 + th\right)\,h^n\,dt.$$

$$(7.47)$$

In the above expansion, $F^{(k)}\left(x_0\right) h^k$ denotes the value of the k-linear operator $F^{(k)}$ at $(h, \ldots, h)$. The relation between the various differentiation concepts for a nonlinear operator F is shown schematically below, where "uniform in h" indicates the validity of the relation $\lim\limits_{h \to 0} \frac{\|F(x+h)-F(x)-\delta F(x;h)\|}{\|h\|} = 0$:

G-Differential $\xrightarrow{\text{uniform in } h}$ F-Differential $\xrightarrow{\text{linear in } h}$ F-Derivative,

G-Differential $\xrightarrow{\text{linear in } h}$ G-Derivative $\xrightarrow{\text{uniform in } h}$ F-Derivative.

8

Appendix C

CONTENTS

8.1 Parameter Identification and Estimation

The *parameter identification* issue can be formulated mathematically as the one-to-one property of mapping from the space of system outputs to the space of parameters. Such a mapping is generally known as an *inverse problem*, in contradistinction with the forward mapping, which maps the space of parameters to the space of outputs. The notion of identifiability addresses the issue of obtaining *unique solutions* of the inverse problem for unknown parameters of interest in a mathematical model, from data collected in the spatial and temporal domains. A working definition of *identifiability* is as follows: "*An unknown parameter is 'identifiable' if it can be determined uniquely in all points of its domain by using the input-output relation of the system and the input-output data.*" The uniqueness of inverse mappings is difficult to establish.

Chavent and Lemonnier [31] presented a definition of identifiability using an output least-squares error criterion. For such a criterion, the parameter is said to be "output least-square identifiable" if and only if a unique solution of the optimization problem exists and the solution depends continuously on the observations. Chavent also recognized that this inverse problem is often ill-posed, characterized by nonuniqueness and instability of the identified parameters, in the sense that small errors in the data will induce large errors in the identified parameters. The ill-posedness of parameter identification problems is illustrated in subsection 8.1.1 by means of a simple example, admitting an analytic solution (see Banks and Kunisch [8]) which clearly displays the nonexistence of a continuous inverse of the parameter-to-observation mapping.

8.1.1 Mathematical Framework for Parameter Identification and Regularization

In simple terms, the usual "forward problem" can be described as the parameter-to-output mapping $\mathbf{\Phi} : \mathbf{Q} \rightarrow Z$, with $\mathbf{\Phi}(\mathbf{q}) = \mathbf{Cu}(\mathbf{q})$, where: $\mathbf{Q}$ denotes the set of parameters which guarantee that a chosen model equation has solution $\mathbf{u}(\mathbf{q})$; $\mathbf{C}$ denotes the observation operator, which is a mapping from the solution space of the model to the observation space, denoted by Z. Conversely, the "inverse problem" is to determine uniquely $\mathbf{Q}$ by inverting the mapping $\mathbf{\Phi}(\mathbf{q}) = \mathbf{Cu}(\mathbf{q})$. Inverse problems are notoriously ill-posed, because the differentiation operator is not continuous with respect to any physically meaningful observation topology. For example, the problem of identifying spatially dependent coefficients appearing in the differential operator of a partial differential equation is usually both nonlinear and ill-posed. This fact can be readily illustrated by considering a simple model admitting a unique, well-defined solution for the forward problem (Banks and Kunisch [8]), but clearly displaying the nonexistence of a continuous inverse of the parameter-to-observation mapping. Thus, consider the model

$$-(qu_x)_x = f, \quad x \in (0,1), \quad u(0) = u(1) = 0, \tag{8.1}$$

where f is assumed to be known and q is the unknown parameter. For any particular point $x_p \in (0,1)$, Eq. (8.1) can be formally integrated to obtain

$$q(x) = \frac{u_x(x_p)}{u_x(x)} q(x_p) - \frac{1}{u_x(x)} \int_{x_p}^{x} f(s)\, ds, \quad x \in [0,1]. \tag{8.2}$$

If $u_x > 0$ (or $u_x < 0$) on [0,1], then q is uniquely determined by Eq. (8.2), provided that $q(x_p)$ is given for some point $x_p \in (0,1)$. If u_x has precisely one root, this root may be taken to be x_p, provided that $u_{xx}(x_p) \neq 0$. It then follows from Eq. (8.1) that

$$q(x_p) = \frac{f(x_p)}{u_{xx}(x_p)}. \tag{8.3}$$

As indicated by the above expression, q is determined uniquely although it has not been specified at any point over its domain. Consider now that $f = 0$ and $q(0) = 1$ are known, such that Eq. (8.1) admits the following sequence of solutions

$$u_k(x) = x + \frac{1}{(2k+1)\pi} sin(2k\pi x), \quad x \in [0,1], \quad k = 1, 2, \dots. \tag{8.4}$$

It follows from Eq. (8.4) that $u_k \to u^*$ with $u^*(x) = x$, in the space $C([0,1])$ of continuous functions of $x \in [0,1]$. It also follows that

$$(u_k)_x = 1 + 2k/(2k+1), \quad u_k(0) = 0, \quad u_k(1) = 1. \tag{8.5}$$

On the other hand, Eq. (8.2) indicates that

$$q_k(x) = \frac{(u_k)_x(0)}{(u_k)_x(x)} = \frac{1 + \frac{2k}{(2k+1)}}{1 + \frac{2k}{(2k+1)}cos(2k\pi x)}, \tag{8.6}$$

which is a divergent series in $C([0,1])$. Thus, the inverse of $\Phi(q) = u(q)$ as a mapping from $C([0,1])$ to $C([0,1])$ is not bounded even in a neighborhood of some u^* satisfying $u_x^* > 0$. This example illustrates the lack of a continuous inverse of the parameter-to-observation mapping in parameter identification problems.

Since inverse problems are generally ill-posed, afflicted by nonuniqueness and instabilities of the identified parameters, a procedure called "regularization" is often employed to address such problems. "*Regularization*" of a problem refers to solving a related problem, called the "*regularized problem*," the solution of which approximates the solution of the original problem, but is more regular, in a well-defined sense, than the solution of the original problem. Thus, regularization is an approach to circumvent the lack of continuous dependence on the data, but in a way such that the regularized problem is a well-posed problem whose solution yields a physically meaningful answer to the given ill-posed problem. The idea of regularization appears to have been first proposed by Tikhonov [197].

The *parameter estimation* problem refers to finding an estimated value $\widehat{Q}$ of the parameter Q from knowledge of data Z, of the parameter to output mapping $\Phi : Q \to Z$ and some "a priori" knowledge (e.g., lower and upper bounds, trends, regularity) on the parameter which is condensed in a set of admissible parameters Q_{ad}. A systematic mathematical effort, starting with the seminal works of Richard Bellman and Kalaba [13], has been devoted to understanding in depth the issues of parameter estimation. The current mathematical formulation of the parameter identification and estimation problem can be stated as follows: consider a mathematical system represented in operator form as

$$\Psi\left(A, U\right) = f, \tag{8.7}$$

where A represents a Banach space of partial differential operators, U represents the solution of Eq. (8.7) that belongs to a Banach space $\Im$, f belongs to a Banach space F, and Ψ is a mapping (in general, nonlinear) from $A \times \Im$ into F. Consider, in addition, that the following assumptions are fulfilled:

(A.1) Ψ belongs to C^k, where, as usual, the vector space denoted by $C^k\left(\Omega, \mathbb{R}^m\right)$ consists of all $\mathbb{R}^m$-valued functions defined on Ω such that all partial derivatives up to and including those of order $k > 0$ of all components are bounded and continuous;

(A.2) There is an open subset A_c of A and an open subset $\Im_c$ of $\Im$ such that for $\forall A \in A_c$, Eq. (8.7) admits a unique solution $U \in \Im_c$;

(A.3) For $\forall A \in A_c$, and $\forall u \in \Im_c$, $\left(\frac{\partial \psi}{\partial u}\right) A\left(u\right)$ is a linear homeomorphism of $\Im$ onto F.

If the above assumptions are fulfilled, the implicit function

$$U = \Phi\left(A\right), \tag{8.8}$$

exists and can be defined as the solution of Eq. (8.7), with Φ belonging to C^k from A_c into $\Im_c$.

Furthermore, consider that A depends on a set of parameters p belonging to the Banach space Λ, and consider that the set of physically admissible parameters p is denoted by Λ_{ad}. Under the assumptions that: (i) the mapping

$\Lambda \to A$ belongs to C^k; (ii) Λ_{ad} is a norm-closed convex subset of Λ; and $A(\Lambda_{ad}) \leq A_c$, the parameter identification problem can be posed as follows: find $p \in \Lambda_{ad}$ so as to satisfy equation Eq. (8.7), knowing the mappings Ψ: $A \times \Im \to F$, $\Lambda \to A$, the element $f \in F$, and a given observation $Z_d \in H$ of U, with H representing the corresponding Banach space.

Thus, the identification problem can be viewed as solving in Λ_{ad} the non-linear operator equation

$$(C \cdot \Phi \cdot A)(p) = Z_d. \tag{8.9}$$

If the operator $(C \cdot \Phi \cdot A){:}\Lambda_{ad} \to H$ has an unique inverse, and the inverse is continuous, the least-squares or the adjoint methods can be used to perform parameter estimation, by minimizing over Λ_{ad} the "least-squares" functional

$$J_{LS}(p) = \|C \cdot \Phi \cdot A(p) - Z_d\|_H^2. \tag{8.10}$$

Often the problem of solving Eq. (8.9) is ill-posed, so the minima of $J_{LS}(p)$ over Λ_{ad} will not depend continuously on the data Z_d. To regularize the parameters p, a Hilbert (rather than Banach) space, R, needs to be introduced for these parameters, such that

(R.1) R is a Hilbert space;
(R.2) R is densely imbedded in Λ;
(R.3) The embedding operator from R into Λ is compact.

By defining $R_{ad} = R \cap \Lambda_{ad}$, such that R_{ad} is a norm-closed convex subset of R, it is possible to introduce the "stabilizing" (regularization) functional

$$J_s(p) = \|p\|_R^2, \quad p \in R_{ad}, \tag{8.11}$$

which is added to the "least-squares" functional to obtain the "smoothing" functional $J_\beta(p) = J_{LS}(p) + \beta J_s(p)$, defined as

$$J_\beta(p) = \|C \cdot \Phi \cdot A(p) - Z_d\|_H^2 + \beta \|p\|_R^2, \quad p \in R_{ad}. \tag{8.12}$$

Identification and estimation of parameters by regularization proceeds now by minimizing the cost functional $J_\beta\,(p)$ in order to determine $p_\beta \in R_{ad}$, given the observations $Z_d \in H$. Several key issues need to be examined, including:

1. whether Eq. (8.9) admits a unique solution ("identifiability"), in which case the optimally estimated parameters could be expected to be close to the true values of the unknown parameters;

2. whether the solution of equation Eq. (8.9) depends continuously on the observations Z_d ("stability"), in which case the estimated parameters are expected to be stable with respect to perturbations in the observations;

3. the relationship between dimensionality of the state, observation, and parameter spaces, which is essential for choosing efficient numerical procedures for computing the "estimated" parameters, in order to avoid:

    (a) possible instabilities that might occur when refining the discretization of parameters;

    (b) possible lack of uniqueness of the estimated parameters;

    (c) the optimization algorithm becoming attracted by and stuck at some local minimum.

More general cost functionals can also be chosen for parameter estimation; in atmospheric sciences, for example, the following functional could be considered:

$$J\,(p) = J_h\,(p) + J_d\,(p) + J_f\,(p) + J_R\,(p)\,, \qquad (8.13)$$

where $J_h\,(p)$ denotes a weighted least-squares term accounting for the differences between "measured" and computed model parameters (these differences are usually called "residuals"), with weights that are related to the sensitivities of, and/or confidence in, the data; $J_f\,(p)$ denotes a weighted least-squares error between "measured" and computed model parameters at the final time of data assimilation, optimizing improvements in data measurements made at later times; $J_d\,(p)$ denotes a weighted prior data-measurements error term, representing prior knowledge about the parameters, with weights representing the confidence in the measured prior data; and $J_r\,(p)$ denotes a Tikhonov regularization term that regularizes instabilities which may affect values of parameter estimates that are close to the noise in measured data. This term smoothes parameter estimates by imposing a penalty on oscillations in parameter estimates.

8.1.2 Maximum Likelihood (ML) Method for Parameter Estimation

The estimation of parameter values using the maximum likelihood (ML) method is performed by minimizing a criterion (cost functional) expressed in terms of "prior errors," which are usually taken to be the residuals from errors arising from the numerical model and measurements. Assuming that all of the model parameters can be represented by a set of discrete values, and following the argument that the prior errors, ε^*, in the model parameters stem from by a variety of uncorrelated causes, ε^* is invariably assumed to be a Gaussian random vector with zero mean and covariance matrix $\sigma_i^2 V_i$, i.e.,

$$\varepsilon^* = N\left(0, \sigma_i^2 \mathbf{V}_i\right), \tag{8.14}$$

where σ_i^2 is either a known or unknown positive scalar, and $\mathbf{V}_i$ is a known symmetric positive-definite matrix. Furthermore, the prior estimates of various parameter types are also assumed to be uncorrelated, with a block diagonal global covariance matrix denoted here as $\mathbf{C}_p$. The true parameter values are denoted by the vector $\mathbf{p}$, while the measured values (i.e., "prior estimates") are denoted by the vector $\mathbf{p}^*$.

Consider now that the model and parameter data are concatenated in the vector $\mathbf{Z}^* = (\mathbf{x}^*, \mathbf{p}^*)$, while the unknown statistical parameters characterizing prior errors are assembled in the vector $\boldsymbol{\theta}$. Furthermore, consider that the column vector $\boldsymbol{\beta} = (\mathbf{p}, \boldsymbol{\theta})$ denotes the vector of all unknown parameters. Then, the likelihood $L(\boldsymbol{\beta} | \mathbf{Z}^*)$ of a hypothesis regarding the value of $\boldsymbol{\beta}$ given $\mathbf{Z}^*$ and a specific model structure (e.g., a numerical method parameterization) is proportional to the probability density of observing $\mathbf{Z}^*$ if $\boldsymbol{\beta}$ was true; this probability is denoted here as $f(\mathbf{Z}^* | \boldsymbol{\beta})$. For a given model structure, the optimal parameter estimates are taken to be those which maximize $L\left(\widehat{\boldsymbol{\beta}} | \mathbf{Z}^*\right)$, where $\widehat{\boldsymbol{\beta}}$ is an estimate of $\boldsymbol{\beta}$. Assuming that all of the available data has been properly transformed to yield Gaussian distributions for the prior errors, the likelihood function becomes itself a Gaussian of the form

$$L\left(\beta\,|\mathbf{Z}^*\right) = f\left(\mathbf{Z}^*\,|\beta\right)$$

$$= \left(2\pi\right)^{-\frac{1}{2}}|\mathbf{C}_Z|^{-\frac{1}{2}}exp\left[\frac{1}{2}(\mathbf{Z}^* - \mathbf{Z})^T\mathbf{C}_Z^{-1}\left(\mathbf{Z}^* - \mathbf{Z}\right)\right], \qquad (8.15)$$

where $\mathbf{C}_Z$ denotes the covariance matrix of the prior errors defined as

$$\mathbf{C}_Z = \begin{bmatrix} \mathbf{C}_x & \mathbf{0} \\ \mathbf{0} & \mathbf{C}_p \end{bmatrix}, \qquad (8.16)$$

with $\mathbf{C}_x$ and $\mathbf{C}_p$ denoting, respectively, the covariance matrices of the prior model and model parameter errors $\left(\mathbf{C}_x = \sigma_x^2\mathbf{V}_x\right)$.

In practice, maximum likelihood estimates are generally obtained by minimizing the "log-likelihood" function $S = -2ln\left[L\left(\beta\,|\mathbf{Z}^*\right)\right]$, which has the practically useful property that the log-likelihood of a hypothesis, given all the data, is the sum of the log-likelihood of the same hypothesis, given each separate set of data. This property facilitates the stepwise introduction of prior information about the parameters into the estimation procedure in order to analyze data representing conditions created by a variety of initial and boundary conditions.

8.1.3 Maximum Total Variation as an L_1-Regularization Method for Estimation of Parameters with Discontinuities

Standard regularization techniques do not admit discontinuous solutions. The "total variation" (TV) regularization alleviates this restriction, thus allowing consideration of parameter estimation where the location and size of discontinuities are important. The term "blocky profile" refers to functions that are piecewise constant and hence have sharply defined edges. Since 1990, there has been a wide interest in the TV-regularization methods for recovering "blocky," possibly discontinuous images, from noisy data.

The TV-regularization method is used for computing a physically meaningful solution, of bounded variation (i.e., possibly discontinuous, rather than smooth), of the ill-posed operator equation

$$\mathbf{AU} = \mathbf{Z}, \tag{8.17}$$

where $\mathbf{A}$ denotes a linear operator from $L^p(\Omega)$ into a Hilbert space Z containing the data vector $\mathbf{Z}$. Both $\mathbf{A}$ and $\mathbf{Z}$ are considered to be affected by uncertainties (e.g., numerical and experimental errors). The TV-regularization method minimizes a TV-functional $J(\mathbf{U})$ of the form

$$\min_{U} J(\mathbf{U}) ; \quad J(\mathbf{U}) \triangleq \frac{1}{2}\|\mathbf{AU} - \mathbf{Z}\|^2 + \alpha \int_\Omega \sqrt{|\nabla \mathbf{U}|^2 + \beta}. \tag{8.18}$$

Under mild conditions on the operator $\mathbf{A}$, this minimization procedure yields unique minimizer $\mathbf{U}^*$ as the optimal solution to Eq. (8.18), such that $\mathbf{U}^*$ depends continuously on the operator $\mathbf{A}$, on the data $\mathbf{Z}$, and on the parameters α and β. The penalty parameter $\alpha > 0$ controls the trade-off between the goodness of fit to data (measured by $\|\mathbf{AU} - \mathbf{Z}\|^2 = \sigma^2$, where σ^2 represents an estimate of the errors in the data) and the variability at jumps in the approximate solution $\mathbf{U}^*$ (as measured by the term $\int_\Omega \sqrt{|\nabla \mathbf{U}|^2 + \beta}$, $\beta \geq 0$, which represents a semi-norm or a "bounded variation" norm of $\mathbf{U}$). The choice of the penalty parameter α is influenced by the magnitude of the errors in the data (e.g., signal-to-noise ratio); the magnitude of α decreases with decreasing errors.

8.1.4 Parameter Estimation by Extended Kalman Filter

The use of the so-called Extended Kalman Filter (EKF) method can be illustrated by considering a time-evolution model of the form

$$\mathbf{W}_k^f = \boldsymbol{\psi}_{k-1}\left(\mathbf{W}_{k-1}^a\right), \tag{8.19}$$

where $\mathbf{W}_{k-1}^a$ denotes the best estimation of the model's state, $\boldsymbol{\psi}_{k-1}$ denotes the state transition function, and the subscript k is associated with the time-advancing scheme (in this case, a simple two-step scheme). The superscripts a, f, and t will be used in this subsection to indicate "analysis," "forecast," and "true," respectively. In reality, the true system actually transits from one state to the next one according to the evolution equation

$$\mathbf{W}_k^t = \boldsymbol{\psi}_{k-1}\left(\mathbf{W}_{k-1}^t\right) + \mathbf{b}_{k-1}^t, \tag{8.20}$$

where $\mathbf{b}_{k-1}^t$ denotes errors that are considered here to be modeled as a Gaussian white-noise sequence, with zero means and covariances $\mathbf{E}[\mathbf{b}_k^t(\mathbf{b}_l^t)]^T = \mathbf{Q}_k\delta_{k,l}$, with $\mathbf{Q}_k$ denoting the model error covariance.

The evolution of the system is measured by observations that are related to the true state of the system through the relation

$$\mathbf{W}_k^o = \mathbf{h}_k\left(\mathbf{W}_k^t\right) + \mathbf{b}_k^o, \tag{8.21}$$

where the observation errors are denoted by $\mathbf{b}_k^o$, which are also assumed to behave like white noise, with zero means and covariances comprised in a covariance matrix $\mathbf{R}_k$.

Although both the dynamical model $\boldsymbol{\psi}_k$ and the observation $\mathbf{h}_k$ are usually nonlinear operators, it is *assumed that the model state can be estimated based on observations using a linear relation* of the form

$$\mathbf{W}_k^a = \mathbf{W}_k^f + \mathbf{K}_k\left(\mathbf{W}_k^o - \mathbf{W}_k^f\right), \tag{8.22}$$

where $\mathbf{K}_k$ is called the *Kalman gain*, representing weights assigned to observations, which are optimized by minimizing the *trace* of the covariance matrix, $\mathbf{P}_k^a$, of analysis errors, defined as

$$trace\left[\mathbf{P}_k^a\right] = E\left[\left(\mathbf{W}_k^a - \mathbf{W}_k^t\right)^T\left(\mathbf{W}_k^a - \mathbf{W}_k^t\right)\right]. \tag{8.23}$$

At each time step, the EKF *linearly approximates* the operators $\boldsymbol{\psi}_k$ and $\mathbf{h}_k$ along the trajectory in phase space, continually updating the model with "observations." When "observations" are available at time-step k, the *EKF* computational algorithm comprises the following sequence of computations:

1. forecast step:

$$\boldsymbol{\Psi}_{k-1}\left(\mathbf{W}_{k-1}^{a}\right) \triangleq \frac{\partial \psi_{k-1}\left(\mathbf{W}\right)}{\partial \mathbf{W}}\bigg|_{\mathbf{W}=\mathbf{W}_{k-1}^{a}},$$

$$\mathbf{W}_{k}^{f} = \boldsymbol{\Psi}_{k-1}\left(\mathbf{W}_{k-1}^{a}\right),$$

$$\mathbf{P}_{k}^{f} = \boldsymbol{\Psi}_{k-1}\left(\mathbf{W}_{k-1}^{a}\right) \mathbf{P}_{k-1}^{a} \boldsymbol{\Psi}_{k-1}^{T}\left(\mathbf{W}_{k-1}^{a}\right) + \mathbf{Q}_{k-1},$$

2. updating step:

$$\mathbf{H}_{k}\left(\mathbf{W}_{k}^{f}\right) = \frac{\partial \mathbf{h}_{k}\left(\mathbf{W}\right)}{\partial \mathbf{W}}\bigg|_{\mathbf{W}=\mathbf{W}_{k}^{f}},$$

$$\mathbf{K}_{k} = \mathbf{P}_{k}^{f}\mathbf{H}_{k}^{T}\left(\mathbf{W}_{k}^{f}\right) \left[\mathbf{H}_{k}\left(\mathbf{W}_{k}^{f}\right)\mathbf{P}_{k}^{f}H_{k}^{T}\left(\mathbf{W}_{k}^{f}\right) + \mathbf{R}_{k}\right]^{-1},$$

$$\mathbf{W}_{k}^{a} = \mathbf{W}_{k}^{f} + \mathbf{K}_{k}\left[\mathbf{W}_{k}^{o} - \mathbf{h}_{k}\left(\mathbf{W}_{k}^{f}\right)\right],$$

$$\mathbf{P}_{k}^{a} = \left[\mathbf{I} - \mathbf{K}_{k}\mathbf{H}_{k}\left(\mathbf{W}_{k-1}^{a}\right)\right]\mathbf{P}_{k}^{f}.$$

In the above computational algorithm, the matrix that needs to be inverted to compute the Kalman gain $\mathbf{K}_k$ has the same dimension as the number of parameters to be estimated. It is also important to note that the operator $\boldsymbol{\Psi}_k$ is a *linear propagator of forecast error covariance* in time, which acts as the de facto state transition matrix within the EKF algorithm. Thus, in the EKF algorithm, the action of the original nonlinear propagator ψ_k is approximated by the linear operator $\boldsymbol{\Psi}_k$. The computations of $\boldsymbol{\Psi}_k$, H_k, and the Kalman gain $\mathbf{K}_k$ all involve linearization of the respective original (nonlinear) operators at each time step along the phase-space trajectory. Therefore, the parameter estimation based on the EKF method depends on the phase-space trajectory along which the various original operators are linearized. The number of observations required for successful optimal parameter estimation is found to be proportional to the size of the errors in the assimilated state fields (when the parameter error is left uncorrected) and also to the sensitivity of the state-estimation errors to the parameter value. Hence, a correct estimation of "important" model parameters contributes significantly towards improving the estimation of the model's state.

Bibliography

[1] E. Aarts and J. Korst. Simulated annealing and Boltzmann machines. New York, NY; John Wiley and Sons Inc., 1988.

[2] A.K. Alekseev and I. M. Navon. The analysis of an ill-posed problem using multi-scale resolution and second-order adjoint techniques. *Computer Methods in Applied Mechanics and Engineering*, 190(15):1937–1953, 2001.

[3] A.K. Alekseev and I.M. Navon. On estimation of temperature uncertainty using the second order adjoint problem. *International Journal of Computational Fluid Dynamics*, 16(2):113–117, 2002.

[4] R.A. Anthes. Data assimilation and initialization of hurricane prediction models. *Journal of Atmospheric Sciences*, 31:702–719, 1974.

[5] D. Auroux, J. Blum, et al. A nudging-based data assimilation method: the back and forth nudging (BFN) algorithm. *Nonlinear Processes in Geophysics*, 15:305–319, 2008.

[6] F. Baer. Adjustment of initial conditions required to suppress gravity oscillations in nonlinear flows. *Beitraege zur Physik der Atmosphaere*, 50(3):350–366, 1977.

[7] N.L. Baker and R. Daley. Observation and background adjoint sensitivity in the adaptive observation-targeting problem. *Quarterly Journal of the Royal Meteorological Society*, 126(565):1431–1454, 2000.

[8] H.T. Banks and K. Kunisch. *Estimation techniques for distributed parameter systems*. Birkhauser Boston, 1989.

[9] J.W. Bao and Y.H. Kuo. On-off switches in the adjoint method: step functions. *Monthly Weather Review*, 123(5):1589–1594, 1995.

[10] G.A. Barnard. Studies in the history of probability and statistics: IX. Thomas Bayes's essay towards solving a problem in the doctrine of chances. *Biometrika,* 45(3/4):293–315, 1958.

[11] S.L. Barnes. A technique for maximizing details in numerical weather map analysis. *Journal of Applied Meteorology,* 3(4):396–409, 1964.

[12] E.M.L. Beale. A derivation of conjugate gradients. *Numerical methods for nonlinear optimization,* pages 39–43, Academic Press, London, 1972.

[13] R.E. Bellman and R.E. Kalaba. Quasilinearization and nonlinear boundary-value problems. *Elsevier, New York,* 1965.

[14] A.F. Bennett. *Inverse methods in physical oceanography.* Cambridge University Press, 1992.

[15] J.O. Berger. *Statistical decision theory and Bayesian analysis,* 2nd edition. *Springer, New York,* 1985.

[16] J. Bernoulli. *Die Werke von Jakob Bernoulli: Bd. 3: Wahrscheinlichkeitsrechnung,* volume 2. Birkhäuser Basel, 1975.

[17] J.F. Bonnans, J.C. Gilbert, C. Lemaréchal, and C. Sagastizábal. *Optimisation Numérique: aspects théoriques et pratiques.* Springer, 1997.

[18] D.G. Bounds. New optimization methods from physics and biology. *Nature,* 329:215–219, 1987.

[19] F. Bouttier and P. Courtier. Data assimilation concepts and methods march 1999. *Meteorological training course lecture series. ECMWF,* 2002.

[20] C.G. Broyden. The convergence of a class of double-rank minimization algorithms 1. general considerations. *IMA Journal of Applied Mathematics,* 6(1):76–90, 1970.

[21] A. Buckley and A. Lenir. QN-like variable storage conjugate gradients. *Mathematical programming,* 27(2):155–175, 1983.

[22] D.G. Cacuci. Sensitivity theory for nonlinear systems. i. nonlinear functional analysis approach. *Journal of Mathematical Physics,* 22:2794, 1981.

[23] D.G. Cacuci. Sensitivity theory for nonlinear systems. II. extensions to additional classes of responses. *Journal of Mathematical Physics*, 22:2803, 1981.

[24] D.G. Cacuci. Global optimization and sensitivity analysis. *Nuclear Science and Engineering*, 104(1):78–88, 1990.

[25] D.G. Cacuci and M.C.G. Hall. Physical interpretation of the adjoint functions for sensitivity analysis of atmospheric models. *Journal of the Atmospheric Sciences*, 40(10):2537–2546, 1983.

[26] D.G. Cacuci and M.C.G. Hall. Systematic analysis of climatic model sensitivity to parameters and processes. *Geophysical Monograph Series*, 29:171–179, 1984.

[27] D.G. Cacuci, M.C.G. Hall, and M.E. Schlesinger. Sensitivity analysis of a radiative-convective model by the adjoint method. *Journal of the Atmospheric Sciences*, 39(9):2038–2050, 1982.

[28] D.G. Cacuci and M. Ionescu-Bujor. On the evaluation of discrepant scientific data with unrecognized errors. *Nuclear Science and Engineering*, 165(1):1–17, 2010.

[29] D.G. Cacuci, M. Ionescu-Bujor, and I.M. Navon. *Sensitivity and Uncertainty Analysis:Applications to Large-Scale Systems*, volume II. CRC, 2005.

[30] M.R. Celis, J.E. Dennis, and Tapia. A trust region strategy for nonlinear equality constrained optimization. In *Proceedings of the SIAM Conference on Numerical Optimization, Edited by Paul T. Boggs, Richard H. Byrd and Robert B. Schnabel*, pages 71–82. SIAM, 1985.

[31] G. Chavent and P. Lemonnier. Identification de la non-linéarité d'une équation parabolique quasilinéaire. *Applied Mathematics & Optimization*, 1(2):121–162, 1974.

[32] X. Chen and I.M. Navon. Optimal control of a finite-element limited-area shallow-water equations model. *Studies in Informatics and Control*, 18(1):41–62, 2009.

[33] S.E. Cohn. An introduction to estimation theory. *Journal of the Meteorological Society of Japan SERIES 2*, 75:147–178, 1997.

[34] A.R. Conn, G.I.M. Gould, and P.L. Toint. *LANCELOT: a Fortran package for large-scale nonlinear optimization (Release A)*. Springer Publishing Company, Inc., 1992.

[35] A.R. Conn, N.I.M. Gould, and P.L. Toint. Trust-region methods: MPS-SIAM series on optimization. *Society of Industrial and Applied Mathematics*, Philadelphia, 2000.

[36] R. Courant and D. Hilbert. *Methods of mathematical physics. Vol. II: Partial differential equations*. Interscience, New York, 1962.

[37] P. Courtier and O. Talagrand. Variational assimilation of meteorological observations with the direct and adjoint shallow-water equations. *Tellus A*, 42(5):531–549, 1990.

[38] P. Courtier, J.N. Thépaut, and A. Hollingsworth. A strategy for operational implementation of 4D-VAR, using an incremental approach. *Quarterly Journal of the Royal Meteorological Society*, 120(519):1367–1387, 1994.

[39] R.T. Cox. Probability, frequency and reasonable expectation. *American Journal of Physics*, 14(1):1–13, 1946.

[40] G.P. Cressman. An operational objective analysis system. *Monthly Weather Review*, 87(10):367–374, 1959.

[41] A. Da Silva, J. Pfaendtner, J. Guo, M. Sienkiewicz, and S.E. Cohn. Assessing the effects of data selection with the daos physical-space statistical analysis system, 1995. In *Proceedings of the second international symposium on the assimilation of observations in meteorology and oceanography, Tokyo, Japan, WMO and JMA*.

[42] D.N. Daescu and I.M. Navon. An analysis of a hybrid optimization method for variational data assimilation. *International Journal of Computational Fluid Dynamics*, 17(4):299–306, 2003.

[43] R. Daley. Atmospheric data analysis, Cambridge atmospheric and space science series. *Cambridge University Press*, page 457, 1991.

[44] R. Daley. The effect of serially correlated observation and model error on atmospheric data assimilation. *Monthly Weather Review*, 120(1):164–177, 1992.

[45] C. Darwin. On the origins of species by means of natural selection. London: Murray, 1859.

[46] B. Das, H. Meirovitch, and I.M. Navon. Performance of enriched methods for large scale unconstrained optimization as applied to models of proteins. *Journal of Computational Chemistry*, 24(10):1222, 2003.

[47] W.C. Davidon. Variable metric method for minimization. *SIAM Journal on Optimization*, 1(1):1–17, 1991.

[48] R. Dawkins. *The blind watchmaker: Why the evidence of evolution reveals a universe without design.* W.W. Norton & Company, 1986.

[49] K.A. De Jong. Analysis of the behavior of a class of genetic adaptive systems. *Doctoral dissertation, University of Michigan*, 1975.

[50] K. Deb. An efficient constraint handling method for genetic algorithms. *Computer Methods in Applied Mechanics and Engineering*, 186(2):311–338, 2000.

[51] D.P. Dee. Bias and data assimilation. *Quarterly Journal of the Royal Meteorological Society*, 131(613):3323–3343, 2005.

[52] R.S. Dembo and T. Steihaug. Truncated-Newton algorithms for large-scale unconstrained optimization. *Mathematical Programming*, 26(2):190–212, 1983.

[53] J.E. Dennis, M. Heinkenschloss, and L.N. Vicente. Trust-region interior-point SQP algorithms for a class of nonlinear programming problems. *SIAM Journal on Control and Optimization*, 36(5):1750–1794, 1998.

[54] J.C. Derber. The variational four-dimensional assimilation of analyses using filtered models as constraints. *Ph.D. Thesis, University of Wisconsin-Madison*, 1985.

[55] J.C. Derber. A variational continuous assimilation technique. *Monthly Weather Review*, 117:2437–2446, 1989.

[56] J.C. Derber, D.F. Parrish, and S.J. Lord. The new global operational analysis system at the national meteorological center. *Weather and Forecasting*, 6(4):538–547, 1991.

[57] J.C. Derber, D.F. Parrish, and J.G. Sela. The SSI analysis system and extensions to 4–D. In *Proceedings of ECMWF Workshop on variational assimilation with emphasis on three-dimensional aspects, Shinfield Park, Reading RG2 9AX, UK*, pages 15–29, 1992.

[58] A. Eliassen. Provisional report on calculation of spatial covariance and autocorrelation of the pressure field. inst. *Weather and Climate Research, Academy of Science*, Oslo, Report No5, 1954.

[59] M. Fisher. Background error covariance modelling. In *Seminar on Recent Development in Data Assimilation for Atmosphere and Ocean*, pages 45–63. European Centre for Medium-Range Weather Forecasts, 2003.

[60] M. Fisher. *Estimation of entropy reduction and degrees of freedom for signal for large variational analysis systems*. European Centre for Medium-Range Weather Forecasts, London, 2003.

[61] M. Fisher. Assimilation algorithms lecture 3: 4–D VAR. *European Centre for Medium-Range Weather Forecasts*, London, 2010.

[62] M. Fisher, J. Nocedal, Y. Trémolet, and S.J. Wright. Data assimilation in weather forecasting: a case study in PDE-constrained optimization. *Optimization and Engineering*, 10(3):409–426, 2009.

[63] R. Fletcher. A new approach to variable metric algorithms. *The Computer Journal*, 13(3):317–322, 1970.

[64] R. Fletcher. A model algorithm for composite nondifferentiable optimization problems. *Nondifferential and Variational Techniques in Optimization, Mathematical Programming Studies*, 17:67–76, Springer,1982.

[65] R. Fletcher, S. Leyffer, and P.L. Toint. On the global convergence of a filter–SQP algorithm. *SIAM Journal on Optimization*, 13(1):44–59, 2002.

[66] R. Fletcher and C.M. Reeves. Function minimization by conjugate gradients. *The Computer Journal*, 7(2):149–154, 1963.

[67] W. Freeden and U. Windheuser. Spherical wavelet transform and its discretization. *Advances in Computational Mathematics*, 5(1):51–94, 1996.

[68] F.H. Fröhner. Assigning uncertainties to scientific data. *Nuclear Science and Engineering*, 126(1):1–18, 1997.

[69] F.H. Fröhner. Evaluation of data with systematic errors. *Nuclear Science and Engineering*, 145(3):342–353, 2003.

[70] L.S. Gandin. *Objective analysis of meteorological fields, Translated from the Russian.* Jerusalem (Israel Program for Scientific Translations), 1965.

[71] G. Gaspari and S.E. Cohn. Construction of correlation functions in two and three dimensions. *Quarterly Journal of the Royal Meteorological Society*, 125(554):723–757, 2006.

[72] D.M. Gay. Computing optimal locally constrained steps. *SIAM Journal on Scientific and Statistical Computing*, 2(2):186–197, 1981.

[73] J.K. Gibson, P. Kallberg, S. Uppala, A. Nomura, E. Serrano, and A. Hernandez. ERA description. ECMWF reanalysis project report 1: project organisation. Technical report, ECMWF, European Centre of Medium-Range Weather Forecasts, Reading, 1997.

[74] J.C. Gilbert and C. Lemaréchal. Some numerical experiments with variable-storage Quasi-Newton algorithms. *Mathematical Programming*, 45(1):407–435, 1989.

[75] P.E. Gill and W. Murray. Conjugate-gradient methods for large-scale nonlinear optimization. *Defense Technical Information Center*, page 64, 1979.

[76] P.E. Gill, W. Murray, and M.H. Wright. *Practical optimization.* Addison-Wesley Professional Academic Press, 1981.

[77] D.E. Goldberg. Genetic algorithms in search, optimization, and machine learning. *Addison-Wesley Professional,* 1st edition:432, 1989.

[78] D. Goldfarb. A family of variable metric methods derived by variational means. *Mathematics of Computation,* 24(109):23–26, 1970.

[79] G.H. Golub and C.F. Van Loan. *Matrix computations.* John Hopkins University Press, 1989.

[80] A. Grammeltvedt. A survey of finite-difference schemes for the primitive equations for a barotropic fluid. *Monthly Weather Review,* 97(5):384–404, 1969.

[81] S. Gratton, A.S. Lawless, and N.K. Nichols. Approximate Gauss-Newton methods for nonlinear least squares problems. *SIAM Journal on Optimization,* 18(1):106–132, 2007.

[82] J.J. Grefenstette. Optimization of control parameters for genetic algorithms. *Systems, IEEE Transactions on Systems, Man and Cybernetics,* 16, 1986.

[83] A. Griewank and P.L. Toint. Partitioned variable metric updates for large structured optimization problems. *Numerische Mathematik,* 39(1):119–137, 1982.

[84] A. Griewank and A. Walther. Evaluating derivatives: principles and techniques of algorithmic differentiation. *Society for Industrial and Applied Mathematics (SIAM),* page 460, 2008.

[85] A.K. Griffith and N.K. Nichols. Accounting for model error in data assimilation using adjoint methods. *Computational Differentiation: Techniques, Applications and Tools,* pages 195–204, 1996.

[86] A.K. Griffith and N.K. Nichols. Adjoint methods in data assimilation for estimating model error. *Flow, Turbulence and Combustion,* 65(3):469–488, 2000.

[87] M.R. Hestenes. Multiplier and gradient methods. *Journal of Optimization Theory and Applications*, 4(5):303–320, 1969.

[88] R.N. Hoffman. A four-dimensional analysis exactly satisfying equations of motion. *Monthly Weather Review*, 114(2):388–397, 1986.

[89] J.H. Holland. *Adaptation in natural and artificial systems: an introductory analysis with applications to biology, control and artificial intelligence*. University of Michigan Press, 1975.

[90] A. Hollingsworth and P. Lönnberg. The statistical structure of short-range forecast errors as determined from radiosonde data. Part I: the wind field. *Tellus A*, 38(2):111–136, 1986.

[91] R.B. Hollstein. Artificial genetic adaptation in computer control systems. Doctoral dissertation, University of Michigan, Dissertation Abstracts International 32.(3), 1510B, 1971.

[92] C. Homescu and I.M. Navon. Optimal control of flow with discontinuities. *Journal of Computational Physics*, 187(2):660–682, 2003.

[93] B. Ingleby. The statistical structure of forecast errors and its representation in the meteorological office global model. *Quarterly Journal of the Royal Meteorological Society*, 124:1783–1807, 2001.

[94] C. Jakob. Ice clouds in numerical weather prediction models–progress, problems and prospects. pages 327–345, In *Cirrus* edited by David K. Lynch et al., Oxford University Press, 327-345, 2002.

[95] M. Janisková, J. Mahfouf, J. Morcrette, and F. Chevallier. Linearized radiation and cloud schemes in the ECMWF model: development and evaluation. *Quarterly Journal of the Royal Meteorological Society*, 128(583):1505–1527, 2002.

[96] M. Janisková, J.N. Thépaut, and J.F. Geleyn. Simplified and regular physical parameterizations for incremental four-dimensional variational assimilation. *Monthly Weather Review*, 127(1):26–45, 1999.

[97] T. Janjic and S.E. Cohn. Treatment of observation error due to unresolved scales in atmospheric data assimilation. *Monthly Weather Review*, 134(10):2900–2915, 2006.

[98] E.T. Jaynes. Prior probabilities. *IEEE Transactions on Systems Science and Cybernetics*, 4(3):227–241, 1968.

[99] E.T. Jaynes. Probability, statistics and statistical physics. *Dordrecht: Reidel*, 1983.

[100] A.H. Jazwinski. Stochastic processes and filtering theory. *Mathematics in Science and Engineering*, 64:376, 1970.

[101] E. Kalnay. *Atmospheric modeling, data assimilation and predictability*. Cambridge University Press, 2003.

[102] N.M.S. Karmitsa and M.M. Mäkelä. Adaptive limited memory bundle method for bound constrained large-scale nonsmooth optimization. *Optimization*, 59(6):945–962, 2009.

[103] N.M.S. Karmitsa, M.M. Mäkelä, and M.M. Ali. Limited memory interior point bundle method for large inequality constrained nonsmooth minimization. *Applied Mathematics and Computation*, 198(1):382–400, 2008.

[104] S. Kirkpatrick. Optimization by simulated annealing: quantitative studies. *Journal of Statistical Physics*, 34, Issues 5–6:975–986, 1984.

[105] S. Kirkpatrick, M.P. Vecchi, et al. *Optimization by simulated annealing*. Research Report RC 9355, IBM (Yorktown Heights, NY), 1982.

[106] K.C. Kiwiel. *Methods of descent for nondifferentiable optimization, lecture Notes in Mathematics*. Number 1133. Springer-Verlag Berlin, 1985.

[107] K.C. Kiwiel. Proximity control in bundle methods for convex nondifferentiable minimization. *Mathematical Programming*, 46(1):105–122, 1990.

[108] A.N. Kolmogorov, W.L. Doyle, and I. Selin. Interpolation and extrapolation of stationary random sequences. *RAND Reports and Bookstore, Research Memoranda*, RM(3090), 1962.

[109] D.G. Krige. *A statistical approach to some mine valuation and allied problems on the Witwatersrand*. University of the Witwatersrand, 1951.

[110] T.N. Krishnamurti, J. Xue, H.S. Bedi, K. Ingles, and D. Oosterhof. Physical initialization for numerical weather prediction over the tropics. *Tellus B*, 43(4):53–81, 1991.

[111] P.S. Laplace. *Théorie analytique des probabilités*. Paris, V. Courcier, 1812.

[112] S. Laroche and P. Gauthier. A validation of the incremental formulation of 4D variational data assimilation in a nonlinear barotropic flow. *Tellus A*, 50(5):557–572, 1998.

[113] A.S. Lawless. A note on the analysis error associated with 3D-FGAT. *Quarterly Journal of the Royal Meteorological Society*, 136(649):1094–1098, 2010.

[114] A.S. Lawless, S. Gratton, and N.K. Nichols. An investigation of incremental 4D-VAR using non-tangent linear models. *Quarterly Journal of the Royal Meteorological Society*, 131(606):459–476, 2005.

[115] A.S. Lawless and N.K. Nichols. Inner-loop stopping criteria for incremental four-dimensional variational data assimilation. *Monthly Weather Review*, 134(11):3425–3435, 2006.

[116] F.X. Le Dimet. Une application des méthodes de contrôle, optimal 'a l'analyse varationelle. *Report, LAMP, Université Blaise-Pascal*, 63170, 1981.

[117] F.X. Le Dimet, I.M. Navon, and D.N. Daescu. Second-order information in data assimilation. *Monthly Weather Review*, 130(3):629–648, 2002.

[118] F.X. Le Dimet and O. Talagrand. Variational algorithms for analysis and assimilation of meteorological observations: theoretical aspects. *Tellus A*, 38(2):97–110, 1986.

[119] D. Legler and I.M. Navon. VARIATM - A Fortran program for objective analysis of pseudo-stress wind fields using large-scale conjugate-gradient minimization. *Computers and Geosciences*, 17(1):1–21, 1991.

[120] D. Legler, I.M. Navon, and J.J. O'Brien. Objective analysis of pseudostress over the Indian Ocean using a direct-minimization approach. *Monthly Weather Review*, 117(8):709–720, 1989.

[121] C. Lemaréchal. Nondifferentiable optimization, subgradient and subgradient methods. *Optimization and Operations Research, Lecture Notes in Economics and Mathematical Systems*, 117:191, 1976.

[122] C. Lemaréchal. Nondifferentiable optimization. *Handbooks in Operations Research and Management Science*, 1:529–572, 1989.

[123] J.M. Lewis, S. Lakshmivarahan, and S. Dhall. *Dynamic data assimilation: a least squares approach*, volume 104. Cambridge University Press, 2006.

[124] Y. Li, I.M. Navon, P. Courtier, and P. Gauthier. Variational data assimilation with a semi-lagrangian semi-implicit global shallow-water equation model and its adjoint. *Monthly Weather Review*, 121(6):1759–1769, 1993.

[125] Z. Li, I.M. Navon, and Y. Zhu. Performance of 4–D VAR with different strategies for the use of adjoint physics with the FSU global spectral model. *Monthly Weather Review*, 128(3):668–688, 2000.

[126] D.C. Liu and J. Nocedal. On the limited memory BFGS method for large scale optimization. *Mathematical Programming*, 45(1):503–528, 1989.

[127] A.C. Lorenc. Analysis methods for numerical weather prediction. *Quarterly Journal of the Royal Meteorological Society*, 112(474):1177–1194, 1986.

[128] D.G. Luenberger. *Linear and nonlinear programming*. Addison-Wesley Publishing Company, 1984.

[129] L. Lukšan and J. Vlček. NDA: Algorithms for nondifferentiable optimization. *Technical Report*, V-797, ICS AS CR, 2000.

[130] P. Lynch and X. Huang. Initialization of the HIRLAM model using a digital filter. *Monthly Weather Review*, 120(6):1019–1034, 1992.

[131] P. Lynch and A. McDonald. A multi-level limited-area slow-equation model: application to initialization. *Quarterly Journal of the Royal Meteorological Society*, 116(493):595–609, 1990.

[132] B. Machenhauer. On the dynamics of gravity oscillations in a shallow water model with applications to normal mode initialization. *Beiträge zur Physik Atmosphäre*, 50:253–271, 1977.

[133] J.F. Mahfouf and F. Rabier. The ECMWF operational implementation of four-dimensional variational assimilation. II: experimental results with improved physics. *Quarterly Journal of the Royal Meteorological Society*, 126(564):1171–1190, 2000.

[134] M.M. Makela and P. Neittaanmäki. *Nonsmooth optimization: analysis and algorithms with applications to optimal control*. World Scientific Singapore, 1992.

[135] N. Metropolis, A.W. Rosenbluth, M.N. Rosenbluth, A.H. Teller, and E. Teller. Equation of state calculations by fast computing machines. *The Journal of Chemical Physics*, 21:1087, 1953.

[136] K. Miyakoda and R.W. Moyer. A method of initialization for dynamical weather forecasting. *Tellus*, 20(1):115–128, 1968.

[137] J.L. Morales and J. Nocedal. Automatic preconditioning by limited memory quasi-Newton updating. *SIAM Journal on Optimization*, 10(4):1079–1096, 1999.

[138] J.L. Morales and J. Nocedal. Algorithm 809: Preqn: Fortran 77 subroutines for preconditioning the conjugate gradient method. *ACM Transactions on Mathematical Software (TOMS)*, 27(1):83–91, 2001.

[139] S.G. Nash. Truncated-Newton methods. *Ph.D. Thesis, Computer Science Departement, Stanford University*, 1982.

[140] S.G. Nash. Newton-type minimization via the Lanczos method. *SIAM Journal on Numerical Analysis*, 21(4):770–788, 1984.

[141] S.G. Nash. Solving nonlinear programming problems using Truncated-Newton techniques. *Numerical Optimization*, pages 119–136, 1984.

[142] S.G. Nash. Truncated-Newton methods for large-scale function minimization. *Application of Nonlinear Programming to Optimization and Control*, page 91, 1984.

[143] S.G. Nash. User's guide for TN. *TNBC: Fortran Routines for Nonlinear Optimization Report*, 397, 1984.

[144] S.G. Nash. Preconditioning of Truncated-Newton methods. *SIAM Journal on Scientific and Statistical Computing*, 6(3):599–616, 1985.

[145] S.G. Nash and J. Nocedal. A numerical study of the limited memory BFGS method and the Truncated-Newton method for large scale optimization. *SIAM Journal on Optimization*, 1(3):358–372, 1991.

[146] S.G. Nash and A. Sofer. Block Truncated-Newton methods for parallel optimization. *Mathematical Programming*, 45(1):529–546, 1989.

[147] S.G. Nash and A. Sofer. Assessing a search direction within a Truncated-Newton method. *Operations Research Letters*, 9(4):219–221, 1990.

[148] S.G. Nash and A. Sofer. *Linear and nonlinear programming*. McGraw-Hill Science/Engineering/Math, 1996.

[149] I.M. Navon, F.B. Brown, and D.H. Robertson. A combined simulated annealing and quasi-Newton-like conjugate-gradient method for determining the structure of mixed argon-xenon clusters. *Computers & Chemistry*, 14(4):305–311, 1990.

[150] I.M. Navon and R. De Villiers. Combined penalty multiplier optimization methods to enforce integral invariants conservation. *Monthly Weather Review*, 111(6):1228–1243, 1983.

[151] I.M. Navon and D. Legler. Conjugate-gradient methods for large-scale minimization in meteorology. *Monthly Weather Review*, 115(4):1479–1502, 1987.

[152] I.M. Navon, P.K.H. Phua, and M. Ramamurthy. Vectorization of conjugate-gradient methods for large-scale minimization in meteorology. *Journal of Optimization Theory and Applications*, 66(1):71–93, 1990.

[153] I.M. Navon, X. Zou, J. Derber, and J. Sela. Variational data assimilation with an adiabatic version of the NMC spectral model. 120(7):1443–1446, 1992.

[154] J. Nocedal. Theory of algorithms for unconstrained optimization. *Acta Numerica*, 1:199–242, 1992.

[155] J. Nocedal and S.J. Wright. *Numerical optimization.* Springer Verlag, 2006.

[156] J.M. Ortega and W.C. Rheinboldt. *Iterative solution of nonlinear equations in several variables.* Academic Press, New York, 1970.

[157] D.F. Parrish and J.C. Derber. The national meteorological center's spectral statistical-interpolation analysis system. *Monthly Weather Review*, 120(8):1747–1763, 1992.

[158] V. Penenko and N.N. Obraztsov. A variational initialization method for the fields of the meteorological elements. *Meteorologiya i Gidrologiya*, 11:1–11, 1976.

[159] N.A. Phillips. On the problem of initial data for the primitive equations. *Tellus*, 12:121–126, 1960.

[160] M. Pincus. Letter to the editor: A Monte Carlo Method for the approximate solution of certain types of constrained optimization problems. *Operations Research*, 18(6):1225–1228, 1970.

[161] E. Polak and G. Ribiere. Note sur la convergence de méthodes de directions conjuguées. *Revue française d'informatique et de recherche opérationnelle*, 16:35–43, 1969.

[162] S. Polavarapu, M. Tanguay, and L. Fillion. Four-dimensional variational data assimilation with digital filter initialization. *Monthly Weather Review*, 128(7):2491–2510, 2000.

[163] M.J.D. Powell. Rank one methods for unconstrained optimization. Technical report, Atomic Energy Research Establishment, Harwell (England), 1969.

[164] M.J.D. Powell. Restart procedures for the conjugate gradient method. *Mathematical Programming*, 12(1):241–254, 1977.

[165] F. Rabier, H. Järvinen, E. Klinker, J.F. Mahfouf, and A. Simmons. The ECMWF operational implementation of four-dimensional variational assimilation. i: Experimental results with simplified physics. *Quarterly Journal of the Royal Meteorological Society*, 126(564):1143–1170, 2000.

[166] M.K. Ramamurthy and F.H. Carr. Four-dimensional data assimilation in the monsoon region. i: experiments with wind data. *Monthly Weather Review*, 115(8):1678–1706, 1987.

[167] M.K. Ramamurthy and I.M. Navon. The conjugate-gradient variational analysis and initialization method: an application to MONEX SOP-2 data. *Monthly Weather Review*, 120(10):2360–2377, 1992.

[168] A. Renyi. Probability calculus, Tankonyvkiado, Budapest, 1954; German transl. *Hochschulbucher fur Mathematik, VEB Deutscher Verlag*, 54, 1962.

[169] A.H.G. Rinnooy Kan and G.T. Timmer. Chapter IX. Global Optimization. *Handbooks in Operations Research and Management Science*, 1:631–662, 1989.

[170] C. Robert, E. Blayo, and J. Verron. Comparison of reduced-order, sequential and variational data assimilation methods in the tropical Pacific Ocean. *Ocean Dynamics*, 56(5):624–633, 2006.

[171] C. Robert, E. Blayo, and J. Verron. Reduced-order 4–D VAR: A preconditioner for the incremental 4–D VAR data assimilation method. *Geophysical Research Letters*, 33:L18609, 2006.

[172] Kirkpatrick S., C.D. Jr. Gerlatt, and M.P. Vecchi. Optimization by simulated annealing. *IBM Research Report*, (9355), 1982.

[173] Y. K. Sasaki. Proposed inclusion of time variation terms, observational and theoretical, in numerical variational objective analysis. *Journal Meteorological Society Japan*, 47(2):115–124, 1969.

[174] Y. K. Sasaki. Numerical variational analysis formulated under the constraints as determined by longwave equations and a low-pass filter. *Monthly Weather Review*, 98(12):884–898, 1970.

[175] Y. K. Sasaki. Numerical variational analysis with weak constraint and application to surface analysis of severe storm gust. *Monthly Weather Review*, 98(12):899–910, 1970.

[176] Y. K. Sasaki. Some basic formalisms in numerical variational analysis. *Monthly Weather Review*, 98(12):875–883, 1970.

[177] Y. K.K. Sasaki. A fundamental study of the numerical prediction based on the variational principle. *Journal Meteorological Society Japan*, 33(6):262–275, 1955.

[178] T. Schlick. Modified Cholesky factorizations for sparse preconditioners. *SIAM Journal on Scientific Computing*, 14(2):424–445, 1993.

[179] T. Schlick and A. Fogelson. Algorithm 702: TNPACK–a Truncated Newton minimization package for large-scale problems: I. algorithm and usage. *ACM Transactions on Mathematical Software (TOMS)*, 18(2):141, 1992.

[180] T. Schlick and A. Fogelson. TNPACK: A Truncated Newton minimization package for large-scale problems: II. implementation examples. *ACM Transactions on Mathematical Software (TOMS)*, 18(1):71–111, 1992.

[181] T. Schlick and M. Overton. A powerful Truncated Newton method for potential energy minimization. *Journal of Computational Chemistry*, 8(7):1025–1039, 1987.

[182] H. Schramm and J. Zowe. A version of the bundle idea for minimizing a nonsmooth function: conceptual idea, convergence analysis, numerical results. *SIAM Journal on Optimization*, 2(1):121–152, 1992.

[183] E. Schrödinger. The foundation of the theory of probability: I. In *Proceedings of the Royal Irish Academy. Section A: Mathematical and Physical Sciences*, volume 51, pages 51–66, 1945.

[184] F.H.M. Semazzi and I.M. Navon. A comparison of the bounded derivative and the normal-mode initialization methods using real data. *Monthly Weather Review*, 114:2106–2121, 1986.

[185] D.F. Shanno. Conjugate gradient methods with inexact searches. *Mathematics of Operations Research*, 3(3):244–256, 1978.

[186] D.F. Shanno. On the convergence of a new conjugate gradient algorithm. *SIAM Journal on Numerical Analysis*, 15(6):1247–1257, 1978.

[187] D.F. Shanno et al. Conditioning of quasi-Newton methods for function minimization. *Mathematics of Computation*, 24(111):647–656, 1970.

[188] D.F. Shanno and K.H. Phua. Remark on algorithm 500-a variable method subroutine for unconstrained nonlinear minimization. *ACM Transactions on Mathematical Software*, 6(4):618–622, 1980.

[189] E. Shannon. A mathematical theory of evidence:. *Bell System Technical Journal*, 27:379–423, 1948.

[190] N.Z. Shor, K.C. Kiwiel (translator), and R. Ruszczynski (translator). Minimization methods for non-differentiable functions. *Springer series in computational mathematics, 3, Berlin*, page 172, 1985.

[191] D.C. Sorensen. Newton's method with a model trust region modification. *SIAM Journal on Numerical Analysis*, 19(2):409–426, 1982.

[192] J.L. Steward, I.M. Navon, M. Zupanski, and N. Karmitsa. Impact of nonsmooth observation operators on variational and sequential data assimilation for a limited-area shallow-water equation model. *Quarterly Journal of the Royal Meteorological Society*, 138(663):323–339, Part B 2012.

[193] O. Talagrand and P. Courtier. Variational assimilation of meteorological observations with the adjoint vorticity equation. I: Theory. *Quarterly Journal of the Royal Meteorological Society*, 113(478):1311–1328, 1987.

[194] J.N. Thépaut and P. Courtier. Four-dimensional variational data assimilation using the adjoint of a multilevel primitive-equation model. *Quarterly Journal of the Royal Meteorological Society*, 117(502):1225–1254, 1991.

[195] P.D. Thompson. Reduction of analysis error through constraints of dynamical consistency. *Journal of Applied Meteorology*, 8:738–742, 1969.

[196] M. Tiedtke. Representation of clouds in large-scale models. *Monthly Weather Review*, 121(11):3040–3061, 1993.

[197] A.N. Tikhonov. Regularization of non-linear ill-posed problems. *Doklady Akademii Nauk*, 49(4), 1963.

[198] A.M. Tompkins and M. Janisková. A cloud scheme for data assimilation: description and initial tests. *Quarterly Journal of the Royal Meteorological Society*, 130(602):2495–2517, 2004.

[199] Y. Trémolet. Diagnostics of linear and incremental approximations in 4-D VAR. *Quarterly Journal of the Royal Meteorological Society*, 130(601):2233–2251, 2004.

[200] Y. Trémolet. Incremental 4-D VAR convergence study. *Tellus A*, 59(5):706–718, 2007.

[201] Y. Trémolet. Computation of observation sensitivity and observation impact in incremental variational data assimilation. *Tellus A*, 60(5):964–978, 2008.

[202] M. Tribus. Rational descriptions, decisions, and designs... *Elsevier Science and Technology xix*, page 478, 1969.

[203] J. Tshimanga, S. Gratton, A.T. Weaver, and A. Sartenaer. Limited-memory preconditioners, with application to incremental four-dimensional variational data assimilation. *Quarterly Journal of the Royal Meteorological Society*, 134(632):751–769, 2008.

[204] T. Tsuyuki. Variational data assimilation in the tropics using precipitation data. part III: Assimilation of SSM/I precipitation rates. *Monthly Weather Review*, 125(7):1447–1464, 1997.

[205] S.P. Uryasev. New variable-metric algorithms for nondifferentiable optimization problems. *Journal of Optimization Theory and Applications*, 71(2):359–388, 1991.

[206] F. Veerse and J.N. Thépaut. Multiple-truncation incremental approach for four-dimensional variational data assimilations. *Quarterly Journal of the Royal Meteorological Society*, 124(550):1889–1908, 1998.

[207] P.A. Vidard, E. Blayo, F.X. Le Dimet, and A. Piacentini. 4–D variational data analysis with imperfect model. *Flow, Turbulence and Combustion*, 65(3):489–504, 2000.

[208] P.A. Vidard, A. Piacentini, and F.X. Le Dimet. Variational data analysis with control of the forecast bias. *Tellus A*, 56(3):177–188, 2004.

[209] J. Vlček and L. Lukšan. Globally convergent variable metric method for nonconvex nondifferentiable unconstrained minimization. *Journal of Optimization Theory and Applications*, 111(2):407–430, 2001.

[210] T. Vukićević and R.M. Errico. Linearization and adjoint of parameterized moist diabatic processes. *NCAR*, 1030(92), 1992.

[211] A. Wald. *Statistical decision functions*. New York, John Wiley, 1950.

[212] Z. Wang, K.K. Droegemeier, L. White, and I.M. Navon. Application of a new adjoint newton algorithm to the 3–D ARPS storm-scale model using simulated data. *Monthly Weather Review*, 125(10):2460–2478, 1997.

[213] Z. Wang, K.K. Droegemeier, M. Xue, S.K. Park, J.G. Michalakes, and X. Zou. Sensitivity analysis of a 3–D compressible storm-scale model to input parameters. In *Preprints, Proceedings of the International Symposium on Assimilation of Observations in Meteorology and Oceanography, Tokyo, Japan*, 1995.

[214] Z. Wang, I.M. Navon, X. Zou, and F.X. Le Dimet. A Truncated Newton optimization algorithm in meteorology applications with analytic hessian/vector products. *Computational Optimization and Applications*, 4(3):241–262, 1995.

[215] A. Weaver and P. Courtier. Correlation modelling on the sphere using a generalized diffusion equation. *Quarterly Journal of the Royal Meteorological Society*, 127(575):1815–1846, 2001.

[216] N. Weiner. Extrapolation, interpolation, and smoothing of stationary time series With Engineering Applications. *MIT Press*, page 166, 1949.

[217] P. Wolfe. Convergence conditions for ascent methods. *SIAM Review*, 11(2):226–235, 1969.

[218] P. Wolfe. Convergence conditions for ascent methods. II: some corrections. *SIAM Review*, 13(2):185–188, 1971.

[219] Q. Xu. Generalized adjoint for physical processes with parameterized discontinuities. part I: basic issues and heuristic examples. *Journal of the Atmospheric Sciences*, 53(8):1123–1142, 1996.

[220] Q. Xu. Generalized adjoint for physical processes with parameterized discontinuities. part IV: problems in time discretization. *Journal of the Atmospheric Sciences*, 54(23):2722–2728, 1997.

[221] Q. Xu, J. Gao, and W. Gu. Generalized adjoint for physical processes with parameterized discontinuities. part V: Coarse-grain adjoint and problems in gradient check. *Journal of the Atmospheric Sciences*, 55(11):2130–2135, 1998.

[222] A.M. Yaglom. *Introduction to the theory of stationary random functions*. Dover, 1962.

[223] S. Zhang, X. Zou, J. Ahlquist, I.M. Navon, and J.G. Sela. Use of differentiable and nondifferentiable optimization algorithms for variational data assimilation with discontinuous cost functions. *Monthly Weather Review*, 128(12):4031–4044, 2000.

[224] K. Zhu, I.M. Navon, and X. Zou. Variational data assimilation with a variable resolution finite-element shallow-water equations model. *Monthly Weather Review*, 122(5):946–965, 1994.

[225] Y. Zhu and I.M. Navon. Impact of parameter estimation on the performance of the FSU global Spectral Model using its full-physics adjoint. *Monthly Weather Review*, 127(7):1497–1517, 1999.

[226] X. Zou. Tangent linear and adjoint of "on-off" processes and their feasibility for use in 4-dimensional variational data assimilation. *Tellus A*, 49(1):3–31, 1997.

[227] X. Zou, A. Barcilon, I.M. Navon, J. Whitaker, and D.G. Cacuci. An adjoint sensitivity study of blocking in a two-layer isentropic model. *Monthly Weather Review*, 121(10):2833–2857, 1993.

[228] X. Zou and Y.H. Kuo. Rainfall assimilation through an optimal control of initial and boundary conditions in a limited-area mesoscale model. *Monthly Weather Review*, 124(12):2859–2882, 1996.

[229] X. Zou, H Liu, J. Derber, J.G. Sela, R. Treadon, I.M. Navon, and B. Wang. Four–dimensional variational data assimilation with a full–physics version of the NCEP spectral model: system development and preliminary results. *Quarterly Journal of the Royal Meteorological Society*, 127(573):1095–1122, 2001.

[230] X. Zou, I.M. Navon, M. Berger, K.H. Phua, T. Schlick, and F.X. Le Dimet. Numerical experience with limited-memory Quasi-Newton and Truncated Newton methods. *SIAM Journal on Optimization*, 3(3):582–608, 1993.

[231] X. Zou, I.M. Navon, and F.X. Le Dimet. Incomplete observations and control of gravity waves in variational data assimilation. *Tellus A*, 44(4):273–296, 1992.

[232] X. Zou, I.M. Navon, and F.X. Le Dimet. An optimal nudging data assimilation scheme using parameter estimation. *Quarterly Journal of the Royal Meteorological Society*, 118(508):1163–1186, 1992.

[233] X. Zou, I.M. Navon, and J.G. Sela. Control of gravitational oscillations in variational data assimilation. *Monthly Weather Review*, 121:272–289, 1993.

[234] X. Zou, I.M. Navon, and J.G. Sela. Variational data assimilation with moist threshold processes using the NMC spectral model. *Tellus A*, 45(5):370–387, 1993.

[235] D. Zupanski. The effects of discontinuities in the Betts–Miller cumulus convection scheme on four-dimensional variational data assimilation. *Tellus A*, 45(5):511–524, 1993.

[236] D. Zupanski. A general weak constraint applicable to operational 4D-VAR data assimilation systems. *Monthly Weather Review*, 125(9):2274–2292, 1997.

[237] M. Zupanski. Regional four-dimensional variational data assimilation in a quasi-operational forecasting environment. *Monthly Weather Review*, 121(8):2396–2408, 1993.

[238] M. Zupanski. Maximum likelihood ensemble filter: theoretical aspects. *Monthly Weather Review*, 133(6):1710–1726, 2005.

[239] M. Zupanski, I.M. Navon, and D. Zupanski. The maximum likelihood ensemble filter as a non-differentiable minimization algorithm. *Quarterly Journal of the Royal Meteorological Society*, 134(633):1039–1050, 2008.

[240] M. Zupanski, D. Zupanski, T. Vukicevic, K. Eis, and T.V. Haar. CIRA CSU four–dimensional variational data assimilation system. *Monthly Weather Review*, 133(4):829–843, 2005.

Index